MW00718814

*Advances in*

# BOTANICAL RESEARCH

incorporating *Advances in Plant Pathology*

VOLUME 23

## Pathogen Indexing Technologies

*Advances in*

# BOTANICAL RESEARCH

incorporating *Advances in Plant Pathology*

*Advances in*

# BOTANICAL RESEARCH

incorporating *Advances in Plant Pathology*

VOLUME 23
Pathogen Indexing Technologies

*Guest editor*

### S.H. De Boer

*Agriculture and Agri-Food Canada,*
*Pacific Agriculture Research Centre,*
*6660 N.W. Marine Drive,*
*Vancouver, B.C. V6T 1X2*

*Volume editors*

### J.H. Andrews *and* I.C. Tommerup

*Department of Plant Pathology,*    *CSIRO Centre for Mediterranean*
*The University of Wisconsin,*    *Agricultural Research,*
*Madison, Wisconsin,*    *Perth, PO Wembley 6014,*
*USA*    *Australia*

*Series editor*

### J.A. Callow

*School of Biomedical Sciences,*
*University of Birmingham,*
*Birmingham, UK*

1996

ACADEMIC PRESS
Harcourt Brace & Company, Publishers
London   San Diego   New York   Boston
Sydney   Tokyo   Toronto

# Contents

# Contributors

K.R. BARKER, *Plant Pathology Department, North Carolina State University, Box 7616, Raleigh, North Carolina 27695-7616, United States of America*

D.A. CUPPELS, *Pest Management Research Centre, Agriculture and Agri-Food Canada, 1391 Sandford Street, London, Ontario, N5V 4T3, Canada*

E.L. DAVIS, *Plant Pathology Department, North Carolina State University, Box 7616, Raleigh, North Carolina 27695-7616, United States of America*

S.H. DE BOER, *Pacific Agriculture Research Centre, Agriculture and Agri-Food Canada, 6660 N.W. Marine Drive, Vancouver, British Columbia, V6T 1X2, Canada*

I.G. DINESEN, *The Danish Plant Directorate, Skovbrynet 20, Lyngby, DK-2800, Denmark*

D.M. EAVES, *Department of Mathematics and Statistics, Simon Fraser University, Burnaby, British Columbia, V5A 1S6, Canada*

D.L. EBBELS, *Central Science Laboratory, Ministry of Agriculture, Fisheries and Food, Hatching Green, Harpenden, Hertfordshire AL5 2BD, United Kingdom*

A.A.J.M. FRANKEN, *Centre for Plant Breeding and Reproduction Research (CPRO-DLO), Droevendaalsesteeg 1, P.O. Box 16, 6700 AA Wageningen, The Netherlands*

R.D. GITAITIS, *Department of Plant Pathology, University of Georgia, Coastal Plain Station, Tifton, Georgia 31793, United States of America*

H. HUTTINGA, *Research Institute for Plant Protection (IPO-DLO), Binnenhaven 5, P.O. Box 9060, 6700 GW Wageningen, The Netherlands*

C.J. LANGERAK, *Centre for Plant Breeding and Reproduction Research (CPRO-DLO), Droevendaalsesteeg 1, P.O. Box 16, 6700 AA Wageningen, The Netherlands*

C.A. LÉVESQUE, *Pacific Agriculture Research Centre, Agriculture and Agri-Food Canada, 6660 N.W. Marine Drive, Vancouver, British Columbia, V6T 1X2, Canada*

I. MASTENBROEK, *General Inspection Service for Agricultural Seeds and Seed Potatoes (NAK), P.O. Box 51, 6710 BB Ede, The Netherlands*

S.A. MILLER, *Plant Pathology Department, Ohio State University, Wooster, Ohio 44691, United States of America*

*Contributors*

A.W. PEMBERTON, *Central Science Laboratory, Ministry of Agriculture, Fisheries and Food, Hatching Green, Harpenden, Hertfordshire AL5 2BD, United Kingdom*

S.A. SLACK, *Department of Plant Pathology, Cornell University, 334 Plant Science Building, Ithaca, New York 14853-4203, United States of America*

D.E. STEAD, *Central Science Laboratory, Ministry of Agriculture, Fisheries and Food, Hatching Green, Harpenden, Hertfordshire AL5 2BD, United Kingdom*

C. SUTULA, *Agdia Inc., 30380 County Road 6, Elkhart, Indiana 46514, United States of America*

G.W. VAN DEN BOVENKAMP, *General Inspection Service for Agricultural Seeds and Seed Potatoes (NAK), P.O. Box 51, 6710 BB Ede, The Netherlands*

R.W. VAN DEN BULK, *Centre for Plant Breeding and Reproduction Research (CPRO-DLO), Droevendaalsesteeg 1, P.O. Box 16, 6700 AA Wageningen, The Netherlands*

A. VAN ZAAYEN, *Inspection Service for Floriculture and Arboriculture (NAKB), Johan de Wittlaan 12, 2517 JR Den Haag, The Netherlands*

# Series Preface

*Advances in Botanical Research* is one of Academic Press' longest standing serials, and has established an excellent reputation over more than 30 years. *Advances in Plant Pathology*, although somewhat younger, has also succeeded in attracting a highly respected name for itself over a period of more than a decade.

The decision has now been made to bring the two serials together under the title of *Advances in Botanical Research incorporating Advances in Plant Pathology*. The resulting synergy of the merging of these two serials is intended to greatly benefit the plant science community by providing a more comprehensive resource under one 'roof'.

John Andrews and Inez Tommerup, the previous editors of *Advances in Plant Pathology*, are now on the editorial board of the new series. Our joint aim is to continue to include the very best articles, thereby maintaining the status of a high impact factor review series.

# *Preface*

This thematic volume in the *Advances in Botanical Research incorporating Advances in Plant Pathology* series evolved from the perception that very significant advancements have been made in pathogen detection technologies during the last decade but that application of these technologies in strategies to control plant diseases was limited. Yet indexing of plants and plant parts for the presence of specific pathogens has been most effective in some instances for avoiding and/or controlling disease. The new technologies for detecting low levels of pathogens will only increase the value of indexing as a tool for plant disease control. Providing an overview of the status of detection technology, this volume is directed not only to scientists and students interested in detection technology, but also to those involved in formulating and implementing disease control and quarantine regulations. A fairly large body of scientific literature has accumulated on pathogen detection but plant health administrators primarily rely on secondary sources of information on which to base their decisions. Thus it is anticipated that this volume will become a standard reference for certification agencies, quarantine laboratories, and research facilities supporting these functions. The book is not meant to be a laboratory manual or description of methods but rather to provide a conceptual framework presenting a review of current scientific literature, assessment of the state-of-the-art, and speculation on future developments and requirements.

The first chapter is introductory to the book, setting the stage for the other chapters, and is followed by chapters that cover the different pathogen groups, namely, bacteria, viruses, fungi, and nematodes. Subsequent chapters review current practices in areas where detection technology has become important and provide perspectives on how indexing technologies can be applied, how well it has worked, and what problems remain. The penultimate chapter is a statistical treatment of detection limits, sampling strategies, and risk assessment. The issues of cost, standardization, and quality control, which are of great significance for decision makers and regulatory bodies, are addressed in the final chapter.

For each of the chapters an attempt has been made to select authors from the international community of plant pathologists who are experts in their field and have practical experience. Hopefully, their efforts in writing chapters for this volume will prove to have been a worthwhile exercise.

Vancouver, 1995                                                                 S.H. De Boer

# 1

# THE VALUE OF INDEXING FOR DISEASE CONTROL STRATEGIES

D.E. Stead, D.L. Ebbels and A.W. Pemberton

*Central Science Laboratory, Ministry of Agriculture, Fisheries and Food, Harpenden, Hertfordshire AL5 2BD, United Kingdom*

|       |                                                              |    |
|-------|--------------------------------------------------------------|----|
| I.    | Introduction                                                 | 1  |
| II.   | Definitions                                                  | 3  |
| III.  | How Valuable Is Indexing?                                    | 4  |
| IV.   | Need for Rapid and Accurate Diagnosis                        | 5  |
|       | A. Specificity                                               | 6  |
|       | B. Sensitivity                                               | 6  |
|       | C. Time Frame and Cost                                       | 8  |
| V.    | Diagnostic Methods and Their Value in Indexing               | 9  |
| VI.   | Automated Indexing Kits and Expert Systems                   | 10 |
| VII.  | Dependency of Control Strategies on Correct Diagnosis        | 11 |
| VIII. | Indexing in Various Contexts                                 | 11 |
|       | A. Indexing to Prevent Spread of Pathogens in International Trade | 11 |
|       | B. Indexing in Eradication Campaigns                         | 15 |
|       | C. Indexing for the Maintenance of Pest Free Areas           | 16 |
|       | D. Indexing in Healthy Stock Schemes                         | 17 |
|       | E. Indexing in Advisory Work                                 | 19 |
| IX.   | International Co-Operation for Indexing                       | 20 |
| X.    | Documentary Procedures and Voucher Specimens                 | 20 |
|       | References                                                   | 21 |

## I. INTRODUCTION

Plant diseases may be controlled by many means, including the use of chemicals, altering cultural conditions, breeding or engineering resistant varieties, and by various administrative methods. One aspect of the administrative approach aims to exclude particular pests and pathogens. Exclusion can take many forms, for example, prohibiting the movement of plants from an infested country into one which is free of a particular pest or disease. Indexing is essential for this purpose and implies that some assessment for the presence of a particular pathogen or pest on a given crop or consignment has been made. In most cases, absence of the pathogen or pest must be demonstrated but in some cases a tolerance may be allowed. For example, a low percentage of infection in a seed crop may be allowed because at that level it is not considered damaging. Above the given level the risks of damage are considered too

Advances in Botanical Research Vol. 23
Incorporating Advances in Plant Pathology
ISBN 0-12-005923-1

high and the crop is rejected.

In its simplest form, indexing may comprise a visual inspection of a sample of the crop for the presence of a pest or for internal or external symptoms of a disease. However, since many pests and diseases cannot be reliably detected in this way, perhaps because of long periods of latent infection, visual inspection may not be adequate. In such cases, laboratory testing is required. Also for many diseases and some pests, simple visual assessment by inspection for symptoms does not allow accurate diagnosis of the pest or pathogen, particularly where there is a need to differentiate between physically similar pests or pathogens, only one of which is being controlled. Thus indexing often involves laboratory tests for diagnosing diseases and/or detecting the pests and pathogens that cause them.

Indexing has particular value in three distinct areas and the methods employed, the sampling required, and the acceptable tolerance levels vary accordingly. These three areas are: (1) Preventing international and national spread of pests and diseases into new areas as a result of commercial movement of plants and plant products. This is particularly important for propagating material but also for ware crops. (2) Maintaining the absence of a given pest or pathogen in the production of propagating material especially at the beginning of the propagation process. These processes are often governed by a certification or classification scheme which specifies methods for indexing and tolerance levels for particular pests and diseases at the various stages of multiplication. Provision of propagating material in commercially acceptable quantities usually requires bulking-up of the initial selection over several years, during which there is the possibility of re-introduction. This multiplication has several stages and examples are given in Table I. At the initial nuclear stock or indexed clone stage, perhaps each plant would be tested individually, whereas at the intermediate propagation stages a sample would be taken and rigorously tested. In all these cases a zero tolerance would be enforced and any infected lots would be discarded or at least downgraded. The end product, certified stock for final crop or ware production, would be sampled perhaps less stringently or with stringency varying according to the standards set. Such certified stock may be allowed for sale with a small level of infection with non-quarantine organisms affecting only the quality of the material. (3) Maintaining the absence of, or acceptable low levels of a pathogen or pest in the marketing of plant produce. This is largely a quality con-

*Table I.* Certification scheme terminology used by the European Community and by the European and Mediterranean Plant Protection Organisation

| Generation | EC Schemes | EPPO Schemes |
|------------|------------|--------------|
| 1 | Indexed clone or line | Nuclear Stock |
| 2 | Pre-basic | Propagation Stock I |
| 3 | Basic[a] | Propagation Stock II |
| 4 | Certified Stock[a] | Certified Stock |

[a]In some schemes there may be more than one generation. In EPPO schemes there is a general principle that material shall be multiplied in as few steps as possible to produce the required volume of Certified Stock

trol exercise and may simply comply with what the trade is prepared to accept. Most of the emphasis in this chapter will concern the first and second areas.

## II. DEFINITIONS

At this stage, to prevent ambiguity, it is necessary to define some of the terms used.

*Indexing* - the process of assessing or screening plant material for the presence of a pest or pathogen. This may involve a visual assessment for a disease or the pathogen/pest, perhaps with the aid of a microscope or a laboratory test, to demonstrate the presence of the pathogen.

*Diagnosis* - the process of determining the cause of a disease, either involving one or more tests to demonstrate the presence of a pathogen in the tissue or involving its isolation and subsequent identification.

*Identification* - the process of placing a pathogen in a given taxon based on visual characteristics or by the use of several well selected tests which show it to be typical of the particular taxon and which differentiate it from closely related taxa. This process is also often referred to as *determination*.

*Detection* - the process of demonstrating the presence of a pathogen directly in a crop or in plant tissue. This may be in symptomatic or asymptomatic tissue where the pathogen is present as a latent infection or as an epiphyte.

*Isolation* - the process of obtaining a pure sample or culture of the pathogen, usually by culturing and sub-culturing until a pure isolate has been selected.

*Certification* - quality assurance for planting material.

*Certification scheme* - a system for producing plant material of an acceptable level of quality and occurrance of a pathogen or disease. Visual inspection is usually used to check for freedom from disease. Some form of testing involving detection and/or diagnosis may be required during certification, since most propagating material is multiplied over several seasons. Most certification schemes also concentrate on assuring higher quality in the early years of multiplication. As a result it is often necessary to have several grades, each with a defined level of quality assurance. Grade standards may also need to take account of age as well as quality since risks of pest or disease re-introduction increase with time.

*Nuclear stock production* - production of a small nucleus of indexed, pest and pathogen free stocks, and maintenance of this in isolation.

*Infection* - the initial entry of a pest or a pathogen into a plant, or the established condition of a pest or pathogen in or on the cells or tissues of a plant.

*Latent infection* - establishment of a pest or pathogen in a plant without expression of symptoms. Each disease has a latent infection period between the initial entry and the expression of symptoms, but for some host-parasite interactions, this period can be extended indefinitely.

*Contamination* - the presence of the pest or pathogen on a plant, in a substance, or on land, not in a situation where infection of a host has occurred.

## III.  HOW VALUABLE IS INDEXING?

Where a pathogen is established, the ultimate objective of indexing is usually eradication.  However, it may not be cost effective to achieve this goal even though it may well be highly cost effective to reduce and maintain the level of a pest/pathogen below that which leads to economic crop loss.  This is probably the main practical aim of most indexing systems.  Success will depend on several factors including: (1) the nature of the plant material to be indexed, (2) the size of sample required, (3) the diagnostic methods available and their specificity and sensitivity, (4) the biology of the plant and the pests and pathogens concerned, (5) the training of the diagnosticians, (6) the management of the certification scheme, if any, in which the indexing is applied, and (7) the possibility for re-introduction after indexing.

In addition to its value in eradication programs, indexing is valuable for production of pathogen-free micropropagated plants.  Micropropagation provides an excellent opportunity for eliminating pests and pathogens to provide a healthy source of plant material for rapid multiplication.  Although meristems are often sterile or at least pathogen free, this is not necessarily the case for some host/pest combinations.  The subject of indexing microplants during micropropagation is reviewed by Cassells (1992).

Although microplants may be killed by microorganisms which are normally saprophytic, infections by true pathogens sometimes remain latent, not causing disease symptoms, during microplant multiplication.  Thus visual examination is not necessarily an effective indexing method for microplants and some form of laboratory test is advisable for the mother plant and the young micropropagated plants, both during multiplication and after transferring to the open environment.  Nevertheless, despite these problems, micropropagation can offer an excellent source of pest/pathogen free material, provided it originates from fully indexed mother stock.

With seed crops, visual inspection during the growing season prior to harvest will not necessarily result in pest or pathogen free seed and testing is advisable, especially in early generations in the multiplication cycle.  The prospects and potential for indexing seed are discussed by Ball and Reeves (1992).  Indexing of seed is often a feature of quarantine legislation and certification.  Minimum seed standards are set by the Food and Agriculture Organisation in its Quality Declared Scheme. International co-operation on seed indexing also occurs through the International Seed Testing Association (ISTA) which develops and publishes standard protocols for sampling and testing (Ball and Reeves, 1992).

Other kinds of propagating material such as bulbs, tubers, cuttings and fruit tree budwood all have their specific disease or pest problems and often there are internationally agreed baseline protocols for indexing for specific pests and diseases, depending on the species.

Reducing the incidence of the pest or pathogen is likely to reduce the amount of disease but because all indexing methods have a threshold of detection below which they give a negative reaction, they are unlikely to guarantee freedom from a specific pathogen in a single indexing exercise.  The extent to which indexing is successful is not commonly reported, although De Haan (1994) claims that in Canada's Prince Edward Island, indexing based on laboratory testing has dramatically decreased incidence of the potato ring rot disease and eradicated the potato spindle tuber viroid. There are also relatively few reports of its failure to control or at least reduce disease incidence. Relatively few diseases have been successfully eradicated, but this is perhaps largely due to re-introduction and not the failure of indexing. Methods of plant multiplication are important in connection with indexing. For example, many diseases can be easily transmitted through cuttings or by cutting tubers prior to planting. Failure to reduce the incidence of potato ring rot in much of North America is partly due to spread of infection through the practice of cutting seed potatoes and the use of pricker planters. Bacterial diseases of carnations and pelargoniums can also be spread by the cutting knife when taking cuttings. Several carnation diseases have been successfully controlled in Europe by indexing and certification, but bacterial blight of pelargonium caused by *Xanthomonas campestris* pv. *pelargonii* still remains an important disease of pelargonium despite the regular use of indexing to control it. This is thought to be due largely to lack of sensitivity of detection methods followed by accidental spread on the cutting knife. When bacterial blight of pea (*Pseudomonas syringae* pv. *pisi*) was first found in the United Kingdom in 1985, laboratory tests were immediately brought into the seed certification process. Ball and Reeves (1992) report that within one year of introducing the testing regime, the frequency of infected seed lots declined from 12% to 4% but in subsequent years it rose again. They argue that this rise was caused more by loopholes in the regulations resulting in some infected seed remaining in the system than as a consequence of the lack of sensitivity of the direct plating method used.

Because of widespread latent infection, potato blackleg is not well controlled by indexing based on growing season inspection. The use of pathogen free micropropagated plants or virus-tested-stem-cutting-seed stock has failed to maintain freedom for more than a couple of years of multiplication, almost certainly because of regular re-introduction and subsequent rapid spread in rain-generated aerosols. Nevertheless, blackleg incidence is related to the level of seed contamination (Wale *et al.*, 1986; Bain *et al.*, 1990) and thus indexing by quantifying the pathogen allows a useful blackleg potential index to be developed (Jones *et al.*, 1994). Stocks with high risk indices should not be planted in conditions suitable for blackleg development. Whether or not such indexing is cost effective remains to be seen.

## IV. NEED FOR RAPID AND ACCURATE DIAGNOSIS

Indexing implies that some assessment for the presence of a particular plant pathogen or pest is made. Even when the initial breeding material is shown to be free of the target pest/pathogen it is usually necessary for some regular check during multiplication to ensure that the target pest/pathogen has not been re-introduced. The need for this may decrease during the period of certification and in the final years of bulking up a small tolerance may be allowed.

The diagnostic methods used must be as cheap and as rapid as possible but most importantly they need to be specific, sensitive, reproducible and often require quantification rather than simple presence or absence.

## A. Specificity

The ability to define the target pest or pathogen is essential and will depend on various taxonomic parameters. For example is the pest/pathogen accurately named? Is it a discrète organism? Has it been well classified? For many organisms, identity at the infra-specific level is required. For example, many plant pathogenic bacteria are classified as pathovars including those of *Pseudomonas syringae* and *Xanthomonas campestris* many of which have recently been transferred to other *Xanthomonas* spp. especially *X. axonopodis* (Vauterin *et al.*, 1995). Thus *P. syringae* pv. *phaseolicola* but not *P. syringae* pv. *syringae* may be the subject of regulation on beans and hence the method for diagnosis must be able to separate them.

Similarly the formae speciales of the fungi in the genus *Fusarium* comprise pathogens which may or may not be subject to control.

Classical methods of diagnosis include isolation of the pest/pathogen and subsequent identification. Such methods are often labour intensive and thus expensive. Less labour intensive methods include isolation and characterisation by genetic or biochemical methods. Molecular methods include the polymerase chain reaction (PCR) and various profiling techniques for DNA fragments. Biochemical methods include analysis of gel electrophoretic protein profiles as well as fatty acid and nutritional profiling which are especially of value for bacteria. Viruses are often good antigens and are best characterised by serological tests or genetic profiles. Serology is also useful for many bacteria and fungi but is usually used for direct detection in plant tissue or sap. Immunofluorescence and ELISA are the most common techniques, but PCR is becoming increasingly common as a detection method.

Monoclonal antibodies may be too specific and may miss some strains (false negative result). Polyclonal antisera often contain antibodies to epitopes common to many related pathogens and hence give cross reactions (false positive result). Antisera for plant viruses, phytoplasmas and other mollicutes may also contain antibodies which react with host cell constituents.

Even when using PCR, finding primers which have the required specificity can be difficult. Table II lists general and more specific references for diagnostic techniques. Table III lists the modern techniques available for each and indicates the pests or pathogens to which they are most appropriate. Hansen and Wick (1993) give a good recent review of diagnostic techniques, while Torrance and Jones (1981) review techniques in relation to large-scale indexing programmes.

## B. Sensitivity

Each diagnostic method has a threshold or limit of detection. There are two major factors which affect this threshold, namely the efficiency of sampling, and the efficiency of the test method to detect presence of the pathogen in selected samples.

Whereas we strive to improve the technology of detection in the final sample we rarely consider the sample size. What is the point of having a method which detects a single virus particle, bacterium or spore per millilitre of plant tissue if the sample taken is not representative of the plant population under test? Sampling efficiency

*Table II.* Summary of diagnostic indexing techniques for plant pathogens.

| Type of organisms | Techniques | Reference |
|---|---|---|
| General | General | Hansen and Wick, 1993; Duncan and Torrance, 1992 |
| Viruses | General | Hill, 1984 |
| | Serological | Torrance, 1992 |
| | Genetic | Salazar and Querci, 1992; Jones, 1992 |
| Phytoplasmas and other mollicutes | General | Whitcomb and Tully, 1989 |
| | Serological | Clark, 1992 |
| | Genetic | Lee *et al.*, 1993 |
| Bacteria | General | Lelliott and Stead, 1987; Stead, 1992; Klement *et al.*, 1990 |
| | Serological | Stead, 1992; Hampton *et al.*, 1990 |
| | Fatty acids | Stead *et al.*, 1992; Stead, 1992 |
| | Genetic | Vivian, 1992 |
| Fungi | General | Dhingra and Sinclair, 1985 |
| | Serological | Dewey, 1992 |
| | Genetic | Coddington and Gould, 1992 |
| Nematodes | General | Southey, 1986; Mai and Lyon, 1975; Zukerman *et al.*, 1990; Fortuner, 1988 |
| | Molecular | Hussey, 1982 |
| | Genetic | Harris *et al.*, 1990 |
| | Electron microscopy | Shepherd and Clark, 1986 |

is often limited by cost and practicability. For example, a large seed sample may represent the bulk of what is available for continued multiplication. Equally, the practical problems of handling a sample of 10,000 potato tubers add to the costs of indexing. A balance must be achieved and each part of the indexing system considered in relation to this balance. For example, when including a new potato cultivar in a certification scheme, micropropagated plants may be derived from individually tested tubers. Each mother plant is then tested before multiplication. Once in culture, regular sampling is maintained and absence of specific pathogens confirmed by indexing. At the commercial grades indexing is normally by visual inspection during the growing season rather than by laboratory testing.

A negative result in an indexing test means only that the target organism was not found. It may not mean that the pathogen was absent from the actual sample or from the crop from which the sample was taken. By contrast, a positive result confirms the presence of the target organism in the original crop, provided false positive results are known not to occur.

The ultimate objective of most indexing is to guarantee freedom from certain pathogens to effect eradication, but because of insufficiently sensitive methods or re-

Table III. Modern diagnostic techniques and their most appropriate targets.

| Technique | Pest or pathogen |
|---|---|
| Genetic | |
| Nucleic acid hybridisation | |
| Dot blot/squash blot | All |
| Tissue print blot | Viruses |
| Colony blot | Bacteria, fungi |
| Polymerase chain reaction | All |
| dsRNA | Viruses |
| | |
| Phenetic | |
| Biochemical | |
| PAGE protein profiles | All |
| Enzyme profiles | All |
| Fatty acid profiles | Bacteria |
| Nutritional profiles | Bacteria |
| Serological | |
| ELISA | Viruses, mollicutes, bacteria, fungi |
| Immunofluorescence | Mollicutes, bacteria, fungi, nematodes |
| Immunomagnetic separation | Viruses, mollicutes, bacteria, fungi, nematodes |
| Immunospecific electron microscopy | Viruses, mollicutes |
| Agglutination | Viruses, mollicutes |
| Morphological | |
| Light microscopy | Fungi, nematodes, insects |
| Electron microscopy | Viruses, mollicutes, fungi, nematodes |
| Bioassay | |
| Inoculations | All |
| Grafting | Viruses, mollicutes |
| Baiting | Bacteria, fungi |

introduction, eradication may not be practical. Latent infections add to these problems and if the methods used are not likely to detect these latent infections it may be necessary to increase sensitivity. This can be done by indexing the plant material after culturing under conditions favourable for multiplication of the pathogen and for disease expression. For example, the efficiency of indexing raspberry roots for the presence of *Phytophthora fragariae* var. *rubi* can be improved by keeping soil conditions dry for approximately 6 weeks, and then flooding and maintaining waterlogged conditions before checking the roots for the presence of the pathogen (Anon., 1994c).

## C. Time Frame and Cost

The two issues here are the length of time it takes to obtain the results of indexing procedures, and the costs due to actual staff input.

Classical indexing of viruses often includes grafting to woody indicators which can take many months for results to be obtained, especially if woody plants must bear fruit before the tests are read. Host tests for some bacteria can take many weeks, e.g. eggplant tests for potato ring rot (Lelliott and Sellar, 1976). Many diagnoses are labour intensive. For example, isolation, purification and identification by traditional bacteriological methods (Lelliott and Stead, 1987) are often expensive in staff costs and the results take a long time to obtain.

Indexing should be reliable, quick, cheap and simple. These desirable criteria are obviously interrelated and few techniques will satisfy all of them. Reliability, which includes both consistency and accuracy, is probably the most important characteristic by which any indexing technique must be judged. Techniques which are not very consistent or accurate are of very limited use and cannot be relied upon, especially for cases liable to be contested in a court of law. A rapid result is not always essential, especially for pathogens which are not able to spread quickly, such as those causing some soil-borne diseases. For example, tests for beet rhizomania in soil at present take eight weeks. In some situations action can be taken when results become available without much loss of efficacy. However, in other circumstances, for example when trying to check outbreaks of aerially disseminated organisms or when suspect perishable produce is awaiting marketing or landing from shipment, speed of diagnosis can be extremely important and may make the difference between substantial financial loss or gain. The cost of indexing is always a factor to consider and must be judged against the value of the material being tested and the potential effects of using other methods or not indexing at all. A high cost for effective indexing may be well worth while if the material is a valuable new cultivar which will subsequently be extensively propagated, or if the consequences of using a cheaper alternative are potentially very damaging to a substantial industry or area of crops. Where other considerations are equal, simple indexing methods are to be preferred to complicated ones. Simple methods obviously have less to go wrong and are usually cheaper, but simplicity often has an inverse relationship with the other desirable criteria mentioned.

## V. DIAGNOSTIC METHODS AND THEIR VALUE IN INDEXING

A review of diagnostic methods available is given by Hansen and Wick (1993). Such methods, listed in Table III, fall into two categories. (1) Methods which determine whether or not the sample contains the target microorganism, such as methods that use specific reagents (e.g. antisera, DNA probes and primers). (2) Methods which identify any pest or pathogen in the sample, for example, profiling techniques which compare the profile of an unknown with a library of profiles, herbaceous indicators for viruses, and isolation on general media for fungi.

Both types of methods have their value in indexing but since indexing is usually for a specific target pest or pathogen, the former is often of greatest value providing the method used has the required specificity and sensitivity. In particular, diagnosticians will be attracted to methods which can handle large numbers of samples. Obvious candidates would be various ELISA formats and the polymerase chain reaction. Although generally excellent for most plant viruses, ELISA has its problems for some organisms. Antiserum specificity can be a problem with antigenically more complex organisms such as bacteria and fungi. Monoclonal antibodies may be too specific and give false negative results with some strains. Polyclonal antibodies may not be specific enough and cross react with other bacteria. The sensitivity of ELISA may also be too low for meaningful indexing. Also ELISA is particularly suited to trapping soluble or small antigens whereas cells and mycelia may easily be lost in washing, hence further reducing the sensitivity of the test. Various potential improvements include flow-through ELISA formats based on microtitre plates with basal membrane filters which physically trap antigen, and poly-L-lysine-glutaraldehyde ELISA in which plates coated with poly-L-lysine are centrifuged to trap the antigen onto the adhesive layer which is then fixed to it with glutaraldehyde. For bacteria, some ELISA formats include an enrichment phase (Alarcon *et al.*, 1994) to increase the populations of the target bacterium to levels which can be detected.

The polymerase chain reaction also has great value for indexing because it has great potential sensitivity and is potentially applicable to large numbers of samples. However, PCR is relatively easily inhibited. Sap of many plant species can contribute to this inhibition and therefore adequate controls must be included to determine whether inhibition is occurring. Inhibition can be decreased or prevented in several ways including purification of the target pathogen by immunoseparation, and by the use of nested primers. The efficiency of DNA extraction from the plant samples can also be improved by phenol chloroform extraction, or by the use of commercially available DNA purification kits. Current PCR protocols are not always well suited to large-scale indexing on numerous samples, but costs are decreasing and reliability is improving.

## VI.  AUTOMATED INDEXING KITS AND EXPERT SYSTEMS

Indexing increasingly relies on fairly sophisticated, expensive equipment and on the supply of reagents such as antibodies, DNA probes and PCR primers. Whereas there are many publications showing that improved diagnosis is possible, the techniques demand technical skills and the reagents are expensive and often not commercially available. Although commercial kits are becoming more common, users perhaps often expect too much of them and need to be reminded that a negative result means that the target pest or pathogen was not detected in the sample rather than its guaranteed absence as discussed in section IV. In a market which requires cheap, rapid results it is tempting for companies to sell kits which in practice are inadequate. It is also tempting for the trade to ignore these inadequacies in meeting the minimum indexing requirements, since official requirements imposed may not always be wholly acceptable to the trade. Also kits usually rely on a single test. When laboratory-based diagnosticians find a particular test or reagent is inadequate, they would rarely rely on it as the sole indexing method.

Kits are now available for a wide range of pathogens including bacteria (potato ring rot, potato brown rot, pelargonium blight, Pierce's disease of grape), mollicutes (*Spiroplasma citri*), a wide range of viruses, and fungi (*Pythium* spp., *Rhizoctonia solani* and various fungal pathogens of turf grass). Most of these kits are based on ELISA but dipstick assays will undoubtedly become increasingly common.

A range of nutritional kits is available for identifying plant pathogenic bacteria, of which, perhaps the Biolog GN Microplate system is currently the market leader. Commercial kits based on DNA technology are not yet widely available although the improvements in non-radioactive labels and in commercially available DNA extraction and purification kits may soon change this. Dot blot DNA hybridisation tests may be the most common format.

Various expert systems are now available for a number of crops including potato (Adams *et al.*, 1990) and tomato (Blancard *et al.*, 1985) but these are of limited use in indexing other than during growing season inspections.

Automation is an important feature in large scale commercial indexing and will become increasingly important. Several techniques are particularly suited to automation. In particular, ELISA can now be fully automated although the computer-operated equipment is expensive. Various profiling techniques especially suited to bacterial identification are semi-automated. Commercial fatty acid profiling systems use gas chromatographs with fully automated sample injection, but sample extraction is still done manually. Nutritional kits such as the Biolog GN Microplate system allow computerised pattern recognition and automated scoring of substrate utilisation patterns.

## VII. DEPENDENCY OF CONTROL STRATEGIES ON CORRECT DIAGNOSES

The success of all disease and pest control programmes depends on the correct identification of the causal agent. Without this, the outcome of any action would be only a matter of luck. This applies equally to all manner of plant protection strategies, whether advisory or statutory, aimed at prevention, control or eradication, and irrespective of the use of chemical, biological or administrative means. Correct identification gives immediate access to information vital to all these areas. It will assist in the choice of a particular strategy and may allow for a confident guess at the prospects for success. It will enable a suitable chemical or biological pesticide to be selected and may indicate a promising means of application.

To be effective, diagnosis relies on good taxonomic classification. For most organisms a phylogenetic system is not only logical but will normally result in physiologically similar organisms being placed taxonomically close together. Hereby the likely behaviour and reactions of poorly known organisms may often be deduced from those of their better known close relatives.

## VIII. INDEXING IN VARIOUS CONTEXTS

### A. Indexing to Prevent Spread of Pathogens in International Trade

National governments have a sovereign right and duty to protect both the managed (agriculture, horticulture, forestry) and natural environments from the introduction and/or spread of potentially harmful and exotic organisms not currently established in their country or, if present, of only limited distribution. These are the principles of the International Plant Protection Convention of 1951 (IPPC) (Anon., 1992e; Anon., 1994a). However, through the Sanitary and Phytosanitary section of the General Agreement on Tariffs and Trade (GATT) (Anon., 1994b) governments also have the obligation to ensure that any measures aimed at preventing such introductions or spread are the minimum necessary to achieve that aim.

The prohibition of a particular trade (i.e. of a particular commodity of a particular origin) is the option of last resort adopted only in circumstances where the assessed risk from potentially harmful organisms cannot be overcome by other measures. It is thus incumbent on regulatory authorities to establish procedures which will reduce to an acceptable level the risk of organisms being introduced or spread. The range of possibilities to achieve such a goal is substantial.

Prohibition in practice is rarely absolute. For research or varietal development, most governments permit the importation of small quantities of otherwise prohibited plant material provided it is first subjected to inspection and testing under quarantine conditions to ensure freedom from harmful organisms. In the European Union, all seed potatoes and potato breeding material imported from non European Union countries must be indexed according to agreed protocols at officially recognised quarantine stations (Anon., 1980; Anon., 1993a; Jeffries *et al.*, 1993) and similar protocols are being developed for other prohibited commodities, especially fruit plants (*Citrus, Vitis, Malus, Prunus*, etc.).

More generally, international and national trade in plants and plant products is amenable to control. Traders generally wish to trade in quality products satisfactory for their intended purpose which do not present their customers with difficulties or problems.

Avoiding the introduction or spread of harmful organisms does not necessarily entail total elimination of organisms from traded commodities. For most trade there is no plant health risk because the ecological conditions at the point of destination are not appropriate for the organism. For example, bananas present no plant health risk to the United Kingdom because the host plant is not grown and the climate is not suitable for the development of tropical or sub-tropical diseases. There is the theoretical possibility that high temperature glasshouse cultures ("stove-house" cultures) could be at risk from organisms carried by banana fruit, but the pathway for the organism from the fruit to the "stove-house" is obscure and such specialist cultures are best protected by good management and hygiene practices.

Commodities for consumption (fruit, vegetables, cut flowers, etc.) can carry diseases but the risk of the disease establishing in a new area is low. There is generally no clear pathway by which the disease organism may move from, say, the imported vegetable to a growing crop. For planting material the converse applies. If no action is taken then the risk of a disease becoming established and spreading is so high as to be almost a certainty, assuming of course the conditions at the point of destination are suitable for development of the disease organism.

Measures to avoid the introduction across national borders and subsequent spread of harmful organisms are necessary and various, as exemplified by the plant health regime of the EU Single Market (Anon., 1977), as amended. Such measures can be applied at the following points:
- at the place of production in the country of origin,

- immediately prior to export from the country of origin,
- in transit between the exporting country and the importing country,
- at the point of entry or destination in the importing country, and
- following arrival at the destination.

Most procedures involved in the plant health control of international trade rely to a greater or lesser extent on visual inspections. For many pest and disease organisms, this is an efficient and reliable indexing procedure. Although a single individual organism may not be detected, most pests and diseases occur in sufficient numbers or cause such significant symptoms as to be detected if inspected at an appropriate time.

Visual inspection however will not detect latent or symptomless infection. Disease indexing which includes specific detection and diagnostic testing can overcome such difficulties by either replacing or, more normally, supplementing visual inspection.

*1. Export measures*

In international regulations, specific disease indexing is normally required to be done at the place of production in the season, or sometimes for several seasons, prior to export. This is a timely period in which to require such examinations.

Although serological and nucleic acid procedures are relatively quick (a matter of hours once samples are in the laboratory), the processes of sampling the plants, packaging, labelling and transport to the laboratory (which will often be at a distance from the growers' premises) all takes time. So do the laboratory procedures if individual tests must be made on large numbers of samples. If each test takes two hours, tests on a sample of 300 plants, even with bulking of samples, can take a number of days to complete. This is perfectly acceptable to the trade at a time when perishable plants are still growing but it may not be acceptable when the plants are in the pack house or are already packed ready for export at the dock or the airport.

Indexing at the producers' premises can provide several layers of security. Indexing of samples from plants destined for export is often a standard requirement, especially for virus diseases and bacteria-like organisms such as phytoplasmas and other mollicutes. Examples include plum pox virus, apple proliferation phytoplasma, and tomato ringspot virus. Some bacteria fall into the same category, such as potato ring rot (*Clavibacter michiganensis* subsp. *sepedonicus*) and fireblight (*Erwinia amylovora*). The level of security varies with the sensitivity of the test and the number of plants in the sample tested. As mentioned earlier, the latter is often a limiting factor. In a consignment of half a million plants, resources may only permit a sample of 300 plants to be tested, which would only enable detection of 1% infection at the 95% level of confidence. Lower levels of infection will commonly be missed. Testing of the mother plants can provide a higher level of confidence that the stock is healthy because smaller numbers are often involved and every plant can be tested. However, the earlier the testing the greater the chance that infection will occur between testing and marketing. Testing alone is thus rarely effective and is normally supplemented by visual inspection at appropriate times.

Conversely, visual inspection should also be supplemented by the diagnostic testing of samples to check on an inspector's visual examination. Leaf symptoms of plum pox virus can easily be confused with a range of common virus diseases such as plum line pattern. In raspberry cane nurseries, raspberry ringspot virus can be confused with eriophyid mite feeding damage. Visual inspections will be made on the plants intended for export and, if required, on the mother plants from which they were derived (as is required for viruses such as plum pox virus). Visual inspection is also commonly required for host plants "in the immediate vicinity" of the plants intended for export. This is a standard requirement for the control of fireblight and other diseases with aerial vectors or aerial dispersal mechanisms.

Detection and diagnostic testing procedures vary with the pathogen and resources available. Serological and nucleic acid techniques are often preferred methods where suitable reagents are available because they are rapid and cost-effective. However, less rapid and cost-effective techniques such as cell culturing, test plant inoculation, grafting, electron microscopy, etc. all continue to have important roles. Although test kits are becoming available which permit "in field" testing, most plant health services still rely on some form of centralised laboratory facility where efficiency of scale, availability of specialised equipment, and diversity of expertise can generally provide a more effective service.

Although normally undertaken at the place of production (i.e. the grower's premises) sampling and testing of a consignment can be undertaken at any point in the export-import chain. It is, however, normally only as an adjunct to visual inspection. An inspection must be done by the plant health service of the exporting country immediately prior to export so that an international phytosanitary certificate can be issued. This certificate which accompanies the consignment states that, to the best of the inspector's knowledge, the consignment is free from quarantine pests and diseases and is believed to conform with the regulations of the importing country.

*2. Import measures*
At the point of entry to the importing country, as part of the Customs' clearance procedure, a plant health inspection may be required and if in doubt the inspector may submit samples for testing. Often it is not convenient for a consignment to be thoroughly inspected at the point of entry (such as an airport) and the inspection may be deferred to the importer's premises where space and time permit a fuller inspection. Once again, samples may be taken for testing, but these will only be check samples and are always limited in their effectiveness by the constraints of time and resources. If plants are in a dormant phase, visual inspection for most systemic disorders is pointless and laboratory testing is the only reliable procedure. However, in the United Kingdom we also rely substantially on our Plant Health Inspectors carrying out follow-up inspections during the growing season or period after import. For dormant trees this can be up to six months after import. We consider this an efficient means to detect newly introduced organisms when containment and eradication measures can be applied to prevent establishment and spread.

These "import" inspections and testing are regarded as no more than checks to ensure that effective measures have been taken in the exporting country. They are intended to check firstly that the required export inspections and tests have been done and secondly that those requirements are appropriate to exclude the diseases which threaten the importing country (i.e. that the regulations are sound). They are not intended as a quarantine exclusion measure in their own right, although through the early detection of new diseases enabling prompt eradicating action, they can have that effect.

## B. Indexing In Eradication Campaigns

Plant Health import regulations can be termed exclusion measures. However, whenever there is movement of transport vehicles, goods, or people there will also be an associated risk of spread of pest and disease organisms. Plant health import regulations aim to minimise such risks. When a pest or disease organism breaches the regulatory barrier the plant health authorities will, where a pest risk analysis deems it justified, embark on an eradication or at least a containment exercise.

To be effective, eradication requires early detection and accurate diagnosis of the organism. The detection method must be effective in finding low levels of the organism at a stage prior to widespread dispersal. The eradication campaign in the United Kingdom against plum pox virus failed because in the early 1970s such methods were not available. Sap inoculations to herbaceous indicators such as *Chenopodium foetidum* and *Nicotiana clevelandii* were insensitive for much of the year. Graft inoculations to seedling peach trees can take nine months to develop characteristic symptoms. Both methods thus permitted ample time for extensive spread via aphid vectors and eradication measures were always lagging behind the infection front. By the time the ELISA technique became available, the disease had already become well established in fruiting orchards and native vegetation so that total eradication was no longer a cost effective option.

A similar fate almost befell the campaign against potato spindle tuber viroid (PSTV). Prior to the development of polyacrylamide gel electrophoresis for the detection of viroids, the main test for PSTV was a tomato cross-protection test to detect a mild strain of the viroid. Test conditions were critical and temperatures in the range of 30-35°C plus high light intensities were needed as was scrupulous attention to hygiene for such a highly contagious pathogen. The campaign was successful due to early detection (it was restricted to just a few breeders' lines), absence of an effective vector (it is mechanically transmitted) and the tight control permitted by the documentary procedures of compulsory seed potato certification (Harris *et al.*, 1979).

In the United Kingdom, the campaign to contain beet rhizomania disease (beet necrotic yellow vein virus) owes its success to the development of sensitive and definitive serological reagents which can be combined with quick and relatively cheap diagnostic systems such as ELISA and immunospecific electron microscopy (ISEM). It is a soil-borne organism so natural spread is relatively slow. Statutory controls combined with codes of conduct for farmers and processors restrict the import and movement of soil and beets. Visual inspection of beet fields undertaken annually by either ground or aerial surveying facilitate early detection. Although symptoms can be confused with other disorders, ELISA and ISEM testing provides sensitive and specific diagnosis. In the absence of indicator beet plants, soil testing using the Gross-Gerau bait plant technique (Buttner and Burckey, 1990) permits out of season checking of field soils where infection may be suspected (Ebbels, 1994).

The United Kingdom campaign against potato brown rot (*Pseudomonas solanacearum*) also relies heavily on sensitive and specific diagnosis. First found in two potato crops on a farm in Oxfordshire in November 1992, the severe rotting symptoms could easily have been confused with a range of other bacterial or fungal tuber rotting organisms. Confirmation of *P. solanacearum* was obtained by isolation, purification, identification by fatty acid profiling, and by host tests. However, these techniques were not well suited to the essential intensive surveys of potato crops and harvested potatoes that followed. A range of serological methods, including ELISA and immunofluorescence, DNA amplification by PCR and use of sensitive semi-selective media were investigated and a combination of the better methods was employed. Later surveys done using these specific and sensitive methods have failed to detect further evidence of the bacterium in potatoes. Testing of soil samples from the affected fields suggests that under conditions in the United Kingdom the organism does not survive more than a few months in the absence of certain *Solanum* host plants. Field controls are thus aimed at eliminating potato groundkeepers and other *Solanum* hosts from affected fields. Controls on the possible movement of soil are also applied as a precaution.

The source of *P. solanacearum* has been traced to infected *Solanum dulcamara* weeds growing on the banks of an adjacent river from which water was used to irrigate the two affected potato crops. *S. dulcamara* is a symptomless carrier of infection and only roots actually immersed in the river water have been found to be infected. Infection in these roots can be confirmed by PCR, ELISA and isolation. The exact source of these infections is not known at present, although waste from imported ware potatoes entering the river via sewage treatment works is being investigated. Techniques developed for indexing plants may not be as useful for detecting the organism in complex substrates such as soil, river water and sewage and alternative approaches are required.

## C.  Indexing for the Maintenance of Pest Free Areas

The guarantee provided by crops being grown in "pest free areas" (PFAs) is of increasing importance in the provision of phytosanitary/plant health assurance to assist international trade. Although the concept has been noted in regulations for many years, international interest in greater specification for the concept has derived from its inclusion in the recent GATT agreement on Sanitary and Phytosanitary measures as a measure which should be recognised by contracting parties. (Anon., 1994b).

Under the auspices of the International Plant Protection Convention of 1951, the FAO Committee of Experts on Phytosanitary Measures is developing an International Standard for Pest Free Areas (Hedley, 1994). This standard will lay down basic procedures for both general surveillance for pests and diseases and specific surveys for particular organisms. In addition to visual inspection, good record keeping and documentation, standard operating procedures and recommended reagents for detection and diagnosis will be critical to the process. The reliability, repeatability, sensitivity and specificity of these reagents and procedures will need to be tested and documented in the various circumstances in which they are to be used.

The European Union has the outline for such a PFA in its regulations relating to its Protected Zones, for example that covering fireblight (Anon., 1992a). Regular surveys must be made within the zone together with the monitoring of host plants moved into the zone. Plants showing signs or symptoms of fireblight must be thoroughly investigated using appropriate laboratory methods such as isolation and identification by serological and nutritional methods. Host plants being moved into the zone must be certified as either originating from a similar protected zone or must originate in an officially designated "buffer zone" (Anon., 1992b). These "buffer zones" are in effect mini-PFAs within an otherwise infested area. The controls are stringent and include an indexing programme. They must cover an area of at least 50 km². Fireblight host plants must be controlled to minimise the risk of the disease spreading. Host plant propagation beds and all host plants within 250 m must be inspected at least twice during the growing season and further inspections carried out in the surrounding 1 km. All incidents or suspected incidents of fireblight must be investigated, with random samples being tested for latent infection or epiphytic presence if appropriate. Any confirmed occurrence of E. *amylovora* within the buffer zone removes the status of the zone for at least one growing season after the infection was detected.

## D. Indexing in Healthy Stock Schemes

Schemes for the production of healthy planting material, especially those covering vegetatively propagated species, require the examination of entered material for evidence of diseases. Stock which meets the stipulated standards for health and other attributes is given a certificate of authentication. Stocks may be examined visually, tested by various means, either in the field or in the laboratory, or both. Higher quality schemes usually require a pedigree of descent from tested starting material or "nuclear stock" (Anon., 1991a). In these cases it is vital that the indexing of such nuclear stock is done efficiently because the health standard of progeny lower down the multiplication chain cannot be better than that of the nuclear stock.

*1. Development of indexing in certification schemes*
Early certification schemes accepted a wide range of material, and health checks were made by visual inspections for disease symptoms during the growing season. Standards varied from tolerating no disease symptoms at all to accepting a few percent of plants showing symptoms. Inspections also varied in frequency and in the proportion of the crop covered. In some cases all plants might be inspected, whereas in lower quality schemes only a limited sample might be examined.

As more sensitive tests were developed, these were quickly incorporated into certification systems. One of the first tests to support visual inspections was the serological microprecipitin or slide agglutination test (Hardie, 1970), which could be done in the field in a few minutes to detect viruses with elongated particles present in high concentration . However, most tests are unsuitable for field application and so plants were normally sent to the laboratory for further detailed examination. This is still current practice in most certification inspections, but normally only relatively few such checks on suspicious plants are necessary. In most cases well trained inspectors will readily be able to allot the symptoms seen to various predetermined categories and in some cases may even be able to identify the specific virus or other pathogen present, if symptoms are sufficiently distinctive.

Back-up tests in support of certification systems do not normally need to be very sophisticated because such schemes are not concerned with rare pathogens. For the common diseases these schemes are designed to combat, examination under the electron microscope or in a routine ELISA usually will be adequate. However, in the future it is anticipated that PCR will play an increasingly important role here. Smooth and rapid throughput of samples at minimum cost is essential and because the ELISA can be largely automated, it is particularly suitable for this application (Torrance and Jones, 1981).

## 2. Indexing for Nuclear Stock Production

The concept of indexed nuclear stock as starting material for healthy stock schemes developed in the late 1940s when clones of virus-tested raspberries were introduced into the Scottish raspberry certification scheme (Harris and Cadman, 1949; Ebbels, 1979). This was followed in the potato sector of the United Kingdom by issue of fully indexed clones of several important potato cultivars in 1950 (Hirst et al., 1970). Tree fruit clones free from all viruses and virus-like organisms then known were issued from 1953 onward by East Malling Research Station, United Kingdom (Posnette, 1962). Testing of these first virus-free clones was based on indicator test plants and on electron microscopy. For tree fruits, woody indicators often had to be used and for some diseases these had to be retained until several fruit crops had been obtained, which was a lengthy and expensive process taking up to four or five years to complete. The development of ELISA for plant viruses by Clark and Adams (1977) was a major break-through in plant virus indexing and revolutionised testing for all viruses for which an adequate antiserum could be prepared. However, test plants still had to be employed for those which were not good antigens, for diseases of unknown aetiology, and to provide a chance of detecting pathogens previously unknown. This is still the case today, especially in the tree fruit sector, although the number of viruses for which there is as yet no good laboratory test has been reduced. More recently, the development of nucleic acid tests has greatly facilitated the detection and identification of pathogens which are difficult or not suitable for serological detection.

In meeting the concept of pathogen-free nuclear stock, specialist diagnosticians need to bear in mind that a wide range of tests will usually be necessary. Tests may need to be done at certain very specific times of the season or may need to be repeated over a lengthy period (perhaps several years) in order that negative results may be viewed with confidence. For example, tests for apple mosaic ilarvirus normally must be done between leafing out in the spring and the decline in virus concentration in summer (Torrance and Dolby, 1984), although the testing season can be expanded by forcing buds to break in early spring (Torrance, 1981). For many narcissus viruses, tests must also be done during a limited period of the growing season. To guard against the possibility that viruses are present in low concentration or other factors which might give false negative results, they should be repeated over several seasons. In many cases fungal or nematode pathogens may be extremely difficult to detect reliably. For example, tests for Phytophthora cactorum in strawberry may often give false negatives and the strawberry leaf nematode, Aphelenchoides fragariae, may pass through micropropagation without inciting symptoms (Flegg, 1983). Nuclear stock must be exemplary in its standard of health and the lengthy efforts needed to free material from systemic infections should not be negated by allowing the resulting stock to acquire other pathogens, even if these are relatively easily eliminated. For example, nuclear stock of black currant should be free from mildew and midge as well as reversion disease.

### 3. *Routine indexing for certification schemes*
High grade certification schemes often incorporate a requirement for the testing of random samples of stock. With potatoes, for example, virus tests on samples of leaves are often required at the highest grades, while at lower grades post-harvest virus tests may be required on tuber samples. Such tests are at present not fully reliable and take several weeks because the tubers need to have their dormancy broken and be sprouted. Also, costs do not usually permit the testing of samples consisting of large numbers of tubers, which in turn prevents their use in verifying very small percentages of infection. However, they are of value for identifying stocks of very poor virus health in areas of high infection risk. New, more rapid and reliable tests for detection of virus in potato plants at the end of the season, shortly before harvest, are being developed and soon may be ready for routine use (Barker *et al.*, 1992). While this is to be welcomed, the introduction of novel indexing techniques to certification systems must be done with care. It is possible that increased sensitivity of detection (or the ability to detect a hitherto undetectable organism) may reveal the presence of undesirable organisms in existing material where previously none were thought to exist. Unless arrangements are made to permit the continued marketing of such material (bearing in mind that it may still be the best available) until new, fully-indexed stock has been propagated, it may lead to serious imbalance of supply and demand and could cause unwarranted losses to propagators. There is also the possibility that infection with a mild, symptomless virus may give useful control of growth which, in its absence, may be excessive (as was found with some virus-free clones of rhubarb, *Rheum rhaponticum*).

## E. Indexing in Advisory Work

Good diagnosis is a prerequisite for good advice. Reliability, speed and cost of diagnoses are all important in advisory work which aims to provide an effective service at an economic cost, sometimes in competition with rival services.

As well as diagnosing the causal agents of symptoms on which advice is sought, it is often necessary to go beyond simple identification. For example, it may be vital to know not only the identity of the pathogen or pest, but whether it is resistant to certain control chemicals or agents, or whether it belongs to a certain strain of the pathogen. For these purposes it is therefore often necessary for diagnosticians to become involved in strain typing or resistance testing in order that effective advice may be given. Specialist diagnosticians also have to be aware of other disciplines to which the problem in hand may relate and that symptoms may be due to the presence of a combination of several pathogens and other factors, any of which may mutually interact.

Although important in all diagnosis, it is especially important in advisory work to keep track of the progress of each sample received in the laboratory, for this to be documented (either electronically or otherwise), and for information on the progress of the diagnostic or indexing procedure to be available at short notice to answer queries from the advisor or the customer.


## IX. INTERNATIONAL CO-OPERATION FOR INDEXING

Success in preventing international spread of foreign pathogens and in maintaining quality in international trade of plant material is dependent on the quality of indexing. Whereas these topics have been discussed in various sub-sections within section VIII, international collaboration and the work of international organisations have led to indexing schemes that have been tested or will be tested in a number of geographic areas. Such collaboration may be between laboratories with similar interests. For example, international co-operation between workers in North America and Europe has led to fairly rapid and extensive testing of new serological methods for detecting potato ring rot (De Boer *et al.*, 1992a; De Boer *et al.*, 1992b; De Boer *et al.*, 1994).

Whereas work of this type depends on personal contacts, there are several organisations which have wide and major influence. For seed indexing, the International Seed Testing Association (ISTA) is a particularly important and active organisation. Much of this work is published in *Seed Science and Technology*. In other areas of the world, regional plant protection organisations have great potential value, and some, e.g. the European and Mediterranean Plant Protection Organisation (EPPO) and North American Plant Protection Organisation (NAPPO), are particularly active. Many of the indexing methods agreed to by EPPO member states and the certification schemes that contain them are published in the *EPPO Bulletin*. EPPO has recently initiated a co-operative programme for developing and publishing agreed indexing protocols for all quarantine organisms recognized by the European Union. Table IV lists some of the recent publications which deal with pest and disease indexing for certifying specific crops.


## X. DOCUMENTARY PROCEDURES AND VOUCHER PROCEDURES

Reproducibility is perhaps the most critical aspect of good indexing practice. For this, there must not be just strict attention to detail but, as already mentioned, excellent record keeping and documentation of procedures.

Even for routine indexing and diagnosis, standard operating procedures with accurately described reagents should be established. This is necessary not just in cases where litigation may be involved but even where certification or advisory samples are processed. Decisions taken as a result of such diagnoses can then be cross-checked. This is important even where the results are satisfactory but is especially so should subsequent events be at variance with the expected result. Obviously the production of standard operating procedures by international organisations as outlined above have great value in this respect.

The conservation of specimens as received and, where appropriate, isolated cultures should be a routine part of any standard procedure. These "voucher specimens" are

*Table IV.* Examples of international certification schemes

| Crop | Certification objective | Reference |
| --- | --- | --- |
| Rubus | Pathogens | Anon., 1994c |
| Ribes | Pathogens | Anon., 1994d |
| Strawberry | Pathogens, nematodes | Anon., 1994e |
| Ornamental plants (vegetatively propagated) | Pathogens | Anon., 1993e |
| Fruit trees and rootstocks (Part I) | Viruses | Anon., 1991a |
| Fruit trees and rootstocks (Part II-IV) | Viruses | Anon., 1992c |
| Chrysanthemums | Pathogens | Anon., 1993b |
| Narcissus | Pathogens | Anon., 1993c |
| Lilies | Viruses, nematodes | Anon., 1993d |
| Pelargoniums | Pathogens | Anon., 1992d |
| Carnations | Pathogens | Anon., 1991b |
| Nursery requirements | | Anon., 1993f |

essential if litigation is a possibility, but may also be a legal requirement, as in the case of the European Community Potato Ring Rot Directive (Anon., 1993a). The method of maintenance must be appropriate to the organism and host material concerned and should take account of the period over which the voucher specimen is likely to be needed, with a reasonable margin of safety included. Refrigeration or deep freezing may be appropriate for short to medium term maintenance but the viability of many organisms is seriously impaired by simple freezing. The use of other techniques for long term preservation, such as freeze drying or the maintenance of cultures under oil with regular sub-culturing, may be necessary and further information may be obtained from Kirsop & Doyle (1991) who discuss preservation methods used by international collections of microorganisms. The viability of conserved specimens should be checked at appropriate and regular intervals so that remedial action such as sub-culturing or host tests may be done before significant loss of viability or virulence occurs.

The role of international culture collections as sources of well-documented reference strains and for deposition of new and noteworthy strains is of vital importance.

## REFERENCES

Adams, S.S., Stevenson, W.R., Delhotal, P. and Fayet, J. (1990). An expert system for diagnosis of post-harvest potato diseases. *EPPO Bulletin* **20**, 341-347.

Alarcon, B., Gorris, M.T., Lopez, M.M. and Cambra, M. (1994). Comparison of immunological methods for detection of *Erwinia carotovora* subsp. *atroseptica* using monoclonal and polyclonal antibodies. *In* 'Plant Pathogenic Bacteria' (LeMattre, M., Freigoun, S., Rupolph, K. and Swings, J.G., eds). pp 423. INRA Editions, Versailles.

Anonymous (1977). Council Directive 77/93/EEC of 21 December 1976 on protective measures against the introduction into the Member States of organisms harmful to plants or plant products. *Official Journal of the European Communities* No. L **26**, 20-54.

Anonymous (1980). Commission Decision of 21 August 1980 authorising the Member States to provide for derogations from certain provisions of Council Directive 77/93/EEC, in respect of potato breeding material. *Official Journal of the European Communities* No. **L 248**, 25-26.

Anonymous (1991a). Certification scheme: Schema de certification. Virus-free or virus-tested fruit trees and rootstocks. Part I. Basic scheme and its elaboration. *EPPO Bulletin* **21**, 267-277.

Anonymous (1991b). Certification scheme: Schema de certification. Pathogen-tested material of carnation. *EPPO Bulletin* **21**, 269-290.

Anonymous (1992a). Commission Directive 92/76/EEC of 6 October 1992 recognising protected zones exposed to particular plant health risks in the Community. *Official Journal of the European Communities* No. **L 305**, 12-15.

Anonymous (1992b). Commission Directive 92/103/EEC of 1 December 1992 amending Annexes I to IV to Council Directive 77/93/EEC on protective measures against the introduction into the Community of organisms harmful to plants or plant products and against their spread within the Community. *Official Journal of the European Communities* No. **L 363**, 1-65.

Anonymous (1992c). Certification scheme: Schema de certification. Virus-free or virus-tested fruit trees and rootstocks. Part II. Tables of viruses and vectors. Part III. Testing methods for viruses of fruit trees present in the EPPO Region. Part IV. Technical appendices and table of contents. *EPPO Bulletin* **22**, 255-283.

Anonymous (1992d). Certification scheme: Schema de certification. Pathogen-tested material of pelargonium. *EPPO Bulletin* **22**, 285-296.

Anonymous (1992e). 'International Plant Protection Convention.' 17 pp. FAO, Rome.

Anonymous (1993a). Council Directive 93/85/EEC of 4 October 1993 on the control of potato ring rot. *Official Journal of the European Communities* No. **L 259**, 1-25.

Anonymous (1993b). Certification scheme: Schema de certification. Pathogen-tested material of chrysanthemum. *EPPO Bulletin* **23**, 239-247.

Anonymous (1993c). Certification scheme: Schema de certification. Pathogen-tested material of narcissus. *EPPO Bulletin* **23**, 225-237.

Anonymous (1993d). Certification scheme: Schema de certification. Pathogen-tested material of lily. *EPPO Bulletin* **23**, 215-224.

Anonymous (1993e). Scheme for the production of classified vegetatively propagated ornamental plants to satisfy health standards. *EPPO Bulletin* **23**, 735-736.

Anonymous (1993f). Nursery requirements - recommended requirements for establishments participating in certification of fruit or ornamental crops. *EPPO Bulletin* **23**, 249-252.

Anonymous (1994a). Plant Quarantine Principles as Related to International Trade. In 'Publication No 1. International Standards for Phytosanitary Measures.' 12 pp. Secretariat of the International Plant Protection Convention, FAO, Rome.

Anonymous (1994b). Agreement on the Application of Phytosanitary Measures; 15 April 1994. In 'General Agreement on Tariffs and Trade.' 15 pp. Agriculture and Commodities Division, World Trade Organisation, Geneva.

Anonymous (1994c). Certification scheme. Pathogen-tested material of *Rubus*. *EPPO Bulletin* **24**, 865-873.

Anonymous (1994d). Certification scheme. Pathogen-tested material of *Ribes*. *EPPO*

*Bulletin* **24**, 857-864.

Anonymous (1994e). Certification scheme. Pathogen-tested strawberry. *EPPO Bulletin* **24**, 875-889.

Bain, R.A., Perombelon, M.C.M., Tsror, L. and Nachmias, A. (1990). Blackleg development and tuber yield in relation to numbers of *Erwinia carotovora* subsp. *atroseptica* on seed potatoes. *Plant Pathology* **39**, 125-133.

Ball, S. and Reeves, J. (1992). Application of rapid techniques to seed health testing. In 'Techniques for the Rapid Detection of Plant Pathogens'. (Duncan, J.M. and Torrance, L., eds) pp 193-207. Blackwell Scientific Publications, Oxford.

Barker, I., Brewer, G. and Hill, S. (1992). Early detection of potato viruses in home-saved seed. *Aspects of Applied Biology* **33**, 71-75.

Blancard, D., Bonnet, A and Coleno, A. (1985). ToM, un systeme expert en maladies des tomatoes. *Revue Horticole* **261**, 7-14.

Buttner, G. and Burckey, K. (1990). Experiments and considerations for the detection of BNYVV in soil samples using bait plants. *Zeitschrift fur Pflanzenkrankheiten und Pflanzenschutz* **97**, 54-56.

Cassells, A.C. (1992). Screening for pathogens and contaminating micro-organisms in micropropagation. In 'Techniques for the Rapid Detection of Plant Pathogens'. (Duncan, J.M. and Torrance, L., eds) pp 179-192. Blackwell Scientific Publications, Oxford.

Clark, M.F. (1992). Immunodiagnostic techniques for plant mycoplasma-like organisms. In 'Techniques for the Rapid Detection of Plant Pathogens'. (Duncan, J.M. and Torrance, L., eds) pp 34-46. Blackwell Scientific Publications, Oxford.

Clark, M. and Adams, A. N. (1977). Characteristics of the microplate method of enzyme-linked immunosorbent assay for the detection of plant viruses. *Journal of General Virology* **34**, 475-483.

Coddington, A. and Gould, D.S. (1992). Use of RFLPs to identify races of fungal pathogens. In 'Techniques for the Rapid Detection of Plant Pathogens'. (Duncan, J.M. and Torrance, L., eds) pp 162-178. Blackwell Scientific Publications, Oxford.

De Boer, S.H., Janse, J.D., Stead, D.E., Van Vaerenbergh, J. and McKenzie, A.R. (1992a). Detection of *Clavibacter michiganensis* subsp. *sepedonicus* in potato stems and tubers grown from seed pieces with various levels of inoculum. *Potato Research* **35**, 207-216.

De Boer, S.H., Van Vaerenbergh, J., Stead, D.E., Janse, J.D. and McKenzie, A.R. (1992b). A comparative study among five laboratories on detection of *Clavibacter michiganensis* subsp. *sepedonicus* in potato stems and tubers. *Potato Research* **35**, 217-226.

De Boer, S.H., Stead, D. E., Alivizatos, A.S., Janse, J.D., Van Vaerenbergh, J., De Haan, T.l., and Mawhinney, J. (1994). Evaluation of serological tests for detection of *Clavibacter michiganensis* subsp. *sepedonicus* in composite potato stem and tuber samples. *Plant Disease* **78**, 725-729.

DeHaan, T.L. (1994). Seed potato certification and diagnostic testing. *Canadian Journal of Plant Pathology* **16**, 156-157.

Dewey, F.M. (1992). Detection of plant-invading fungi by monoclonal antibodies. In 'Techniques for the Rapid Detection of Plant Pathogens'. (Duncan, J.M. and Torrance, L., eds) pp 47-62. Blackwell Scientific Publications, Oxford.

Dhingra, O.D., and Sinclair, J.B. (1985). 'Basic Plant Pathology Methods'. Chemical

Rubber Company Press, Boca Raton, Florida.

Duncan, J.M. and Torrance, L., (eds) (1992). 'Techniques for the Rapid Detection of Plant Pathogens'. Blackwell Scientific Publications, Oxford.

Ebbels, D.L. (1979). A historical review of certification schemes for vegetatively-propagated crops in England and Wales. *ADAS Quarterly Review* **32**, 21-58.

Ebbels, D.L. (1994). The rhizomania situation in autumn 1993. *British Sugar Beet Review* **62**, 10-11.

Flegg, J.J.M. (1983). Nematodes. *East Malling Research Station Report for 1982*, 110-111.

Fortuner, R. (1988). 'Nematode Identification and Expert System Technology'. Plenum Press, New York and London.

Hampton, R., Ball, E. and De Boer, S. (Eds) (1990). 'Serological Methods for Detection and Identification of Viral and Bacterial Plant Pathogens : a Laboratory Manual'. American Phytopathological Society Press, St Paul, Minnesota.

Hansen, M.A. and Wick, R.L. (1993). Plant disease diagnosis: present status and future prospects. *Advances in Plant Pathology* **10**, 65-125.

Hardie, J.L. (1970). 'Potato Grower's Guide to Clonal Selection'. Department of Agriculture for Scotland, Edinburgh.

Harris, R.V. and Cadman, C.H. (1949). Can the health of raspberry stocks be improved? *Scottish Agriculture* **28**, 194-197.

Harris, P.S., Miller-Jones. D.N. and Howell, P.J. (1979). Control of spindle tuber viroid: the special problems of a disease in plant breeders material. *In* 'Plant Health. The scientific basis for administrative control of plant diseases and pests.' (D.L. Ebbels and J.E. King, eds), pp 231-237. Blackwell Scientific Publications, Oxford.

Harris, T.S., Sandall, L.J. and Powers, T.O. (1990). Identification of single *Meloidogyne* juveniles by polymerase chain reaction amplification of mitochondrial DNA. *Journal of Nematology* **22**, 518-524.

Hedly, J. (1994). Trade and plant quarantine. *In* 'Brighton Crop Protection Conference, Pests and Diseases - 1994'. pp 153-158. British Crop Protection Council, Farnham.

Hill, S.A. (1984). 'Methods in Plant Virology'. Blackwell Scientific Publications, Oxford.

Hirst, J.M., Hide, G.A., Griffith, R.L. and Stedman, O.J. (1970). Improving the health of seed potatoes. *Journal of the Royal Agricultural Society of England* **131**, 87-106.

Hussey, R.S. (1982). Molecular approaches to taxonomy of Heteroderoidea. *In* 'Nematology in the Southern United States' (R.D. Riggs, ed.) pp 50-53. Southern Cooperative Series Bulletin 276, Fayetteville, Arkansas.

Jeffries C.J., Chard J.M. and Brattey C. (1993). Coping with plant health risks posed by gene bank collections of potato. In 'Plant health and the European single market.' (D.L. Ebbels, (ed.), *BCPC Monograph* No. **54**. pp. 145-156. British Crop Protection Council, Farnham.

Jones, A.T. (1992). Application of double-stranded RNA analysis of plants to detect viruses, virus-like agents, virus satellites and subgenomic viral RNAs. In 'Techniques for the Rapid Detection of Plant Pathogens'. (Duncan, J.M. and Torrance, L., eds) pp 115-128. Blackwell Scientific Publications, Oxford.

Jones, D.A.C., Hyman, L.J., Tumeseit, M., Smith, P. and Perombelon, M.C.M. (1994).

Blackleg potential of potato seed : determination of tuber contamination by *Erwinia carotovora* subsp. *atroseptica* by immunofluorescence colony staining and stock and tuber sampling. *Annals of Applied Biology* **124**, 557-568.

Kirsop, B.E. and Doyle, A. (1991). 'Maintenance of Micro-organisms and Cultured Cells: A Manual of Laboratory Methods'. 2nd edition. Academic Press, London.

Klement, Z., Rudolph, K and Sands, D.C. (1990). 'Methods in Phytobacteriology'. Akademiai Kiado, Budapest.

Lee, I.M., Hammond, R.W., Davis, R.E. and Gundergen, D.E. (1993). Universal amplification and analysis of pathogen 16S rDNA for classification and identification of mycoplasmalike organisms. *Phytopathology* **83**, 834-842.

Lelliott, R.A. and Sellar, P. W. (1976) The detection of latent ring rot (*Corynebacterium sepedonicum* (Spieck et Koth.) (Skapt. et Burkh.)) in potato stocks. *EPPO Bulletin* **6**, 101-106.

Lelliott, R.A. and Stead, D.E. (1987). 'Methods for the Diagnosis of Bacterial Diseases of Plants'. Blackwell Scientific Publications, Oxford.

Mai, W.F. and Lyon, H.H. (1975). 'Pictorial Key to Genera of Plant-Parasitic Nematodes-, 4th edition. Comstock Publishing Associates, Cornell University Press, Ithaca and London.

Posnette, A.F. (1962). The Mother Tree Scheme. *East Malling Research Station Report for 1961*, 125-127.

Salazar, L.F. and Querci, M. (1992). Detection of viroids and viruses by nucleic acid probes. In 'Techniques for the Rapid Detection of Plant Pathogens' (Duncan, J.M. and Torrance, L., eds) pp 129-144. Blackwell Scientific Publications, Oxford.

Shepherd, A.M. and Clark, S.A. (1986). Preparation of nematodes for electron microscopy. *In* 'Laboratory Methods for work with Plant and Soil Nematodes', 6th edition, (J.F. Southey, ed.) pp. 121-131. Ministry of Agriculture, Fisheries and Food Reference Book 402, Her Majesty's Stationery Office, London.

Southey, J.F. (ed) (1986). 'Laboratory Methods for Work with Plant and Soil Nematodes', 6th edition. Ministry of Agriculture, Fisheries and Food Reference Book 402. Her Majesty's Stationery Office, London.

Stead, D.E. (1992). Techniques for detecting and identifying plant pathogenic bacteria. In 'Techniques for the Rapid Detection of Plant Pathogens'. (Duncan, J.M. and Torrance, L., eds) pp 76-114. Blackwell Scientific Publications, Oxford.

Stead, D.E., Sellwood, J.E., Wilson, J. and Viney, I. (1992). Evaluation of a commercial microbial identification system based on fatty acid profiles for rapid, accurate identification of plant pathogenic bacteria. *Journal of Applied Bacteriology* **72**, 315-321.

Torrance, L. (1981). Use of forced buds to extend the period of serological testing in surveys for fruit trees viruses. *Plant Pathology* **30**, 213-216.

Torrance, L. (1992). Serological methods to detect plant viruses : production and use of monoclonal antibodies. In 'Techniques for the Rapid Detection of Plant Pathogens'. (Duncan, J.M. ad Torrance, L., eds) pp 7-33. Blackwell Scientific Publications.

Torrance, L and Dolby, C.A. (1984). Sampling conditions for reliable routine detection by enzyme-linked immunosorbent assay of three ilarviruses in fruit trees. *Annals of Applied Biology* **104**, 267-276.

Torrance, L. and Jones, R.A.C. (1981). Recent developments in serological methods suited for use in routine testing for plant viruses. *Plant Pathology* **30**, 1-24.

Vauterin, L., Hoste, B., Kersters, K. and Swings, J. (1995). Reclassification of *Xanthomonas*. *International Journal of Systematic Bacteriology* **45**, 472-489.

Vivian, A. (1992). Identification of plant pathogenic bacteria using nucleic acid technology. In 'Techniques for the Rapid Detection of Plant Pathogens'. (Duncan, J.M. and Torrance, L., eds) pp 145-161. Blackwell Scientific Publications, Oxford.

Wale, S.J., Robertson, K., Fisher, G. (1986). Studies on the relationship of contamination of seed potato tubers with *Erwinia* spp. to blackleg incidence and large scale hot water dipping to reduce contamination. *Aspects of Applied Biology* **13**, 285-291.

Whitcomb, R.F. and Tully, J.G. (1989). 'The Mycoplasmas' Vol. 5. Academic Press, New York.

Zuckerman, B.M., Mai, W.F. and Krusberg, L.R. (1990). 'Plant Nematology Laboratory Manual'. Agricultural Experiment Station, University of Massachusetts, Amherst, Massachusetts.

# 2

# DETECTING LATENT BACTERIAL INFECTIONS

S.H. De Boer[1], D.A. Cuppels[2] and R.D. Gitaitis[3]

[1]*Pacific Agriculture Research Centre, Vancouver, British Columbia, Canada*
[2]*Pest Management Research Centre, London, Ontario, Canada*
[3]*Coastal Plain Experiment Station, University of Georgia, Tifton,
Georgia, United States*

## I. INTRODUCTION

Bacterial infections of plants in the absence of overt symptoms of disease are
considered latent (Hayward 1974). Latent infections may persist for long periods of
time, or until conditions become favorable for bacterial multiplication and disease
development. Although there are instances where relatively large numbers of
bacteria have been found in asymptomatic tissue, the bacterial populations associated
with latent infections are usually small. Bacterial plant pathogens have been found
in ostensibly healthy tissue from virtually every plant part, including seeds and
vegetative propagules. Because they often go undetected, latent infections are an
important inoculum source for several serious bacterial diseases of greenhouse and
agricultural crops, such as bacterial canker of tomato and bacterial ring rot of potato.
Hence development of rapid and accurate indexing procedures to detect these

Advances in Botanical Research Vol. 23
Incorporating Advances in Plant Pathology
ISBN 0-12-005923-1

infections is an important component of some disease control strategies. Once identified, consignments of infected plants, plant parts, or seeds may be treated or destroyed, thereby eliminating or substantially reducing the threat of a disease outbreak. Although classical cultural techniques have been used for indexing, serological procedures tend to be better suited for assaying large numbers of samples, and DNA-based procedures provide a remarkable degree of specificity.

## II. PROCEDURES BASED ON ISOLATION

Often, phytopathogenic bacteria will represent the dominant and possibly the only species of the total microflora of a latent infection. Since symptom development and subsequent cellular breakdown have not yet occurred, there is little competition from opportunistic saprophytes. There should be fewer saprophytic bacteria in latent infected tissues than often is encountered in older lesions. However, eventually one will encounter samples that contain populations of nontarget organisms that will interfere with efficient recovery of the target organism. The source of a sample can also play a major role in the effectiveness of an isolation medium for isolation of bacteria from plant samples. For example, a semi-selective medium (T5) developed for *Pseudomonas viridiflava*, causal agent of bacterial streak and bulb rot of onion, was most effective in preliminary tests for isolation of the bacterium in areas containing Dothan loamy sand that had no history of onion production. When the same medium was used to detect natural populations in areas with a long history of onion production with different soil types, the medium failed to reduce the background microflora sufficiently (Gitaitis *et al.*, 1992). Similar attempts to detect *Clavibacter michiganensis* subsp. *michiganensis* on the semi-selective medium D2anx and SCM resulted in the recovery of strains that appeared morphologically different (Stephens *et al.*, 1988). The original colony type was described as grey and fluidal (Fatmi and Schaad, 1988), however, after screening several samples, a few suspect colony types with atypical morphology (black, butyrous and circular) proved to be *C. m. michiganensis* upon further characterization. Thus, the greater the diversity of samples from wide geographical areas, the greater the likelihood of encountering strains of saprophytic bacteria that grow in large numbers, or strains of target bacteria whose growth and appearance are inhibited or greatly modified by the semi-selective medium.

Isolation usually is a simple matter of diluting the sample, plating it on a nutritional medium and selecting colonies representative of the target. Ideally, the function of a semi-selective medium is to manipulate growth conditions so the target organism will predominate on the medium. The primary means of manipulating growth conditions are using a single carbon source, a nitrogen source, inhibiting agents of nontarget organisms, and adjusting the incubation temperature. Selection of carbon and nitrogen source is based on growth efficiency of the target and the percent inhibition of nontarget organisms. Inhibition agents such as dyes, antibiotics, or other chemical-inhibiting agents are incorporated into the medium based on their ability to inhibit nontarget organisms with minimal negative effect on growth of the target. Another useful trait of some media components is the production of a differential characteristic, such as pitting of crystal violet pectate medium by *Erwinia carotovora* (Cuppels and Kelman, 1974) and CMC-E by most *Xanthomonas campestris* pathovars (Gitaitis *et al.*, 1991); precipitates in tween-containing media by xanthomonads and *Acidovorax avenae* subsp. *citrulli* (Frankle, 1992; McGuire *et al.*, 1986), and in lactalysate medium by *A. avenae* subsp. *avenae* (Gitaitis *et al.*, 1978); or zones of starch hydrolysis produced by *X. c. campestris* (Randhawa and Schaad, 1984). Differential qualities are most useful when selecting 'suspect' colonies from a plate with mixed bacterial populations.

If the goal is to detect latent infections under controlled circumstances, a laboratory strain with an antibiotic resistance marker can be used with media supplemented with a broad spectrum antibiotic. However, if the intended use is in a quality assurance, regulation-certification program, or any system that would encounter populations of the wildtype, such antibiotic-amended media would be of no value.

Perhaps the biggest problems encountered are determining what tissues to sample and the sample size, as well as the logistics of having sufficient labour, materials, and space for processing the samples. When dealing with a small number of samples, tissues can be triturated in an appropriate diluent, such as 0.01 M phosphate-buffered 0.85% saline (pH 6.8-7.4), and aliquants either streaked with a loop or serially diluted and aliquots of each dilution spread with a bent rod on the surface of agar medium. Large numbers of samples often are processed in bulk by shaking tissues in an appropriate diluent prior to dilution and plating. Processing individual samples by either method can result in a large and unmanageable number of plates very quickly and using bulk samples could result in a less sensitive assay.

Under certain circumstances, the type of disease may permit the use of unique methods resulting in an assay that is sensitive, rapid and capable of accommodating a large number of samples. For example, bacterial canker of tomato basically is a wilt disease, with the bacteria colonizing conductive tissues. The use of cross-sections of cut tomato stems pressed on semi-selective agar media was used to detect latent infections (Gitaitis *et al.*, 1991). Exterior portions of the stems were surface disinfected with ethanol and 1 cm sections were excised from stems. The excised stem pieces were gently squeezed as the freshly cut surface was pressed firmly on the surface of modified-CNS agar. Plates were incubated 7-10 days at 30 C. Yellow fluidal colonies, typical of *C. m. michiganensis* were characterized to confirm bacterial identity. This technique was rapid and could accommodate up to 50 samples per plate, thus reducing labour, materials, and space required compared to the spread-plate technique.

Indexing plant material for latent bacterial infections by isolation techniques often tend to be time-consuming, laborious and lacking in adequate sensitivity. Serological procedures, therefore, have been used more widely. DNA-based detection protocols may also become more important in the future because of their potential high level of sensitivity, if simplified techniques for DNA extraction and processing can be developed.

## III. SEROLOGICAL PROCEDURES

## A. Introduction

Serological detection procedures are particularly well suited for bacterial pathogens because these microorganisms possess an assortment of specific antigenic determinants that are readily accessible to antibody probes. The specificity of serological tests, as in any detection or indexing procedure, is of utmost importance. In microbiology, serological specificity is generally described in terms of the degree to which an antibody preparation reacts with all strains of a target bacterium but not with non-target microorganisms. Although antibodies can bind non-specifically to certain molecular moieties, specificity is largely determined by the occurrence and distribution of complementary epitopes. Thus an antiserum may react specifically with only one molecular entity, but have low specificity because the molecule is widespread in the microbial world. Antibodies to the lipopolysaccharide of *Erwinia chrysanthemi*, for example, cross-reacted with some strains of *Pseudomonas fluorescens* having the same antigenic determinants in their lipopolysaccharide (van der Wolf *et al.*, 1992). The enterobacterial common antigen is another example of an antigen that occurs in many different species (Peters *et al.*, 1985). It is also possible that the same epitope, or antibody reaction site, is present on structurally very different antigens (Peters *et al.*, 1985).

Because polyclonal antisera contain a mixture of antibodies directed toward various epitopes of a bacterium, its specificity is usually lower than monoclonal antibody preparations in which all antibody molecules are identical and specific for a single epitope. Polyclonal antisera to *C. michiganensis* subsp. *sepedonicus*, for instance, generally cross-reacts with various related and unrelated bacteria (Miller, 1984; Crowley and De Boer, 1982; De Boer, 1982; Calzolari et al, 1982), whereas monoclonal antibodies could be selected with high specificity for the subspecies (De Boer and Wieczorek, 1984; De Boer *et al.*, 1988). Specificity of polyclonal antisera can be enhanced by dilution to decrease the concentration of antibodies that react with common epitopes (Miller, 1984), or cross-absorbed to remove cross-reacting antibodies (Jones *et al.*, 1993; Roberts, 1980; Vruggink and Maas Geesteranus, 1975).

Serological tests with polyclonal antisera tend to give a more intense response than tests conducted with monoclonal antibodies. Higher values in ELISA and brighter fluorescence in immunofluorescence, for example, are obtained with polyclonal antisera because, in contrast to monoclonal antibodies, they react with multiple epitopes. Nevertheless, monoclonal antibodies have the highest probability of being specific for the bacterium of interest.

"Monospecific" polyclonal antisera can also be generated to specific antigens unique to a target bacterium. Niepold (1992) produced an antiserum specific to *Pseudomonas syringae* pv. *syringae* by targeting a 31 Kd protein identified as being unique to the pathogen by Western blot analysis of bacterial proteins subjected to SDS-polyacrylamide gel electrophoresis. Monospecific polyclonal antisera differs from monoclonal antibodies in that the antiserum still contains a population of different antibodies complementary to different epitopes of the specific antigen, whereas each monoclonal antibody will only react with one of many epitopes on the antigen molecule.

To some extent the usefulness of an antibody preparation for indexing procedures is dependent on the immunoglobulin isotype, but more importantly on the nature of its complementary antigen. One of the primary antigens of Gram negative bacteria is the lipopolysaccharide. The lipopolysaccharide is a structural component of the outer membrane and is readily released into the surrounding medium so that antibodies directed toward it are useful in several different serological indexing procedures. Lipopolysaccharide antigens have been specifically targeted for detection of *Agrobacterium tumefaciens, Erwinia carotovora* subsp. *atroseptica, Xanthomonas campestris* subsp. *citri, X. c. citrumelo, X. c. campestris, X. c. begoniae, X. c. pelargonii,* and *Pseudomonas syringae* (Alvarez *et al.*, 1991; Benedict *et al.*, 1990; Bouzar *et al.*, 1988; De Boer and McNaughton, 1987; Franken *et al.*, 1992; Guillorit and Samson 1993; Murray *et al.*, 1990). Antibodies directed toward a soluble product may work well in an ELISA test but fail in the immunofluorescence test which requires antibodies directed to cell wall components. In contrast, antibodies to a somatic or membrane antigen are ideal for the immunofluorescence test which relies on visualization of intact bacterial cells, but these cannot be used for ELISA unless the cell walls of target bacteria are extracted or solubilized prior to the ELISA test. Whole cells are trapped poorly on microtiter plates in double antibody sandwich configurations of ELISA. The monoclonal antibody specific for the extracellular polysaccharide of *C. m. sepedonicus* cannot be used for immunofluorescence, for example, while it works well for ELISA (De Boer *et al.*, 1988). Another monoclonal antibody that reacts in immunofluorescence with a cell wall antigen of *C. m. sepedonicus* does not work in ELISA unless the antigen is extracted and concentrated, or whole cells are fixed to the solid support (De Boer and Wieczorek, 1984). Antibodies directed to induced metabolites, such as pectate lyase of *E. carotovora*, are not good targets on which to base detection protocols because bacterial cells in asymptomatic infections may be metabolically inactive and fail to secrete the antigen (Ward and De Boer, 1989; Klopmeyer and Kelman, 1988).

Monoclonal antibodies have been produced to diverse bacterial plant pathogens (Table 1). Although for many of the monoclonals, the specific antigens to which the antibodies are directed have not been identified, several involving Gram negative bacteria are directed to the lipopolysaccharide. Of two monoclonals to *C. m. sepedonicus* one was directed toward an uncharacterized somatic antigen and the other to the extracellular polysaccharide. Only a few of the monoclonal antibodies have yet been used for indexing of plant material on a commercial scale. However, since monoclonal antibody-producing hybridoma cells lines are readily preserved and regenerated, specific reagent-grade antibodies can be made available in unlimited quantities. Several of them can undoubtedly be introduced with advantage for indexing of seed and vegetative propagules for bacterial pathogens where more lengthy procedures are currently being used. The use of monoclonal antibodies for testing seed potato lots for the presence of the bacterial ring rot pathogen has been invaluable (chapter 8).

## B. Enzyme-linked Immunoassays

ELISA tests are usually carried out in 96-well microtiter plates which permits testing

*Table I.* Plant pathogenic bacteria and mollicutes to which monoclonal antibodies have been produced.

| Bacterium | Reference |
|---|---|
| *Clavibacter michiganensis* subsp. *sepedonicus* | De Boer and Wieczorek 1984; De Boer *et al.*, 1988 |
| *Erwinia amylovora* | Lin *et al.*, 1987 |
| *Erwinia carotovora* | Klopmeyer and Kelman, 1988; Vernon-Shirley and Burns, 1992; Ward and De Boer, 1989 |
| *Erwinia carotovora* subsp. *atroseptica* | De Boer and Mc Naughton, 1987; Gorris *et al.*, 1994 |
| *Erwinia stewartii* | Lamka *et al.*, 1991 |
| *Pseudomonas andropogonis* | Li *et al.*, 1993 |
| *Pseudomonas cepacia* | Takahashi *et al.*, 1990 |
| *Pseudomonas syringae* pv. *glycinea* | Wingate *et al.*, 1990 |
| pv. *phaseolicola* | Wong, 1990 |
| *Spiroplasma citri* | Lin and Chen, 1985; Jordan *et al.*, 1985 |
| *Xanthomonas campestris* pv. *begoniae* | Benedict *et al.*, 1990 |
| pv. *campestris* | Alvarez *et al.*, 1985 |
| pv. *cerealis* | Bragard and Verhoyen, 1993 |
| pv. *citri* | Alverez *et al.*, 1991 |
| pv. *citrumelo* | Alverez *et al.*, 1991; Permar and Gottwald, 1989 |
| pv. *dieffenbachiae* | Lipp *et al*, 1992 |
| pv. *hordei* | Bragard and Verhoyen, 1993 |
| pv. *oryzae* | Benedict *et al.*, 1989; Jones *et al.*, 1989 |
| pv. *oryzicola* | Benedict *et al.*, 1989 |
| pv. *pelargonii* | Benedict *et al.*, 1990 |
| pv. *phaseoli* | Wong, 1990 |
| pv. *secalis* | Bragard and Verhoyen, 1993 |
| pv. *translucens* | Bragard and Verhoyen, 1993; Duveiller and Bragard, 1992 |
| pv. *undulosa* | Bragard and Verhoyen, 1993; Duveiller and Bragard, 1992 |
| *Xylella fastidiosa* | Wells *et al.*, 1987 |
| Mycoplasma aster yellows | Lin and Chen, 1985a |
| clover phyllody | Garnier *et al.*, 1990 |
| corn stunt spiroplasma | Davis *et al.*, 1987; Jordan *et al.*, 1985; Lin and Chen, 1985b |
| grapevine flavescence doree | Schwartz *et al.*, 1989 |
| maize bushy stunt | Chen and Jiang, 1988 |
| peach eastern X | Jiang *et al.*, 1989 |
| sweet potato witches' broom | Shen and Lin, 1993 |
| tomato big bud | Hsu *et al.*, 1990 |
| tomato stolbur | Garnier *et al.*, 1990 |
| walnut witches' broom | Konai and Chang, 1990 |

of many samples simultaneously and makes the test ideal for routine indexing on a large scale. Various ELISA formats have been devised, but since indexing usually involves detection of low bacterial populations, trapping of bacterial antigens out of a plant extract or homogenate onto the test plate is required to obtain adequate test sensitivity. Trapping is usually accomplished with an antibody coated onto the plate prior to application of the sample. Polyclonal antibody is generally used for coating since specificity is not of paramount importance and tend to work better than monoclonal antibodies (Lamka *et al.*, 1991) perhaps due to their greater affinity and reactivity with multiple antigenic determinants. Subsequent detection of trapped antigen is done with an enzyme conjugate of the same or different antibody preparation in what is often called the double antibody sandwich (DAS) format. If the second antibody is from a different animal source than the trapping antibody, the indirect detection procedure with an anti-antibody enzyme conjugate can be used. The advantage of the indirect DAS-ELISA is that the same enzyme conjugate can be used for antibodies to different pathogens. The advantage of the direct DAS-ELISA, in which the pathogen-specific antibody is conjugated with the enzyme, is that fewer steps are involved in the procedure. Depending on the specific protocol used, indirect DAS-ELISA methods can be completed in 4-6 hr (Alvarez and Lou, 1985; Gitaitis *et al.*, 1991), while at least one direct ELISA protocol could be preformed in 1 hr (Anderson and Nameth 1990). Other protocols call for overnight incubation of samples in the microtiter plates (De Boer *et al.*, 1992a). Details of ELISA protocols useful for bacterial plant pathogens have been published elsewhere (e.g. Duncan and Torrance, 1992; Hampton *et al.*, 1990).

Nitrocellulose and nylon membranes can be used instead of microtiter plates as the solid phase in enzyme immunoassays. Such "dot blot" assays and direct blotting of tissue onto the membrane are useful for small numbers of samples and does not require the initial serological trapping of antigen (Lazarovits *et al.*, 1987; Comstock, 1992). The process of probing samples in dot blot assays with specific antibodies in a direct or indirect procedure is essentially the same as for ELISA carried out on plates. A precipitating rather than a soluble substrate is used in the final development to visualize enzyme activity. Gold-labelled antibodies have also been used in an immunogold silver staining procedure as an alternative to enzyme-linked antibodies (Anderson and Nameth, 1990). Dot blot results can be quantified by use of a scanner and digitization of the image but is not as easily performed as reading absorbance in microtiter plates with commercial ELISA plate readers.

## C. Sensitivity and Specificity of ELISA

ELISA protocols have already been developed and used for indexing various plants and plant parts for the presence of bacterial pathogens. Alvarez and Lou (1985) developed an ELISA procedure with polyclonal antibodies to confirm the presence of *X. c. campestris* in cabbage leaves with atypical symptoms of crucifer black rot. Sensitivity and specificity of the assay, at 94.7 and 97.5%, respectively, exceeded success of direct isolation onto culture media. Similar, sensitivities and specificities were obtained with ELISA for detecting *C. m. sepedonicus*, causal agent of bacterial ring rot, in composite samples of potato stems and tubers (De Boer et al, 1994). A tissue immunoassay technique that was 98.9% effective for detecting X. albilineans in symptomatic sugar cane stalks, was also considered useful for detecting latent infections in asymptomatic stalks, but sensitivity and specificity for asymptomatic infections .were not determined (Comstock, 1992).

Sensitivity of ELISA cannot be accurately expressed in terms of bacterial concentration because the test is based primarily on detection of soluble antigens (Barzic and Trigalet, 1982; De Boer and McCann, 1989). The amount of soluble antigen produced by a given population of bacteria may not be directly correlated with cell density but rather may vary with substrate composition and other growing conditions. In some cases soluble antigen is detected in the complete absence of bacterial cells. For example, soluble antigens of Erwinia salicis that were transported through the xylem vessels of willow trees were detected by ELISA in leaf tissue in which no E. salicis cells could be found by immunofluorescence (De Kam, 1982).

Despite the difficulty in correlating antigen concentration with cell numbers, the minimum bacterial concentration that can be detected is estimated to be in the range of $10^5$ to $10^6$ cells/ml (Table II) although detection as low as $10^4$ cells/ml have been claimed for P. s. phaseolicola in contaminated seed and leaves of bean, and C. m. michiganensis in tomato (Barzic and Trigalet 1982; Stephens et al, 1988). In direct comparisons between ELISA and immunofluorescence, ELISA is usually shown to be about 10X less sensitive than immunofluorescence (Muratore et al., 1986). In contrast, however, Gudmestad et al. (1991) found that ELISA was generally more sensitive than immunofluorescence for detecting C. m. sepedonicus in potato tissue, perhaps because they used a monoclonal antibody to an extracellular polysaccharide which is produced in copious amounts by many strains of the bacterium on appropriate substrates.

Increased sensitivity was described for an ELISA inhibition assay, named 'selected antibody enzyme immunoassay' by the authors, and developed for detection of X. c. citri in citrus plant tissue (Kitagawa et al., 1992). The detection limit was calculated to be 50 colony forming units (cfu) per assay and cross-reactions with non-target bacteria were minimized.

## D. Immunofluorescence

Paton (1964) was the first to use the immunofluorescence technique in phytobacteriology when he used it to detect P. syringae in turnip tissue. Although the procedure was subsequently used by Hockenhull (1979) for in situ detection of E. amylovora in symptomless hawthorn plants, it was not until after Allan and Kelman's (1977) work on E. c. atroseptica, that it became recognized as a useful tool for diagnostic work.

The immunofluorescence procedure involves staining of bacterial cells on microscope slides with antibodies conjugated with a fluorescent dye. Usual protocols involve extraction of a bacterial fraction from tissue of individual or composite plants that is fixed to a well of a multi-well glass microscope slide by flaming or acetone treatment. Acetone fixation is particularly important when monoclonal antibodies are used to prevent heat-denaturation of the one complementary epitope type. Fixed preparations can be stained directly by conjugated antibody or by the indirect

Table II. Detection limits of serological tests for indexing plant samples for the presence of pathogenic bacteria.

| Bacterium | Plant Sample | Detection Limit | Reference |
|---|---|---|---|
| **Immunofluorescence** | | | |
| *Clavibacter michiganensis* | | | |
| subsp. *michiganensis* | tomato stems | $7 \times 10^3$ cells/ml | VanVaerenbergh and Chauveau, 1987 |
| subsp. *sepedonicus* | potato stems | $\sim 10^4$ ifu[a]/g | Gudmestad et al., 1991 |
| | potato stems | $1 \times 10^6$ ifu/g | De Boer and McCann, 1989 |
| *Erwinia chrysanthemi* | composite carnation stems | $8 \times 10^4$ cells/sample | Muratore et al., 1986 |
| *Pseudomonas caryophylli* | composite carnation stems | $3 \times 10^4$ cells/sample | Muratore et al., 1986 |
| **ELISA** | | | |
| *Clavibacter michiganensis* | | | |
| subsp. *michiganensis* | tomato | $1 \times 10^6$ cfu[b]/ml | Stephens et al., 1988 |
| subsp. *sepedonicus* | potato | $\sim 10^5$ cfu/g | Gudmestad et al., 1991 |
| | potato | $\sim 10^5$ cells/g | Corbiere et al., 1987 |
| | potato | $5 \times 10^5$ cells/ml | Underberg and Sander, 1991 |
| | potato | $5 \times 10^6$ cells/ml | Zielke and Kalinina, 1988 |
| *Erwinia chrysanthemi* | composite carnation stems | $8 \times 10^5$ cells/sample | Muratore et al., 1986 |
| *Erwinia stewartii* | corn seed | $6 \times 10^6$ cfu/g | Lamka et al., 1991 |
| *Pseudomonas caryophylli* | composite carnation stems | $3 \times 10^6$ cells/sample | Muratore et al., 1986 |
| *Pseudomonas phaseolicola* | bean seed | $\sim 10^4$ cells/ml | Barzic and Trigalet, 1982 |
| *Xanthomonas campestris* | | | |
| pv. *campestris* | cabbage leaves | $2 \times 10^6$ cells/ml | Alvarez and Lou, 1985 |
| **Dot Immunoassay** | | | |
| *Xanthomonas campestris* | | | |
| pv. *undulosa* | wheat seed | $10^6$ cfu/ml | Duveiller and Bragard, 1992 |
| | wheat seed | $10^5$ cfu/ml | Claflin and Ramundo, 1987 |
| pv. *vesicatoria* | pepper and tomato leaves | $1-2 \times 10^3$ cfu/dot | Lazarovits et al., 1987 |

[a]immunofluorescing units
[b]colony forming units

procedure in which the primary antibody is not conjugated but a secondary conjugated antibody is bound to the primary one. Bacteria can also be trapped on filters rather than fixed on microscope slides to provide a more quantitative test (De Boer, 1984). Bacterial cells can be stained with conjugated antibody prior to or after entrapment. The trapping procedure was used to detect *X. c. citrumelo* and *X. c. citri* on symptomless citrus leaves (Brlansky *et al.*, 1990), *E. amylovora* in apple blossom and fruit tissue (Lin *et al.*, 1987), and the then unnamed coryneform bacterium to diagnose the ratoon stunting disease of sugar cane (Davis and Dean, 1984).

One disadvantage of the immunofluorescence technique for some diagnostic laboratories is the requirement for a relatively costly fluorescence microscope. All major microscope manufacturers have models made specifically for fluorescence microscopy in which high intensity radiation from a high-pressure mercury or xenon lamp is transmitted to the subject in the epi-illumination mode via the objective lens. Primary and secondary filter combinations and a dichroic mirror direct the appropriate wavelengths of radiation to achieve excitation of the dye and permit fluorescence emission to reach the eye-piece. Total magnification of 1000x is usually recommended for observing immunofluorescently-labelled preparations in order to clearly distinguish bacterial cells from other fluorescing debris and examining cell morphology, although lower magnifications have been used successfully.

Immunofluorescence is frequently the most sensitive of the serological tests for detecting bacteria in plant tissue with a sensitivity approaching $10^4$ cells/ml (Table II). Greater sensitivity ($10^2$ cfu/ml) was obtained for *P.s. phaseolicola* on bean seed but required scanning of 500 microscope fields at 500x magnification (van Vuurde *et al.*, 1991). The greater sensitivity has also been obtained with pure cultures and artificial mixtures of bacteria (Slack *et al.*, 1979), but sensitivity is usually limited by presence of plant tissue and physical restrictions of light microscopy. Theoretical sensitivity of immunofluorescence is a function of sample concentration, volume of sample applied to the microscope slide, number of microscope fields observed, and the magnification employed. A single positive cell in 20 microscope fields at 1000x magnification represented $1 \times 10^3$ cells of *X. c. undulosa* per gram of wheat seed in the configuration used by Duveiller *et al.* (1992). In the immunofluorescence procedure for detecting *C. m. sepedonicus* in potato tissue, one cell per microscope field represented $1.3 \times 10^5$ cells/ml (De Boer, unpublished data). Gudmestad *et al.* (1991) could detect *C. m. sepedonicus* populations as low as 10 cfu/g fresh weight of plant tissue, but sensitivity of the test itself was not acceptable at levels below $10^6$ cfu/g. Specificity of antisera determines actual test sensitivity because the presence of cross-reacting bacterial cells necessitates setting a positive/negative threshold level higher than the background reading. For polyclonal antisera, background fluorescing coryneform-like cells at $2.5-5 \times 10^3$ cells/ml was considered normal in tests for bacterial ring rot (Miller, 1984).

Van Vuurde *et al.* (1991) estimated that immunofluorescence was about 100-fold more sensitive than dilution plating for detection of *P. s. phaseolicola* on bean seed. They estimated that test sensitivity for immunofluorescence was 100% compared to 78% for dilution plating. Specificity was estimated at 89 and 100% for immunofluorescence and dilution plating, respectively. In a collaborative study involving six laboratories, De Boer *et al.* (1994) reported that both sensitivity and specificity was 100% in some laboratories for detecting *C. m. sepedonicus* in composite potato stem and tuber samples spiked with diseased tissue. Due to lower efficiency in other laboratories, overall sensitivity was 90.9 and 97.3% for stem and tuber samples, respectively, and specificity was 98.5 and 97.8%, respectively (see also Chapter 8). Attempts to increase sensitivity of immunofluorescence by immunological capture of cells directly on microscope slides have been only partially successful (van Vuurde and van Henten, 1983), probably because trapping antibody sites were saturated with soluble antigen components.

### E. Confirmation of Serological Test Results

When serological tests are used to index ostensibly healthy plants or plant parts for the presence of bacterial pathogens, confirmatory testing of positive samples is often required to verify that the result was not due to cross-reaction with non-target bacteria. In Europe, positive immunofluorescence tests for bacterial ring rot of potato need to be confirmed by bioassay on eggplant (Anon, 1987). Similarly, dilution plating is recommended for confirming the presence of the halo blight bacterium in immunofluorescence-positive samples of bean seed (van Vuurde *et al.*, 1991). When isolation of the pathogen is not an option, two serological tests specific for different antigenic determinants may provide acceptable specificity. This approach was used for reliable detection of asymptomatic ring rot infections by using two different monoclonal antibodies which detect two entirely different antigenic determinants of the causal bacterium (De Boer *et al.*, 1994). Alternatively, methods based on the nucleotide sequence of bacterial DNA such as dot blot hybridization with a labelled probe, amplification of specific DNA fragments by the polymerase chain reaction, and *in situ* hybridization can be used to confirm or reject serological test results.

An immunofluorescence colony-staining procedure was developed to allow serological testing and subsequent isolation of serologically-positive bacteria (van Vuurde, 1987). The procedure involves multiplying bacteria from a plant sample in an agar film, staining of the film by an immunofluorescence procedure, visualization of fluorescing micro-colonies with a fluorescence microscope under low magnification, and finally isolation of bacteria from fluorescing micro-colonies. Isolated cultures can be identified by classical methods or, more rapidly, by PCR (van der Wolf *et al.*, 1994). Sensitivity of recovery, a function of the target bacterium and selectivity of the agar medium employed, was in the range of 64-82% for pectolytic erwinia added to slurries of cattle manure (van Vuurde and Roozen, 1990).

### IV. NUCLEIC ACID-BASED METHODS

### A. Introduction

Over the past decade, considerable progress has been made in the development of nucleic acid probe-based methods for detecting and identifying microorganisms (Sayler and Layton, 1990; Tenover and Unger, 1993; Pickup, 1991; Falkinham, 1994; Shibata, 1992). These methods are a considerable improvement over the more traditional culture-based assays. Not only are they less time-consuming and more sensitive but they also make it possible to detect bacteria that have not been successfully grown in culture. Several commercial probe kits, primarily for the sexually-transmitted human diseases and slow-growing respiratory pathogens, were released in the 1980s (Tenover and Unger, 1993). Although DNA probe-based methods can be expensive and may not have the sensitivity required for certain types of clinical and environmental samples, their use by diagnostic microbiology laboratories, including those concerned with plant pathogens, continues to grow.

A diagnostic nucleic acid probe may be defined as a nucleotide sequence, labelled with a reporter molecule, that is able to identify a target microorganism within a test sample by selectively hybridizing to the complementary sequence present within that microorganism's DNA. Probes, which can be either DNA or RNA, range in size from 15 to several thousand base pairs (bp). Hybridization protocols usually require the probe or target sequence to be immobilized on a solid surface but the reaction can also occur in solution or *in situ*. The probe-target hybrids are visualized using autoradiography, colorimetric assays, or chemiluminescence. Techniques, such as the polymerase chain reaction (PCR), that can be used to amplify either probe or target DNA (Saiki *et al.*, 1985), have increased the sensitivity of nucleic acid probe-based methods to the point where it is now possible to detect one target cell in a background microbial population of $10^9$ cells (Steffan and Atlas, 1991). Coupled with the advent of simple yet sensitive nonradioactive labelling procedures and better approaches to selecting pathogen-specific probes, this technology should revolutionize the manner in which plant tissue samples are monitored for the presence of bacterial plant pathogens.

The remainder of this section will provide a brief description of nucleic acid hybridization and amplification technology, present applications of this technology to bacterial plant pathogens, and discuss future directions for diagnostic molecular phytobacteriology.

## B. Probe Selection

Although there are ssDNA and RNA probes, the majority of probes developed for organisms of environmental interest are dsDNA (Sayler and Layton, 1990). A probe may be several thousand bases long, but the shorter oligonucleotide probes (50 bp or less), which can be chemically-synthesized in large amounts, offer greater sensitivity and specificity and a shorter target-probe hybridization time. A probe may be (1) a unique plasmid sequence, (2) part of a sequence controlling toxin or antigen synthesis, (3) a unique genomic sequence identified using differential hybridization, or (4) a rRNA sequence. Probes for some of the bacterial phytopathogens have been found through selective hybridization and a technique known as subtraction hybridization (Ward and DeBoer, 1990; Cook and Sequeira, 1991; Seal *et al.*, 1992b; Darrasse *et al.*, 1994). Unfortunately, this method is quite labour-intensive. One of the more efficient means of designing a DNA sequence unique to a target bacterium is to amplify variable rDNA sequences using universal primers from the conserved regions of the genes for 16S or 23S rRNA (Barry *et al.*, 1990; Persing, 1993; Schleifer *et al.*, 1991). Oligonucleotide probes based on rRNA sequences are very sensitive because of the high numbers of ribosomal sequences present in a bacterial cell. DeParasis and Roth (1990) determined the potential of this approach for phytopathogenic bacteria by comparing the sequences of a variable 16S rRNA region (34-bp) from three *Erwinia* spp. and several *X. campestris* and *P. syringae* pathovars. This region clearly identified the genera but could not distinguish the pathovars or *Erwinia* subspp. However, there may be other variable regions in the 16S rRNA molecule that are unique to a specific pathovar or subspecies. Perhaps an alternative would be the spacer or intergenic region between the 16S and 23S rRNA genes (Barry *et al.*, 1991).

## C. Hybridization Formats

Nucleic acid hybridizations may be performed in solution or on a solid support. The most commonly-used formats, the dot/slot blot, the Southern blot, and the colony blot, require that nucleic acids be bound to a nitrocellulose or nylon membrane. When multiple samples are to be processed, solid support assays are the most convenient, although there may be problems with background and/or faint signals from the probe-target hybrids, particularly if samples are to be applied directly without prior DNA extraction. Background problems can be reduced somewhat by using capture probes in a sandwich hybridization assay (Steffan and Atlas, 1991). In this assay, an immobilized 'capture' probe binds target DNA which in turn hybridizes to a labelled 'signalling' probe. Because this technique has less background and can analyze larger volumes of environmental samples, it generally tends to be more sensitive than the dot or slot blot assays. Furthermore, because it uses microtitre plates, it is amenable to automation.

Another hybridization assay with a solid format is the *in situ* hybridization assay. This assay, in which cells or tissue sections that have been fixed to a microscope slide are exposed to probe DNA, is used primarily to detect human viruses (Tenover and Unger, 1993). Although this assay is limited by the accessibility of target nucleic acid within individual cells, it holds great potential. Not only is the assay automatable but the results are amenable to digital image analysis, which will not only improve sensitivity but permit the simultaneous visualization of multiple probes (Reid *et al.*, 1992; Celeda *et al.*, 1992). Just recently *in situ* hybridization assays have been applied to studies involving plants and bacteria (Madan and Nierzwicki-Bauer, 1993; Meyerowitz, 1987).

Hybridization assays, in which the probe and target DNA bind in solution, are much more efficient than those requiring a solid support. Furthermore, solution hybridizations may provide greater sensitivity because larger quantities of DNA sample can be processed (Pickup, 1991). The company Gen-Probe has introduced a novel solution hybridization format in which rRNA target sequences that have hybridized to a chemiluminescent-labelled DNA probe in solution are separated from non-hybridized probe by means of magnetic beads. The labelled hybrids are then measured with a luminometer. The Gen-Probe kits are used primarily as culture confirmation assays for selected respiratory and sexually-transmitted pathogens of humans (Tenover and Unger, 1993).

## D. Detection Systems

The first diagnostic DNA probes developed for bacteria were labelled with radioisotopes, primarily $^{32}$P. Although radioactive labels provide very sensitive probes, there are a number of limitations that reduce their effectiveness. $^{32}$P-labelled probes have a short half-life, present a safety hazard, and require costly disposal procedures.

The nonradioactive methods presently available for labelling DNA probes are as sensitive as methods using $^{32}$P with a specific activity of $1 \times 10^8$ cpm/µg but they give quicker results and are not as hazardous. Nonradioactive probes may be labelled either enzymatically or chemically. Although enzymatically-labelled probes are very sensitive, the procedures involved are complex and may give poor reproducibility. The most widely-used modified nucleotide is biotin-11-dUTP. However, biotin is a poor choice when working with plant samples since plants normally contain a certain amount of endogenous biotin. The DIG system (Boehringer Mannheim), which employs the steroid hapten digoxigenin, obviates the non-specific binding seen with biotin/strepavidin. DNA may be labelled using the random-priming method, nick translation, or the 3′ end tailing method (Keller and Manak, 1989). Oligonucleotides are most efficiently labelled by tailing with a modified dNTP. Chemical labelling methods are simple, versatile, and inexpensive. Sulfonation of cytosine residues by bisulfite was one of the first hapten-based labelling systems. The probe is detected using a monoclonal anti-sulfonate antibody followed by an alkaline phosphatase-conjugated anti-mouse IgG antibody. A long developing step is part of this method. One of the more popular chemical labelling methods involves photobiotin, a compound in which biotin is bound to aryl azide by a linker arm. In the presence of light, aryl azide is converted into an aryl nitrene moiety that reacts with nucleic acid. Oligonucleotides are most efficiently labelled chemically by a reactive primary amine. A detectable group, eg., biotin, alkaline phosphatase, or horseradish peroxidase, is added to the amino-oligonucleotide after synthesis.

## E. PCR-Based Methodology

The technology behind *in vitro* nucleic acid amplification is diverse and expanding at a very rapid pace. Amplification may be directed at the target nucleic acid, the probe, or the signal (Persing, 1993). In PCR, the most frequently-used of the amplification techniques, a thermostable DNA polymerase (usually the *Thermus aquaticus* enzyme Taq) is used to exponentially increase target DNA sequences through a series of thermal cycles (Atlas and Bej, 1994). There are three steps in a thermal cycle: (1) the sample DNA is denatured into single strands by high temperature incubation; (2) oligonucleotide primers are annealed at a lower temperature to regions of sample DNA that flank the desired target DNA sequence; and (3) the DNA between the primer binding sites is then extended by the thermostable enzyme. Critical to the success of a PCR protocol is the selection of suitable primers. Length, genome location, melting temperature, and sequence composition are all important considerations (Persing, 1993). If primer pairs have similar annealing temperatures and target DNA lengths, then it is possible to run several amplifications at once in the same amplification reaction (multiplex PCR). Thus it is technically feasible to assay plant samples for several different pathogens at the same time.

The typical PCR procedure, which consists of twenty-five to forty thermal cycles and takes only a few hours to complete, will, under optimal conditions, increase target DNA one million-fold (Atlas and Bej, 1994). For soil microorganisms, the probe detection limit without PCR is approximately $10^3$ to $10^4$ cells/g of soil against a background microbial population of $10^9$ cells/g of soil; with PCR amplification, the detection limit is one target bacterium/g of soil (Stefan and Atlas, 1991). Further improvements in the specificity and amplification efficiency of the PCR can be achieved by using nested primers. In this procedure, PCR proceeds for 15 to 30 cycles with the first primer set and then for another 15 to 30 cycles with a second primer set that recognizes internal sequences of the amplified product of the first set (Persing, 1993). When first described, this procedure involved open transfer from one tube to another and frequently lead to contamination. Recent changes to the nested amplification protocol have reduced the risk of contamination by making it possible to perform the reaction in a single tube (Yourno, 1992).

Until recently the amount of PCR product could not be used to give a reliable estimate of the amount of target DNA present in a test sample. However, it is now known that, under carefully-controlled conditions, PCR protocols can be made quantitative (Gilliland *et al.*, 1990; Ferre 1992; Seibert and Larrick, 1993). Usually quantification is achieved by including known amounts of internal DNA standards that are co-amplified with the target DNA. There are as yet no published reports on using quantitative PCR to detect latent bacterial infections of plants. However, this technique has been applied to *Verticillium* biomass measurements in infected potato plants (Hu *et al.*, 1993).

## F. Sample preparation

Critical to the success of nucleic acid probe-based detection methods is sample preparation (Greenfield and White, 1993). Most environmental samples are a mixture of several different organisms and organic matter. If the target microorganism is present in low numbers, these other sample components may hinder its detection. The most widely-used detection method consists of plating samples directly onto a selective medium followed by colony transfer to a filter and nucleic acid hybridization. Although more sensitive and specific than the standard plate assay, this protocol is limited by the effectiveness of the selective medium; the ratio of target population to background population on the selective medium must not exceed 1:1000 (Sayler and Layton, 1990; Cuppels and Elmhirst, unpublished data). An alternative to colony hybridization is to apply samples directly onto filters and then to proceed with a dot blot hybridization. To maximize the chance of detecting low numbers of the target microbe, one must collect the largest volume sample that is feasible and then concentrate the target cells. A common method for concentrating bacterial cells is to pass the liquid sample through a nylon or nitrocellulose filter.

Another approach is to extract nucleic acid directly from plant tissue (Lee and Davis, 1988; Chen *et al.*, 1993; Prosen *et al.*, 1993). For PCR-based methods the DNA may have to be purified away from potential inhibitors using phenol-chloroform extractions, cesium chloride centrifugation, ethanol (or isopropanol) precipitation, and/or treatment with polyvinylpyrrolidone (Steffan and Atlas, 1991). DNA extraction procedures for infected plants often contain a CTAB (cetyltrimethyl-ammonium bromide) extraction step to remove complex carbohydrates (Murray and Thompson, 1980).

## G. Probe-Based Methods for Plant Pathogenic Bacteria

Mycoplasmalike organisms (MLOs) are wall-less, non-culturable prokaryotes associated with a number of important plant diseases. Prior to 1987 and the development of the first MLO DNA probes (Kirkpatrick *et al.*, 1987), disease diagnosis, which was dependent on electron microscopy and symptomology, was both labour-intensive and time-consuming. Over the last few years, a number of nucleic acid probe-based methods have been developed for detecting MLO DNA in plant tissue (Table III). The majority of these methods are based on dot blot hybridizations in which DNA extracted from infected plants is bound to probes prepared from purified MLO genomic DNA. Because the number of MLOs present in infected plant tissue is usually quite low, DNA must be concentrated before being applied to the filter. The dot blot detection limit with these probes varies with the type of plant and plant tissue sampled but is routinely about 2-30 ng of total DNA from infected plants (Deng and Hiruki, 1990; Bonnet *et al.*, 1990; Daire *et al.*, 1992; Chen *et al.*, 1993). Biotinylated probes may be used provided the protocol for DNA isolation has at least one phenol-chloroform extraction step (Lee *et al.*, 1990). Although they do not bind to healthy plant DNA, many of the MLO probes do react with DNA from other MLOs and thus are being used to study the genetic relatedness of this group of phytopathogens (Lee and Davis, 1988; Bertaccini *et al.*, 1990; Kuske *et al.*, 1991).

MLO disease diagnosis has been facilitated by the recent introduction of PCR technology (Deng and Hiruki, 1991; Ahrens and Seemüller, 1992). As an example, DNA probes are able to recognize grapevine yellows in a 10-ng sample of total DNA from infected periwinkle tissue whereas a PCR-based method using primers derived from one of these probes can detect the pathogen when only $10^{-2}$ pg of total DNA is assayed (Chen *et al.*, 1993). The majority of the primers used for MLOs are derived from 16S rDNA (Table IV). Lee and co-workers have designed a 'universal' primer pair for the specific amplification of 16S rDNA from several different MLOs (Lee *et al.*, 1993). This primer pair has been used to diagnose grapevine yellows disease in both the U.S. and Italy (Prince *et al.*, 1993). Strain classification was based on restriction fragment length polymorphism (RFLP) analysis of the resulting PCR products. An alternative to RFLP-based classification is to use the newly-developed method called recycled PCR (RPCR), in which mollicutes-specific and different MLO group-specific sequences are amplified in the same tube (Namba *et al.*, 1993). Primers specific for the aster yellows MLO cluster are also now available (Davis and Lee, 1993).

A number of DNA probes have been reported for *C. m. sepedonicus*, the causal agent of bacterial ring rot of potato (Verreault et al., 1988; Johansen *et al.*, 1989; Mirza *et al.*, 1993; Drennan *et al.*, 1993). This pathogen, which may persist in symptomless plant tissue for long periods of time, is very difficult to detect by standard cultural and immunological methods. Recently, a 1.1-kb multicopy repeated sequence (RS) from *C. m. sepedonicus*, that had been cloned and characterized by Mogen *et al.* (1990), was used as a probe in direct blotting of potato plant tissue (Drennan *et al.*, 1993). Labelled with $^{32}$P it had a detection limit of approximately $10^5$ cfu, which is similar to the detection limit reported for pEJ79, another bacterial ring rot probe that

*Table III.* Plant pathogenic bacteria and mollicutes detected by nucleic acid probes.

| Bacterium | Target Sequence | Hybrid-ization | Reference |
|---|---|---|---|
| *Agrobacterium tumefaciens* | T-DNA genes | Col | Burr *et al.* 1990 |
| *Clavibacter michiganensis* | | | |
| subsp. *michiganensis* | genomic | Dot | Thompson *et al.*1989 |
| subsp. *sepedonicus* | 16S rDNA | Dot | Mirza *et al.*, 1993. |
| | genomic | Dot | Verreault *et al.*, 1988. |
| | genomic | Dot | Johansen *et al.*, 1989. |
| | repeated-sequence | Dir | Drennan *et al.*, 1993. |
| *Erwinia amylovora* | plasmid DNA | Col | Falkenstein *et al.*, 1988. |
| *Erwinia carotovora* | genomic | Dot | Ward and DeBoer, 1990. |
| | genomic | Dot | Darrasse *et al.*, 1994. |
| *Erwinia herbicola* | plasmid DNA | Dot | Manulis *et al.*, 1991. |
| *Erwinia stewartii* | genomic | Sou | Blakemore *et al.*, 1992. |
| *Pseudomonas syringae* | | | |
| pv. *morsprunorum* | unknown* | Col | Paterson and Jones, 1991. |
| pv. *phaseolicola* | toxin DNA | Col | Schaad *et al.*, 1989. |
| pv. *tomato* | unknown | Dot | Denny, 1988. |
| | toxin DNA | Col | Cuppels *et al.*, 1990. |
| *Pseudomonas solanacearum* | unknown | Sou | Cook and Sequeira, 1991. |
| *Xanthomonas campestris* | | | |
| pv. *citri* | plasmid DNA | Dot | Hartung, 1992. |
| pv. *phaseoli* | plasmid DNA | Col/Dot | Gilbertson *et al.*, 1989. |
| Mycoplasmas | | | |
| Western X-disease | genomic | Dot/Dir | Kirkpatrick *et al.*, 1987. |
| aster yellows | genomic | Dot | Lee and Davis, 1988. |
| periwinkle little leaf | | | Davis *et al.*, 1990. |
| clover proliferation | genomic | Dot | Deng and Hiruki, 1990. |
| apple proliferation | genomic | Dot/Sou | Bonnet *et al.*, 1990. |
| walnut witches' broom | extrachrom. DNA | Sou | Chen *et al.*, 1992 |
| ash yellows | genomic | Dot | Davis *et al.*, 1992 |
| grape flavescence doree | genomic | Dot | Daire *et al.*, 1992. |
| palm leaf yellows | genomic | Dot | Harrison *et al.*, 1992. |
| grapevine yellows | genomic | Dot | Chen *et al.*, 1993. |

Col= colony hybridization; Dot = dot blot; Dir = direct blot of tissue; Sou = Southern hybridization.
* The probe developed by Denny for pv. *tomato* was used to screen for pv. *morsprunorum* by Paterson and Jones.

Table IV. Plant pathogenic bacteria and mollicutes detected using PCR amplification.

| Bacterium | Target Sequence | Reference |
|---|---|---|
| Agrobacterium tumefaciens | T-DNA | Dong et al. 1992 |
| Clavibacter michiganensis | | |
| subsp. sepedonicus | plasmid DNA | Schneider et al., 1993. |
| Erwinia amylovora | plasmid DNA | Bereswill et al., 1992. |
| Pseudomonas solanacearum | genomic DNA | Seal et al., 1992 |
| | 16S rDNA | Seal et al., 1993 |
| Pseudomonas syringae | | |
| pv. phaseolicola | toxin DNA | Prosen et al., 1993 |
| Xanthomonas campestris | | |
| pv. citri | plasmid DNA | Hartung et al., 1993 |
| Mycoplasmalike organisms | | |
| General | 16S rDNA | Deng and Hiruki, 1991 |
| | 16S rDNA | Ahrens and Seemüller, 1992. |
| | genomic | Schaff et al., 1992. |
| | 16S rDNA | Lee et al., 1993. |
| Groups I-III | 16S rDNA | Namba et al., 1993 |
| grapevine yellows | genomic DNA | Chen et al., 1993 |
| grapevine yellows | genomic DNA | Prince et al., 1993 |
| apple proliferation | 16S rDNA | Firrao et al., 1993 |
| clover phyllody | 16S rDNA | Firrao et al., 1993 |

contains a repeated element (Johansen et al., 1989). Although the RS probe appeared specific to bacterial ring rot in field sample testing, it was able to hybridize to DNA extracts of the closely-related alfalfa pathogen C. m. insidiosus. Another recently developed C. m. sepedonicus probe, a 24mer oligonucleotide that targets the V6 region of 16S rDNA, also binds to closely-related subspecies but Mirza et al. (1993) were able to increase its specificity by using more stringent filter washing conditions (60 C). The bacterial ring rot pathogen also can be detected using a PCR method which is based on the amplification of a 258-bp fragment of the C. m. sepedonicus plasmid pCS1. Much more sensitive than the probe-based methods, it has a detection limit of approximately 100 cfu/g of potato tissue (Schneider et al., 1993).

Plasmid sequences have been targeted by diagnostic DNA probes or PCR-based assays for several other bacterial phytopathogens, including X. c. citri, X. c. phaseoli, E. amylovora, E. herbicola pv. gypsophilae and A. tumefaciens (Table III and IV). A PCR assay for the fireblight pathogen E. amylovora utilizes two 17mer oligonucleotide primers to amplify a 0.9-kb fragment of the E. amylovora plasmid pEA29 and ethidium bromide-stained agarose gels to visualize the PCR product (Bereswill et al., 1992). Being able to detect 500 E. amylovora cells out of a population of 50,000 plant-associated bacteria, it should be useful in screening pome fruit orchards and nurseries for latent infections. Another serious fruit disease and the target of international quarantine measures is bacterial citrus canker caused by X. c. citri. Hartung et al. (1993) have developed a PCR-based method for detecting this pathogen that targets a 222-bp fragment of plasmid DNA. Successful amplification of this fragment requires a pH 9.0 buffer supplemented with 1% Triton X-100 and 0.1% gelatin. The detection limit for pure cultures was approximately 10 cfu/ml whereas that for viable three-week-old citrus bacterial canker lesions was 100-800 cfu. The PCR product also was recovered from water extracts of dry, nonviable seven-month-old lesions.

Probe-based methods of screening symptomless plant tissue have been developed for the bean pathogen *P. s. phaseolicola*, the tomato pathogen *P. s. tomato*, and the potato pathogen *P. solanacearum*. The *P. s. tomato* population of symptomless field tomato leaves was determined by plating leaf washes on a semi-selective medium and then hybridizing colony blots with a nonradioactively-labelled probe (Cuppels *et al.*, 1990). The probe was a 4-kb fragment of the *P. s. tomato* gene cluster controlling phytotoxin (coronatine) synthesis. Although this method was highly specific, its sensitivity dropped significantly when there was a large population of saprophytic bacteria present. Similar results were obtained with the toxin gene-based probe developed for *P. s. phaseolicola* (Schaad *et al.*, 1989). Recently the *P. s. phaseolicola* probe assay was modified to one involving PCR-mediated amplification of a 1.9-kb fragment of the probe target (Prosen *et al.*, 1993). This assay, which was used to screen commercial seed lots, had a detection threshold of 10-20 cfu/ml of seed soak water. For the bacterial wilt pathogen *P. solanacearum*, Seal *et al.* (1992ab, 1993) developed a PCR-based method of screening symptomless seed potatoes that employs one of two sets of primers, OL11/Y2 or PS96-H/PS96-I. Although OL11-Y2, which was constructed using 16S rRNA sequence data, was the more sensitive of the two, it was not as specific, producing a PCR product with two closely-related pathogens, *P. syzygii* and the blood disease bacterium. The PCR detection assay with primers PS96-H/PS96-I, which were selected using genomic subtraction, had a detection limit of 10-100 cells per assay and could be completed in just five hours. Additional information on nucleic acid-based assays for the bacterial plant pathogens may be found in recent reviews (Vivian, 1992; Goodwin and Nassuth, 1993; Henson and French, 1993).

## V. FUTURE DIRECTIONS

Although there are several probes and PCR protocols available for the phytopathogenic bacteria, relatively few have been tested extensively with plant tissue. The immediate challenge will be to incorporate these new assay techniques into cost-effective methods of screening large numbers of plant samples for the presence of phytopathogens. Labour costs should not be a deterrent once the automated analyzers currently under development reach the marketplace (Godsey *et al.*, 1994).

There have been several recent advances in diagnostic microbiology that should be of value to researchers concerned with latent bacterial infections of plants. A number of these new technologies, including multiplex PCR, quantitative PCR, and digital image analysis of *in situ* hybridizations using fluorescent probes, have already been discussed. Also of interest are methodologies that combine antibody-based detection systems with PCR. Immuno-PCR was developed by Sano *et al.* (1992) as a means of detecting very small numbers of antigen molecules that had been immobilized in microtitre plate wells. This procedure is similar to conventional ELISA except that the enzyme-conjugated secondary antibody is replaced with a biotinylated DNA molecule which has been bound via a streptavidin-protein A chimera to the antigen-monoclonal antibody complex. This marker DNA, whose sequence is arbitrary, is then PCR-amplified and the PCR product visualized by agarose gel electrophoresis. This simple procedure, which is amendable to automation, is reported to be approximately $10^5$ times more sensitive than ELISA. Hedrum *et al.* (1992) also have developed a technique based on the ELISA format and suitable for large scale screening programs. The target microbe is captured and concentrated using magnetic beads coated with target-specific antibody. The DNA of the captured microbes is amplified by nested PCR using one inner primer that has been labelled with biotin and another that has the *E. coli lac* operator sequence. Captured using streptavidin-coated magnetic beads, the labelled, amplified DNA is detected through a colorimetric signal.

## ACKNOWLEDGEMENTS

DAC wishes to thank D. Drew, Agriculture and Agri-Food Canada librarian, for her help in literature searches and the procurement of journal articles.

## REFERENCES

Ahrens, U. and Seemuller, E. (1992). Detection of DNA of plant pathogenic mycoplasmalike organisms by a polymerase chain reaction that amplifies a sequence of the 16S rRNA gene. *Phytopathology* **82**, 828-832.

Allan, E. and Kelman, A. (1977). Immunofluorescent stain procedures for detection and identification of *Erwinia carotovora* var. *atroseptica*. *Phytopathology* **67**, 1305-1312.

Alvarez, A. M., Benedict, A. A., Mizumoto, C. Y., Pollard, L. W. and Civerolo, E. L. (1991). Analysis of *Xanthomonas campestris* pv. *citri* and *X.˙ c. citrumelo* with monoclonal antibodies. *Phytopathology* **81**, 857-865.

Alvarez, A. M., Benedict, A. A. and Mizumoto, C. Y. (1985). Identification of xanthomonads and grouping of *Xanthomonas campestris* pv. *campestris* with monoclonal antibodies. *Phytopathology* **75**, 722-728.

Alvarez, A. M. and Lou, K. (1985). Rapid identification of *Xanthomonas campestris* pv. *campestris* by ELISA. *Plant Disease* **69**, 1082-1086.

Anderson, M. J. and Nameth, S. T. (1990). Development of a polyclonal antibody--based serodiagnostic assay for the detection of *Xanthomonas campestris* pv. *pelargonii* in geranium plants. *Phytopathology* **80**, 357-360.

Anonymous (1987). 'Scheme for the Detection and Diagnosis of the Ring Rot Bacterium *Corynebacterium sepedonicum* in Batches of Potato Tubers'. EUR 11288. Office for Official Publications of the European Communities, Luxembourg.

Atlas, R. and Bej, A. K. (1994). Polymerase chain reaction. *In* 'Methods for General and Molecular Bacteriology'(P. Gerhardt, R. G. E. Murray, W. A. Wood and N. R. Krieg, eds.) pp. 418-435, American Society for Microbiology, Washington, D.C.

Barry, T., Gollera, G., Glennon, M., Dunican, L. K. and Gannon, F. (1991). The

16s/23s ribosomal spacer region as a target for DNA probes to identify eubacteria. *PCR Methods and Applications* **1**, 51-56.

Barry, T., Powell, R. and Gannon, F. (1990). A general method to generate DNA probes for microorganisms. *Biotechnology* **8**, 233-236.

Barzic, M. R. and Trigalet, A. (1982). Detection de *Pseudomonas phaseolicola* (Burkh.) Dowson par la technique ELISA. *Agronomie* **2**, 389-398.

Benedict, A. A., Alvarez, A. M., Berestecky, J., Imanaka, W., Mizumoto, C. Y., Pollard, L. W., Mew, T. W. and Gonzalez, C. F. (1989). Pathovar-specific monoclonal antibodies for *Xanthomonas campestris* pv. *oryzae* and for *Xanthomonas campestris* pv. *oryzicola*. *Phytopathology* **79**, 322-328.

Benedict, A. A., Alvarez, A. M. and Pollard, L. W. (1990). Pathovar-specific antigens of *Xanthomonas campestris* pv. *begaoniae* and *X. campestris* pv. *pelargonii* detected with monoclonal antibodies. *Applied and Environmental Microbiology* **56**, 572-574.

Bereswill, S., Pahl, A., Bellemann, P., Zeller, W. and Geider, K. (1992). Sensitive and species-specific detection of *Erwinia amylovora* by polymerase chain reaction analysis. *Applied and Environmental Microbiology* **58**, 3522-3526.

Bertaccini, A., Davis, R.E., Hammond, R.W., Vibio, M., Bellardi, M.G. and Lee, I.-M. (1992). Sensitive detection of mycoplasmalike organisms in field-collected and *in vitro* propagated plants of *Brassica, Hydranea*, and *Chrysanthemum* by polymerase chain reaction. *Annals of Applied Biology* **121**, 593-599.

Bertaccini, A., Davis, R. E., Lee, I. M., Conti, M., Dally, E. L. and Douglas, S. M. (1990). Detection of chrysanthemum yellows mycoplasmalike organism by dot hybridization and southern blot analysis. *Plant Disease* **74**, 40-43.

Blakemore, E. J. A., Reeves, J. C. and Ball, S. F. L. (1992). Polymerase chain reaction used in the development of a DNA probe to identify *Erwinia stewartii* a bacterial pathogen of maize. *In* 'Plant Pathogenic Bacteria' (M. Lemattre, S. Freigoun, K. Rudolph and J.G. Swings, eds). pp. 361-363, Institut National de la Recherche Agronomique, Versailles Cedex, France

Bonnet, F., Saillard, C., Kollar, A., Seemüller, E. and Bove, J. M. (1990). Detection and differentiation of the mycoplasmalike organism associated with apple proliferation disease using cloned DNA probes. *Molecular Plant-Microbe Interactions* **3**, 438-443.

Bouzar, H., Moore, L. W. and Schaup, H. W. (1988). Lipopolysaccharide from *Agrobacterium tumefaciens* B6 induces the production of strain-specific antibodies. *Phytopathology* **78**, 1237-1241.

Bragard, C. and Verhoyen, M. (1993). Monoclonal antibodies specific for *Xanthomonas campestris* bacteria pathogenic on wheat and other small grains, in comparison with polyclonal antisera. *Journal of Phytopathology* **139**, 217-228.

Brlansky, R. H., Lee, R. F. and Civerolo, E. L. (1990). Detection of *Xanthomonas campestris* pv. *citrumelo* and *X. citri* from citrus using membrane entrapment immunofluorescence. *Plant Disease* **74**, 863-868.

Burr, T. J., Morelli, J. L., Katz, B. H. and Bishop, A. L. (1990). Use of Ti plasmid DNA probes for determining tumorigenicity of *Agrobacterium* strains. *Applied and Environmental Microbiology* **56**, 1782-1785.

Calzolari, A., Bazzi, C. and Mazzucchi, U. (1982). Cross- reactions between *Corynebacterium sepedonicum* and *Arthrobacter polychromogenes* in immunofluorescence staining. *Potato Research* **25**, 239-236.

Celeda, D., Bettag, U. and Cremer, C. (1992). PCR amplification and simultaneous digoxigenin incorporation of long DNA probes for fluorescence *in situ* hybridization. *Biotechnology* **12**, 98-102.

Chen, J., Chang, C. J. and Jarret, R. L. (1992). DNA probes as molecular markers to monitor · the seasonal occurrence of walnut witches'-broom mycoplasmalike organism. *Plant Disease* **76**, 1116- 1119.

Chen, K. H., Guo, J. R., Wu, X. Y., Loi, N., Carraro, L., Guo, Y. H., Chen, Y. D., Osler, R., Person, R. and Chen, T. A. (1993). Comparison of monoclonal antibodies, DNA probes, and PCR for detection of the grapevine yellows disease agent. *Phytopathology* **83**, 915-922.

Chen, T. A. and Jiang, X. F. (1988). Monoclonal antibodies against the maize bushy stunt agent. *Canadian Journal of Microbiology* **34**, 6-11.

Claflin, L. E. and Ramundo, B. A. (1987). Evaluation of the dot-immunobinding assay for detecting phytopathogenic bacteria in wheat seeds. *Journal of Seed Technology* **11**, 52-61.

Corbiere, R., Hingand, L. and Jouan, B. (1987). Application des methodes ELISA et immunofluorescence pour la detection de *Corynebacterium sepedonicum*: reponses varietales de la pomme de terre au fletrissement bacterien. *Potato Research* **30**, 539-549.

Comstock, J. C. (1992). Detection of the sugarcane leaf scald pathogen, *Xanthomonas albilineans*, using tissue blot immunoassay, ELISA, and isolation techniques. *Plant Disease* **76**, 1033-1035.

Cook, D. and Sequeira, L. (1991). The use of subtractive hybridization to obtain a DNA probe specific for *Pseudomonas solanacearum* race 3. *Molecular and General Genetics* **227**, 401-410.

Crowley, C. F. and De Boer, S. H. (1982). Non-pathogenic bacteria associated with potato stems cross-react with *Corynebacterium sepedonicum* antisera in immunofluorescence. *American Potato Journal* **59**, 1-8.

Cuppels, D. and Kelman, A. (1974). Evaluation of selective media for isolation of soft-rot bacteria from soil and plant tissue. *Phytopathology* **64**, 468-475.

Cuppels, D. A., Moore, R. A. and Morris, V. (1990). Construction and use of a nonradioactive DNA hybridization probe for detection of *Pseudomonas syringae* pv. *tomato* on tomato plants. *Applied and Environmental Microbiology* **56**, 1743-1749.

Daire, X., Boudon-Padieu, E., Berville, A., Schneider, B. and Caudwell, A. (1992). Cloned DNA probes for detection of grapevine flavescence doree mycoplasma-like organisms (MLO). *Annals of Applied Biology* **121**, 95-103.

Darrasse, A., Kotujansky, A. and Bertheau, Y. (1994). Isolation by genomic subtraction of DNA probes specific for *Erwinia carotovora* subsp. *atroseptica*. *Applied and Environmental Microbiology* **60**, 298-306.

Davis, M. J. and Dean, J. L. (1984). Comparison of diagnostic techniques for determining incidence of ratoon stunting disease of sugarcane in Florida. *Plant Disease* **68**, 896-899.

Davis, R.E., and Lee, I.-M. (1993). Cluster-specific polymerase chain reaction amplification of 16S rDNA sequences for detection and identification of mycoplasmalike organisms. *Phytopathology* **83**, 1008-1011.

Davis, R. E., Lee, I.-M., Jordan, R. and Konai, M. (1987). Detection of corn stunt spiroplasma in plants by ELISA employing monoclonal antibodies and by isolation of the pathogen and its phages in serum free medium. *Current Plant Science, Biotechnology and Agriculture* **4**, 306-312.

Davis, R. E., Lee, I.-M. and Douglas, S. M. and Dally, E. L. (1990). Molecular cloning and detection of chromosomal and extrachromosomal DNA of the mycoplasmalike organism (MLO) associated with little leaf disease of periwinkle (*Catharanthus roseus*). *Phytopathology* **80**, 789-793.

Davis, R. E., Sinclair, W. A., Lee, I.-M. and Dally, E. L. (1992). Cloned DNA probes specific for detection of a mycoplasmalike organism associated with ash yellows. *Molecular Plant-Microbe Interactions* **5**, 163-169.

De Boer, S. H. (1982). Cross-reaction of *Corynebacterium sepedonicum* antisera with *C. insidiosum*, *C. michiganense*, and an unidentified coryneform bacterium. *Phytopathology* **72**, 1474-1478.

De Boer, S. H. (1984). Enumeration of two competing *Erwinia carotovora* populations in potato tubers by membrane filter- immunofluorescence procedure. *Journal of Applied Bacteriology* **57**, 517-522.

De Boer, S. H., Janse, J. D., Stead, D. E., Van Vaerenbergh, J. and McKenzie, A. R. (1992). Detection of *Clavibacter michiganensis* subsp. *sepedonicus* in potato stems and tubers grown from seed pieces with various levels of inoculum. *Potato Research* **35**, 207-216.

De Boer, S. H. and McCann, M. (1989). Determination of population densities of *Corynebacterium sepedonicum* in potato stems during the growing season. *Phytopathology* **79**, 946-951.

De Boer, S. H. and McNaughton, M. E. (1987). Monoclonal antibodies to the lipopoly-saccharide of *Erwinia carotovora* subsp. *atroseptica* sergroup I. *Phytopathology* **77**, 828-832.

De Boer, S. H., Stead, D. E., Alivizatos, A. S., Janse, J. D., Van Vaerenbergh, J., De Haan, T. L. and Mawhinney, J. (1994). Evaluation of serological tests for detection of *Clavibacter michiganensis* subsp. *sepedonicus* in composite potato stem and tuber samples. *Plant Disease* **78**, 725-729.

De Boer, S. H., Wieczorek, A. and Kummer, A. (1988). An ELISA test for bacterial ring rot of potato with a new monoclonal antibody. *Plant Disease* **72**, 874-878.

De Boer, S. H. and Wieczorek, A. (1984). Production of monoclonal antibodies to *Corynebacterium sepedonicum*. *Phytopathology* **74**, 1431-1434.

De Kam, M. (1982). Detection of soluble antigens of *Erwinia salicis* in leaves of *Salix alba* by enzyme-linked immunosorbent assay. *European Journal of Forest Pathology* **12**, 1-6.

Deng, S. J. and Hiruki, C. (1990). Molecular cloning and detection of DNA of the mycoplasmalike organism associated with clover proliferation. *Canadian Journal of Plant Pathology* **12**, 385-388.

Denny, T. P. (1988). Differentiation of *Pseudomonas syringae* pv. *tomato* from *P. s. syringae* with a DNA hybridization probe. *Phytopathology* **78**, 1186-1193.

DeParasis, J. and Roth, D. A. (1990). Nucleic acid probes for identification of phytobacteria: identification of genus-specific 16s rRNA sequences. *Phytopathology* **80**, 618-621.

Dong, L. C., Sun, C. W., Thies, K. L., Luthe, D. S. and Graves, C. H. (1992). Use of polymerase chain reaction to detect pathogenic strains of *Agrobacterium*.

*Phytopathology* **82**, 434-439.

Drennan, J. L., Westra, A. A. G., Slack, S. A., Delserone, L. M., Collmer, A., Gudmestad, N. C. and Oleson, A. E. (1993). Comparison of a DNA hybridization probe and ELISA for the detection of *Clavibacter michiganensis* subsp. *sepedonicus* in field-grown potatoes. *Plant Disease* **77**, 1243-1247.

Duncan, J. M., and Torrance, L. (1992). 'Techniques for the Rapid Detection of Plant Pathogens'. Blackwell Scientific Publications, Oxford.

Duveiller, E. and Bragard, C. (1992). Comparison of immunofluorescence and two assays for detection of *Xanthomonas campestris* pv. *undulosa* in seeds of small grains. *Plant Disease* **76**, 999-1003.

Falkenstein, H., Bellemann, P., Walter, S., Zeller, W. and Geider, K. (1988). Identification of *Erwinia amylovora*, the fireblight pathogen, by colony hybridization with DNA from plasmid pEA29. *Applied and Environmental Microbiology* **54**, 2798-2802.

Falkinham, J. O. (1994). Nucleic acid probes. *In* 'Methods for General and Molecular Bacteriology'(P. Gerhardt, R. G. E. Murray, W. A. Wood and N. R. Krieg, eds.) pp. 701-710, American Society for Microbiology, Washington, D.C.

Fatmi, M. and Schaad, N. W. (1988). Semiselective agar medium for isolation of *Clavibacter michiganense* subsp. *michiganense* from tomato seed. *Phytopathology* **78**,

Ferre, F. (1992). Quantitative or semi-quantitative PCR: reality or myth. *PCR Methods and Applications* **2**, 1-9.

Firrao, G., Gobbi, E. and Locci, R. (1993). Use of polymerase chain reaction to produce oligonucleotide probes for mycoplasmalike organisms. *Phytopathology* **83**, 602-607.

Franken, A. A. J. M., Zilverentant, J. F., Boonekamp, P. M. and Schots, A. (1992). Specificity of polyclonal and monoclonal antibodies for the identification of *Xanthomonas campestris* pv. *campestris*. *Netherlands Journal of Plant Pathology* **98**, 81-94.

Frankle, W. G. (1992). 'Bacterial Fruit Blotch of Watermelon: The Relationship of the Causal Bacterium to the Fruit and the Development of a Diagnostic Medium'. Master's Thesis. Graduate School, University of Florida, Gainesville, Florida.

Garnier, M., Martin-Gros, G., Iskra, M. L., Zriek, L., Gandar, J., Fos, A. and Bove, J. M. (1990). Monoclonal antibodies against the MLOs associated with tomato stolbur and clover phyllody. *Zentralblatt fur Bacteriologie, Supplement* **20**, 263-269.

Gilbertson, R. L., Maxwell, D. P., Hagedorn, D. J. and Leong, S. A. (1989). Development and application of a plasmid DNA probe for detection of bacteria causing common bacterial blight of bean. *Phytopathology* **79**, 518-525.

Gilliland, G., Perrin, S., Blanchard, L. and Bunn, H. F. (1990). Analysis of cytokine mRNA and DNA: detection and quantitation by competitive polymerase chain reaction. *Proceedings of the National Academy of Sciences of the United States of America* **87**, 2725-2729.

Gitaitis, R. D., Beaver, R. W. and Voloudakis, A. E. (1991). Detection of *Clavibacter michiganensis* subsp. *michiganensis* in symptomless tomato transplants. *Plant Disease* **75**, 834-838.

Gitaitis, R. D., Chang, C. J., Sijam, K. and Dowler, C. C. (1991). A differential medium for semiselective isolation of *Xanthomonas campestris* pv. *vesicatoria* and other cellulolytic xanthomonads. *Plant Disease* **75**, 1274-1278.

Gitaitis, R. D., Stall, R. E. and Strandberg, J. O. (1978). Dissemination and survival of *Pseudomonas alboprecipitans* ascertained by disease distribution. *Phytopathology* **68**, 227-231.

Gitaitis, R. D., Summer, D., Smittle, D., Gay, B., Maw, Y., Hung, Y. and Tollner, B. (1992). A semiselective agar medium for identification and isolation of *Pseudomonas viridiflava*, causal agent of bacterial blight of onion. *In* 'Proceedings National Onion Research Conference' pp. 23-25, Savannah, Georgia.

Godsey, J. H., Vanden Brink, K., DiMichele, L. J., Beninsig, L. A., Peterson, R. and Sherman, D. G. (1994). Commercialization of nucleic acid probe technology: Current status. *In* 'Antimicrobial susceptibility testing: Critical Issues for the 90's'(J. Poupard, L. R. Walsh and B. Kleger, eds.) pp. 121-129, Plenum Press, New York.

Goodwin, P. and Nassuth, A. (1993). Detection and characterization of plant pathogens. *In* 'Methods in Plant Molecular Biology and Biotechnology'(B. Glick and J. E. Thompson, eds.) pp. 303-319, CRC Press, Inc., Boca Raton, Florida.

Gorris, M. T., Alarcon, G., Lopez, M. M. and Cambra, M. (1994). Characterization of monoclonal antibodies specific for *Erwinia carotovora* subsp. *atroseptica* and comparison of serological methods for its sensitive detection on potato tubers. *Applied and Environmental Microbiology* **60**, 2076-2085.

Greenfield, L. and White, T. J. (1993). Sample preparation methods. *In* 'Diagnostic Molecular Microbiology: Principles and Applications'(D. H. Persing, T. F. Smith, F. Tenover and T. J. White, eds.) pp. 122-137, American Society for Microbiology, Washington, D.C.

Gudmestad, N. C., Baer, D. and Kurowski, C. J. (1991). Validating immunoassay test performance in the detection of *Corynebacterium sepedonicum* during the growing season. *Phytopathology* **81**, 475- 480.

Guillorit, C. and Samson, R. (1993). Serological specificity of the lipopolysaccharides, the major antigens of *Pseudomonas syringae*. *Journal of Phytopathology* **137**, 157-171.

Hampton, R., Ball, E., De Boer, S. (1990). 'Serological Methods for Detection and Identification of Viral and Bacterial Plant Pathogens'. APS Press, St. Paul, Minnesota.

Harrison, N. A., Bourne, C. M., Cox, R. L., Tsai, J. H. and Richardson, P. A. (1992). DNA probes for detection of mycoplasmalike organisms associated with lethal yellowing disease of palms in Florida. *Phytopathology* **82**, 216-224.

Hartung, J. S. (1992). Plasmid-based hybridization probes for detection and identification of *Xanthomonas campestris* pv. *citri*. *Plant Disease* **76**, 889-893.

Hartung, J. S., Daniel, J. F. and Pruvost, O. P. (1993). Detection of *Xanthomonas campestris* pv. *citri* by the polymerase chain reaction method. *Applied and Environmental Microbiology* **59**, 1143-1148.

Hayward, A. C. (1974). Latent infections by bacteria. *Annual Review of Phytopathology* **12**, 87-97.

Hedrum, A., Lundeberg, J., Pahlson, C. and Uhlen, M. (1992). Immunomagnetic recovery of *Chlamydia trachomatis* from urine with subsequent colorimetric DNA detection. *PCR Methods and Applications* **2**, 167-171.

Henson, J. M. and French, R. (1993). The polymerase chain reaction and plant disease diagnosis. *Annual Review of Phytopathology* **31**, 81-109.

Hockenhull, J. (1979). In situ detection of *Erwinia amylovora* antigen in symptomless petiole and stem tissue by means of the fluorescent antibody technique. *Den Kgl.*

*Veterinaer- og Landbohojskole Aarsskrift* **1979**, 1-14.

Hsu, H. T., Lee, I.-M., Davis, R. E. and Wang, Y. C. (1990). Immunization for generation of hybridoma antibodies specifically reacting with plants infected with a mycoplasmalike organism (MLO) and their use in detection of MLO antigens. *Phytopathology* **80**, 946-950.

Hu, X., Nazar, R. N. and Robb, J. (1993). Quantification of *Verticillium* biomass in wilt disease development. *Physiological and Molecular Plant Pathology* **42**, 23-26.

Jiang, Y. P., Chen, T. A., Chiykowski, L. N. and Sinha, R. C. (1989). Production of monoclonal antibodies of peach eastern X- disease and their use in disease detection. *Canadian Journal of Plant Pathology* **11**, 325-331.

Johansen, I. E., Rasmussen, O. F. and Heide, M. (1989). Specific identification of *Clavibacter michiganense* subsp. *sepedonicum* by DNA-hybridization probes. *Phytopathology* **79**, 1019-1023.

Jones, D. A. C., Mcleod, A., Hyman, L. J. and Perombelon, M. C. M. (1993). Specificity of an antiserum against *Erwinia carotovora* subsp. *atroseptica* in indirect ELISA. *Journal of Applied Bacteriology* **74**, 620-624.

Jones, R. K., Barnes, L. W., Gonzalez, C. F., Leach, J. E., Alvarez, A. M. and Benedict, A. A. (1989). Identification of low-virulence strains of *Xanthomonas campestris* pv. *oryzae* from rice in the United States. *Phytopathology* **79**, 984-990.

Keller, G. H., and Manak, M. M. (1989). 'DNA Probes'. Stockton Press, New York.

Kirkpatrick, B. C., Stenger, D. C., Morris, T. J. and Purcell, A. H. (1987). Cloning and detection of DNA from a nonculturable plant pathogenic mycoplasma-like organism. *Science* **238**, 197-200.

Kitagawa, T., Hu, J. G., Ishida, Y., Yoshiuchi, H., Kuhara, S., Koizumi, M. and Matsumoto, R. (1992). A new immunoassay for *Xanthomonas campestris* pv. *citri* and its application for evaluation of resistance in citrus plants. *Plant Disease* **76**, 708- 712.

Klopmeyer, M. J. and Kelman, A. (1988). Use of monoclonal antibodies specific for pectate lyase as serological probes in the identification of soft rot *Erwinia* spp. *Phytopathology* **78**, 1430-1434.

Konai, M. and Chang, C. J. (1990). Detection of a mycoplasma-like organism associated with walnut witches' broom disease by enzyme linked immunosorbent assay (ELISA). *Zentralblatt fur Bakteriologie, Supplement* **20**, 308-312.

Kuske, C.R., Kirkpatrick, B.C. and Seemüller. (1991). Differentiation of virescence MLOs using western aster yellows mycoplasma-like chromosomal DNA probes and restriction fragment length polymorphism analysis. *Journal of General Microbiology* **137**, 153-159.

Lamka, G. L., Hill, J. H., Mcgee, D. C. and Braun, E. J. (1991). Development of an immunosorbent assay for seedborne *Erwinia stewartii* in corn seeds. *Phytopathology* **81**, 839-846.

Lazarovits, G., Zutra, D. and Bar-Joseph, M. (1987). Enzyme- linked immunosorbent assay on nitrocellulose membranes (dot- ELISA) in the serodiagnosis of plant pathogenic bacteria. *Canadian Journal of Microbiology* **33**, 98-103.

Lee, I.-M. and Davis, R. E. (1988). Detection and investigation of genetic relatedness among aster yellows and other mycoplasmalike organisms by using cloned DNA and RNA probes. *Molecular Plant-Microbe Interactions* **1**, 303-310.

Lee, I.-M., Davis, R.E. and DeWitt, N.D. (1990). Nonradioactive screening method of isolation of disease-specific probes to diagnose plant diseases caused by mycoplasmalike organisms. *Applied and Environmental Microbiology* **56**, 1471-1575.

Lee, I.-M., Hammond, R. W., Davis, R. E. and Gunderson, D. E. (1993). Universal amplification and analysis of pathogen 16S rDNA for classification and identification of mycoplasmalike organisms. *Phytopathology* **83**, 834-842.

Li, X., Wong, W. C. and Hayward, A. C. (1993). Production and use of monoclonal antibodies to *Pseudomonas andropogonis*. *Journal of Phytopathology* **138**, 21-30.

Lin, C. P., Chen, T. A., Wells, J. M. and van der Zwet, T. (1987). Identification and detection of *Erwinia amylovora* with monoclonal antibodies. *Phytopathology* **77**, 376-380.

Lin, C. P. and Chen, T. A. (1985a). Monoclonal antibodies against the aster yellows agent. *Science* **227**, 1233-1235.

Lin, C. P. and Chen, T. A. (1985b). Monoclonal antibodies against corn stunt spiroplasma. *Canadian Journal of Microbiology* **31**, 900- 904.

Lin, C. P. and Chen, T. A. (1985c). Production of monoclonal antibodies against *Spiroplasma citri*. *Phytopathology* **75**, 848-851.

Lipp, R. L., Alvarez, A. M., Benedict, A. A. and Berestecky, J. (1992). Use of monoclonal antibodies and pathogenicity tests to characterize strains of *Xanthomonas campestris* pv. *dieffenbachiae* from aroids. *Phytopathology* **82**, 677-682.

Madan, A. P. and Nierzwicki-Bauer (1993). *In situ* detection of transcripts for ribulose-1,5-bisphosphate carboxylase in cyanobacterial heterocysts. *Journal of Bacteriology* **175**, 7301- 7306.

Manulis, S., Gafni, Y., Clark, E., Zutra, D., Ophir, Y. and Barash, I. (1991). Identification of a plasmid DNA probe for detection of strains of *Erwinia herbicola* pathogenic on *Gypsophila paniculata*. *Phytopathology* **81**, 54-57.

McGuire, R. G., Jones, J. B. and Sasser, M. (1986). Tween media for semiselective isolation of *Xanthomonas campestris* pv. *vesicatoria* from soil and plant materials. *Plant Disease* **70**, 887-891.

Meyerowitz, E. M. (1987). *In situ* hybridization to RNA in plant tissue. *Plant Molecular Biology Reporter* **5**, 242-250.

Miller, H. J. (1984). Cross-reactions of *Corynebacterium sepedonicum* antisera with soil bacteria associated with potato tubers. *Netherlands Journal of Plant Pathology* **90**, 23-28.

Mirza, M. S., Rademaker, J. L. W., Janse, J. D. and Akkermans, A. D. L. (1993). Specific 16S ribosomal RNA targeted oligonucleotide probe against *Clavibacter michiganensis* subsp. *sepedonicus*. *Canadian Journal of Microbiology* **39**, 1029-1034.

Mogen, B. D., Olson, H. R., Sparks, R. B., Gudmestad, N. C. and Oleson, A. E. (1990). Genetic variation in strains of *Clavibacter michiganensis* subsp. *sepedonicum*: polymorphisms in restriction fragments containing a highly repeated sequence. *Phytopathology* **80**, 90-96.

Muratore, M. G., Mazzucchi, U., Gaspernini, C. and Fiori, M. (1986). Detection of latent infection of *Erwinia chrysanthemi* and *Pseudomonas caryophylli* in carnation. *EPPO Bulletin* **16**, 1-12.

Murray, J., Fixter, L. M., Hamilton, I. D., Perombelon, M. C. M., Quinn, C. E. and Graham, D. C. (1990). Serogroups of potato pathogenic *Erwinia carotovora* strains: identification by lipopolysaccharide electrophoretic patterns. *Journal of Applied Bacteriology* **68**, 231-240.

Murray, M. G. and Thompson, W. F. (1980). Rapid isolation of high molecular weight plant DNA. *Nucleic Acids Research* **8**, 4321-4325.

Namba, S., Kato, S., Iwanami, S., Oyaizu, H., Shiozawa, H. and Tsuchizaki, T. (1993). Detection and differentiation of plant-pathogenic mycoplasmalike organisms using polymerase chain reaction. *Phytopathology* **83**, 786-791.

Niepold, F. (1992). Development of a method to obtain monospecific antibodies directed against a 31-kD protein of *Pseudomonas syringae* pv. *syringae*. *Journal of Phytopathology - Phytopathologische Zeitschrift* **136**, 137-146.

Paterson, J. M. and Jones, A. L. (1991). Detection of *Pseudomonas syringae* pv. *morsprunorum* on cherries in Michigan with a DNA hybridization probe. *Plant Disease* **75**, 893-896.

Paton, A. M. (1964). The adaptation of the immunofluorescence technique for use in bacteriological investigations of plant tissue. *Journal of Applied Bacteriology* **27**, 237-243.

Permar, T. A. and Gottwald, T. R. (1989). Specific recognition of a *Xanthomonas campestris* Florida citrus nursery strain by a monoclonal antibody probe in a microfiltration enzyme immunoassay. *Phytopathology* **79**, 780-783.

Persing, D. H. (1993). In vitro nucleic acid amplification techniques. *In* 'Diagnostic Molecular Microbiology: Principles and Applications'(D. H. Persing, T. F. Smith, F. Tenover and T. J. White, eds.) pp. 88-104, American Society for Microbiology, Washington, D.C.

Peters, H., Jurs, M., Jann, B., Jann, K., Timmis, K. N. and Bitter-Suermann, D. (1985). Monoclonal antibodies to enterobacterial common antigen and to *Escherichia coli* lipopolysaccharide outer core: Demonstration of an antigenic determinant shared by enterobacterial common antigen and *E. coli* K5 capsular polysaccharide. *Infection and Immunity* **50**, 459-466.

Pickup, R.W. (1991). Development of molecular methods for the detection of specific bacteria in the environment. *Journal of General Microbiology* **56**, 1009-1019.

Prince, J. P., Davis, R. E., Wolf, T. K., Lee, I.-M., Mogen, B. D., Dally, E. L., Bertaccini, A., Credi, R. and Barba, M. (1993). Specific detection of *Pseudomonas syringae* pv. *phaseolicola* DNA in bean seed by polymerase chain reaction-based amplification of a phaseolotoxin gene reaction. *Phytopathology* **83**, 1130-1137.

Prosen, D., Hatziloukas, E., Schaad, N. W. and Panopoulos, N. (1993). Specific detection of *Pseudomonas syringae* pv. *phaseolicola* DNA in bean seed by polymerase chain reaction-based amplification of a phaseolotoxin gene region. *Phytopathology* **83**, 965-970.

Reid, T., Baldini, A., Rand, T. C. and Ward, D. C. (1992). Simultaneous visualization of seven different DNA probes by *in situ* hybridization using combinatorial fluorescence and digital imaging microscopy. *Proceedings of the National Academy of Sciences of the United States of America* **89**, 1388-1392.

Randhawa, P. S. and Schaad, N. W. (1984). Selective isolation of *Xanthomonas campestris* pv. *campestris* from crucifer seeds. *Phytopathology* **74**, 268-272.

Roberts, P. (1980). Problems encountered during immunofluorescent diagnosis of fireblight. *Plant Pathology* **29**, 93-97.

Saiki, R. K., Scharf, S., Fallona, F., Mullis, K. B., Horn, G. T., Erlich, H. A. and Arnheim, N. (1985). Enzymatic amplification of β-globulin genomic sequences and restriction site analysis for the diagnosis of sickle-cell anemia. *Science* **230**, 1350-1354.

Sano, T., Smith, C. and Cantor, C. R. (1992). Immuno-PCR: very sensitive antigen detection by means of specific antibody-DNA conjugates. *Science* **258**, 120-122.

Sayler, G. S. and Layton, A. C. (1990). Environmental application of nucleic acid hybridization. *Annual Review of Microbiology* **44**, 625-648.

Schaad, N. W., Azad, H., Peet, R. C. and Panopoulos, N. (1989). Identification of *Pseudomonas syringae* pv. *phaseolicola* by a DNA hybridization probe. *Phytopathology* **79**, 903-907.

Schleifer, K. L., Amann, R., Ludwig, W., Rothemund, C., Springer, N. and Dorn, S. (1991). Nucleic acid probes for the identification and *in situ* detection of pseudomonads. *In* '*Pseudomonas*: Molecular Biology and Biotechnology'(E. Galli, S. Silver and B. Witholt, eds.) pp. 127-134, American Society for Microbiology, Washington, D.C.

Schneider, J., Zhao, J. L. and Orser, C. (1993). Detection of *Clavibacter michiganensis* subsp. *sepedonicus* by DNA amplification. *FEMS Microbiology Letters* **109**, 207-212.

Schwartz, Y., Boudon-Padieu, E., Grange, J., Meignoz, R. and Caudwell, A. (1989). Monoclonal antibodies to the mycoplasma-like organism (MLO) responsible for grapevine flavescence doree. *Research in Microbiology* **140**, 311-324.

Seal, S. E., Elphinstone, J., Skogllund, L. and Berrios, D. (1992b). Detection of *Pseudomonas solanacearum* latent infections in seed potatoes during their multiplication in Burundi. *ACIAR Bacterial Wilt Newsletter* **8**, 2-3.

Seal, S. E., Jackson, L. A., Young, J. P. W. and Daniels, M. J. (1993). Differentiation of *Pseudomonas solanacearum, Pseudomonas syzygii, Pseudomonas pickettii*, and the blood disease bacterium by partial 16S rRNA sequencing: construction of oligonucleotide primers for sensitive detection by polymerase chain reaction. *Journal of General Microbiology* **139**, 1587-1594.

Seal, S. E., Jackson, L. A. and Daniels, M. J. (1992a). Isolation of a *Pseudomonas solanacearum*-specific DNA probe by subtraction hybridization and construction of species-specific oligonucleotide primers for sensitive detection by the polymerase chain reaction. *Applied and Environmental Microbiology* **58**, 3751- 3758.

Shen, W. C. and Lin, C. P. (1993). Production of monoclonal antibodies against a mycoplasmalike organism associated with sweet potato witches' broom. *Phytopathology* **83**, 671-675.

Shibata, D. K. (1992). The polymerase chain reaction and the molecular genetic analysis of tissue biopsies. *In* 'Diagnostic Molecular Pathology: A Practical Approach'(C. S. Herrington and J. O. McGee, eds.) pp. 85-111, Oxford University Press, Oxford, England.

Siebert, P. D. and Larrick, J. W. (1993). PCR MIMICS: competitive DNA fragments for use as internal standards in quantitative PCR. *Biotechniques* **14**, 244-249.

Slack, S. A., Kelman, A. and Perry, J. B. (1979). Comparison of three serodiagnostic assays for detection of *Corynebacterium sepedonicum. Phytopathology* **69**, 186-189.

Steffan, R. J. and Atlas, R. M. (1991). Polymerase chain reaction: applications in environmental microbiology. *Annual Review of Microbiology* **45**, 137-161.

Stephens, C., Stebbins, T. and Fulbright, D. (1988). Detection of *Clavibacter michiganensis* subsp. *michiganensis* using four selective media and pathoscreen ELISA assay (Agdia). *In* 'Proceedings of the 4th Bacterial Canker Workshop' pp. 6-14, Windsor, Ontario.

Takahashi, Y., Tsuchiya, K., Ahohara, K. and Suzui, T. (1990). Production and utilization of monoclonal antibodies against *Pseudomonas cepacia*. *Annals of the Phytopathological Society of Japan* **56**, 229-234.

Tenover, F. C. and Unger, E. R. (1993). Nucleic acid probès for detection and identification of infectious agents. *In* 'Diagnostic Molecular Microbiology: Principles and Applications'(D. H. Persing, T. F. Smith, F. Tenover and T. J. White, eds.) pp. 3-25, American Society for Microbiology, Washington, D.C.

Thompson, E., Leary, J. V. and Chun, W. W. C. (1989). Specific detection of *Clavibacter michiganense* subsp. *michiganense* by a homologous DNA probe. *Phytopathology* **79**, 311-314.

Underberg, H. A. and Sander, E. (1991). Detection of *Corynebacterium sepedonicum* with antibodies raised in chicken egg yolks. *Journal of Plant Diseases and Protection* **98**, 188-196.

van der Wolf, J. M., van Beckhoven, J. R. C. M., De Boef, E. and Roozen, N. J. M. (1993). Serological characterization of fluorescent *Pseudomonas* strains cross-reacting with antibodies against *Erwinia chrysanthemi*. *Netherlands Journal of Plant Pathology* **99**, 51-60.

van der Wolf, J. M., van Beckhoven, J. R. C. M., de Vries, Ph. M., Raaijmakers, J. M., Bakker, P. A. H. M., Bertheau, Y. and Van Vuurde, J. W. L. (1995). Polymerase chain reaction for verification of fluorescent colonies of *Erwinia chrysanthemi* and *Pseudomonas putida* WCS358 in immunofluorescence colony staining. *Journal of Applied Bacteriology* **79**, 569-577.

Van Vaerenbergh, J. P. C. and Chauveau, J. F. (1987). Detection of *Corynebacterium michiganense* in tomato seed lots. *EPPO Bulletin* **17**, 131-138.

Van Vuurde, J. W. L. (1987). New approach in detecting phytopathogenic bacteria by combined immunoisolation and immunoidentification assays. *EPPO Bulletin* **17**, 139-148.

Van Vuurde, J. W. L., Franken, A. A. J. M., Birnbaum, Y. and Jochems, G. (1991). Characteristics of immunofluorescence microscopy and of dilution-plating to detect *Pseudomonas syringae* pv *phaseolicola* in bean seed lots and for risk assessment of field incidence of halo blight. *Netherlands Journal of Plant Pathology* **97**, 233-244.

Van Vuurde, J. W. L. and Roozen, N. J. M. (1990). Comparison of immunofluorescence colony-staining in media, selective isolation on pectate medium, ELISA and immunofluorescence cell staining for detection of *Erwinia carotovora* subsp. *atroseptica* and *E. chrysanthemi* in cattle manure slurry. *Netherlands Journal of Plant Pathology* **96**, 75-89.

Van Vuurde, J. W. L. and Van Henten, C. (1983). Immunosorbent immunofluorescence microscopy (ISIF) and immunosorbent dilution- plating (ISDP): New methods for the detection of plant pathogenic bacteria. *Seed Science and Technology* **11**, 523-533.

Vernon-Shirley, M. and Burns, R. (1992). The development and use of monoclonal antibodies for detection of erwinia. *Journal of Applied Bacteriology* **72**, 97-102.

Verreault, H., Lafond, M., Asselin, A., Banville, G. and Bellemare, G. (1988). Characterization of two DNA clones specific for identification of *Corynebacterium sepedonicum*. *Canadian Journal of Microbiology* **34**, 993-997.

Vivian, A. (1992). Identification of plant pathogenic bacteria using nucleic acid technology . *In* 'Techniques for the Rapid Detection of Plant Pathogens'(J. M.

Duncan and L. Torrance, eds.) pp. 145-161, Blackwell Scientific Publications, London, England.

Vruggink, H. and Maas Geesteranus, H. P. (1975). Serological recognition of *Erwinia carotovora* var. *atroseptica*, the causal organism of potato blackleg. *Potato Research* **18**, 546-555.

Ward, L. J. and De Boer, S. H. (1989). Characterization of a monoclonal antibody against active pectate lyase from *Erwinia carotovora*. *Canadian Journal of Microbiology* **35**, 651-655.

Ward, L. J. and De Boer, S. H. (1990). A DNA probe specific for serologically diverse strains of *Erwinia carotovora*. *Phytopathology* **80**, 665-669.

Wells, J. M., Raju, B. C.,. Hung, H.-H., Weisburg, W. C., Mandelco-Paul, L. and Brenner, D. J. (1987). *Xylella fastidiosa* gen. nov., sp. no: gram-negative, xylem-limited, fastidious plant bacteria related to *Xanthomonas* spp. *International Journal of Systematic Bacteriology* **37**,

Wingate, V. P. M., Norman, P. M. and Lamb, C. J. (1990). Analysis of the cell surface of *Pseudomonas syringae* pv. *glycinea* with monoclonal antibodies. *Molecular Plant-Microbe Interactions* **3**, 408-416.

Wong, W. C. (1991). Methods for recovery and immunodetection of *Xanthomonas campestris* pv. *phaseoli* in navy bean seed. *Journal of Applied Bacteriology* **71**, 124-129.

Yourno, J. (1992). A method for nested PCR with single closed reaction tubes. *PCR Methods and Applications* **2**, 60-65.

Zielke, R. and Kalinina, I. (1988). Ein beitrag zum nachweis von *Clavibacter michiganensis* subsp. *sepedonicus* (Spieckermann & Kotthoff) Davis et al. im pflanzengewebe mit dem mikroliter ELISA verfahren. *Zentralblatt Mikrobiologie* **143**, 5-16.

# 3

# SENSITIVITY OF INDEXING PROCEDURES FOR VIRUSES AND VIROIDS

## H. Huttinga

*DLO Research Institute for Plant Protection,
Wageningen, The Netherlands*

## I. INTRODUCTION

Viruses and viroids are easily transmitted along with vegetative plant propagules, be it tubers, cuttings, bulbs, scions or budwood for grafting, when these originate from infected plants. The number of plant viruses that can be transmitted by seed is more than one hundred (Bos, 1977). In tropical legume crops alone, 36 viruses have been listed as seed borne (Frison *et al.*, 1990). Some viroids can also be transmitted by seed and pollen (Diener, 1979). Viruses and viroids also are spread in agricultural products that are sold over long distances, in an ever increasing world market under the General Agreement on Tariffs and Trade (GATT). In this way they have been spread all over the world (Bos, 1992).

The only way to prevent unlimited and uncontrolled spread of pathogens is indexing of agricultural products, especially propagating material. A prerequisite for good indexing procedures is basic knowledge of the pathogens of concern. Viruses have to be identified, characterized and classified, to finally enable the development of reliable and sensitive detection methods that can routinely be used in large-scale indexing.

Advances in Botanical Research Vol. 23
Incorporating Advances in Plant Pathology
ISBN 0-12-005923-1

Plant viruses and viroids can, in principle, be studied by two approaches. One way is to study their effects as biological agents in relation to plants; the other way is to study their intrinsic properties, like antigenic properties, size and molecular mass. Both approaches lead to definition of characteristics that may also be used to detect them in indexing procedures.

Some plant viruses and viroids can be detected by methods based on their interaction with plants such as by inoculating an extract of the material to be tested onto an appropriate host plant. This may be the natural host or an alternate host that is sensitive to the infectious agent. In principle one particle of a one-component virus, when inoculated to an infection site, is capable of starting a new infection causing the test plant to produce diagnostic symptoms. From such sympomatic test plants it can be determined whether the material from which the extract used for inoculation was made, was infected by a virus. If specific symptoms are produced, they may give a clue as to which virus is involved.

Detection of viruses and viroids by an intrinsic property means that it first has to be proven that the intrinsic property is uniquely linked with the phytopathological characteristics of the pathogen. That being so, the intrinsic property can be used to detect these pathogens in plants and sometimes also in other material such as aphid and nematode vectors. Detection methods based on intrinsic properties often have the advantage that they can easily be automated for application in large-scale indexing.

In general one should realize that detecting the presence of a pathogen in a diseased plant is not identical to diagnosing a disease. In addition to the pathogen found by a specific detection method, there may be other factors that help or may be entirely responsible for the disease.

## II. INDEXING BASED ON TEST PLANTS

Viruses and viroids may be identified and/or detected by their specific symptoms on natural hosts or specific test plants. This is a very basic method, because it uses the pathogenic properties of the pathogen. Bos (1978) compiled a consistent nomenclature enabling the correct description of the different symptoms of virus diseases in plants, thus promoting mutual understanding among workers in this field.

Sometimes viruses produce very specific symptoms in the plants that have to be indexed. In such cases, like the indexing for the shallot strain of onion yellow stripe virus in shallot corms, the corm lots are sampled and the selected corms grown to see whether or not they will develop symptoms. In many cases, however, specific test plants have to be produced. This may be a burden when the plants have to be in a specific physiological state to be able to react to the virus or the viroid and the test has to be performed throughout the year in different seasons. It may then take specialized facilities to grow the plants prior to inoculation and also to provide optimal conditions for symptom development after infection. Growing such test plants is expensive and time-consuming. Furthermore, tests plants have to be inoculated with extracts from individual samples. Preparing and inoculating many samples is laborious and costly. Finally, the judging of symptoms often requires a specialist's eye and experience.

If plant viruses or viroids do not produce specific symptoms, but instead cause symptoms that are similar to those of other diseases, detection and indexing by symptoms is only partially useful, and diagnosis has to be combined with another method. Latent pathogens, which produce no symptoms at all, cannot be indexed by inoculation to host plants.

Detection by bioassay on indicator plants can be very sensitive because a pathogen in a host multiplies and is amplified. In systemic infections the whole plant will become infected through cell-to-cell movement of the virus, resulting in symptoms in almost all plant parts. In theory one can detect $80 \times 10^{-18}$ g of potato virus Y if one manages to bring one of the potyvirus particles to a site in the plant where it can start a new infection. In reality much higher numbers of infectious units are needed to start an infection. Sänger *et al.* (1976) found that about 10% of tomato plants became infected when they were inoculated with a solution containing 50-100 potato spindle tuber viroid (PSTVd) molecules per plant.

The amplification of pathogens in physiologically active plant material is often used to increase the amount of pathogen present in plant material in order to increase the signal to noise ratio for a subsequent test. An example of that is the sprouting of dormant seed potato tubers. In the dormant tubers some viruses cannot be detected using ELISA. However, after breaking of dormancy these viruses can be detected in the sprouts and/or the leaflets, using ELISA (De Bokx, 1987; Vetten *et al.*, 1983). Morris and Smith (1977) used inoculation of an intermediate tomato plant to amplify PSTVd to increase the analytical sensitivity of their detection method based on polyacrylamide-gel electrophoresis.

## III. INDEXING BASED ON INTRINSIC PROPERTIES

Most plant viruses consist of a relatively simple structure of nucleic acid protected by a coat of multiple repeated protein molecules. In fact both constituents can be used to detect viruses. Approximately 90% of the plant viruses are single-stranded RNA (ssRNA) viruses, 4% contain double-stranded RNA (dsRNA), and 6% of the plant viruses contain DNA (Francki *et al.*, 1991). Because ssRNA is hard to extract from plants, it is not surprising that in the past no methods for detection have been directed towards ssRNA. However, in plants infected by RNA-containing plant viruses one can find detectable amounts of replicative dsRNAs and these can be used to show that plants are infected by an RNA virus. The number and size of the dsRNAs may suggest what kind of virus is involved (Valverde *et al.*, 1990).

## A. Serology

It is quite logical that in the past most of the detection methods have been directed towards the intact particle and/or the coat protein. Many plant viruses are relatively easy to purify, and purified plant viruses with their repetitive coat-protein subunits form ideal antigens. The best known detection methods based on intrinsic properties are serological protocols using specific antibodies directed against antigenic sites on the protein coat of virus particles. The antibodies can, in general, be raised from vertebrates that are injected with antigens. In most cases rabbits are used but when in the past large quantities were needed, sheep and goats and even horses were used (Maat and Huttinga, 1987). Also chickens have been injected with antigens, and the antibodies collected from the egg yolks (Bar-Joseph and Malkinson, 1980). All antibodies produced in animals are so-called polyclonal antibodies.

Köhler and Milstein (1975) were the first to produce so-called monoclonal antibodies. The latter are directed against only one epitope and thus can be more specific than polyclonal antibodies and that is often an advantage. Sometimes when a detection method has to detect all strains of a virus monoclonal antibodies may be too specific.   However, Jordan and Hammond (1991) produced monoclonal antibodies with various specificities, one of them reacts specifically with almost all potyviruses.

Antibodies react specifically with the homologous antigen. The reaction however is not easy to detect. Therefore many protocols have been described to visualize the antibody-antigen reaction. In the first serological detection tests, the blood serum of the immunized animals (antiserum) was used as such. For example, Van Slogteren (1955) mixed small droplets of centrifuged plant sap with droplets of antiserum and monitored the aggregation of virus and antibodies with a binocular microscope. Ouchterlony (1968) described a method in which antibodies and antigens were allowed to approach each other by diffusion in agar. A precipitin band, usually visible to the naked eye, forms at the leading edge were diffusing antigen and antibody molecules meet. The main disadvantage of these methods is the low sensitivity. They are about 100-1000 times less sensitive than a modern ELISA-test. The latex-agglutination test, in which antibodies are first adsorbed to the much bigger latex particles, more or less equals ELISA in sensitivity but has the disadvantage that it is less suited for large-scale application.

Antibodies can also be used in combination with electron microscopy, in procedures known as immunoelectron microscopy (IEM) (Roberts et al., 1982). IEM combines the sensitivity of the electron microscope with the specificity of serological reactions. Many protocols have been developed for IEM (Derrick, 1973; Milne and Luisoni, 1977).

In double antibody sandwich enzyme-linked immunosorbent assays (DAS-ELISA) the antigen is trapped to a solid phase with antibodies and then coated by antibodies coupled to an enzyme. If the homologous antigen is present the enzyme is bound to the solid phase and can be detected by having a substrate cleaved by the enzyme, forming a colored product. DAS-ELISA has become extremely popular for testing viruses because the method is easy to perform in large numbers by relatively untrained personnel at a low cost per test (Clark and Adams, 1977). In the Netherlands alone, about $11 \times 10^6$  ELISA tests are performed annually on propagating material, which includes about $5 \times 10^6$ tests for indexing of seed potatoes and about $6 \times 10^6$ tests for indexing ornamental-plant material.

## B. Nucleic Acid-Based Tests

For viroids, having no coat proteins, serology does not work. However, here another intrinsic property was used to design a detection method, namely the high degree of base pairing within the circular RNA molecule (Sänger *et al.*, 1976). The bi-directional polyacrylamide-gel electrophoresis test makes use of this property and is now used routinely to detect viroids in potato, chrysanthemum, and other crop and ornamental plants (Schumacher *et al.*, 1983; Huttinga *et al.*, 1986; Schumacher *et al.*, 1986). In this test nucleic acid extracts of plants are analyzed in two electrophoretic runs. In the first run under native conditions the viroid molecules are separated from the bulk of the plants' nucleic acids. In the second run under denaturing conditions (high temperature) the viroid molecules having lost their internal base-pairing behave as larger molecules and can be separated from the nucleic acids that co-migrated in the first run. Using silver staining to visualize the viroid band, c. 5 ng of viroid per slot can be detected with this test, which corresponds to c. 10 ng of viroid per gram of leaf material (Huttinga *et al.*, 1986). The test is routinely used by the Dutch Inspection Service for Floriculture and Arboriculture to test for chrysanthemum stunt viroid.

When new molecular-biological methods helped to characterize the genomes of plant viruses and viroids, it became tempting to use molecular hybridization for detecting plant viruses (Sela *et al.*, 1984; Saldarelli *et al.*, 1994) and viroids (Owens and Diener, 1981). When complete sequences of different viruses became available, it was possible to select unique regions of specific viruses to define specific probes, and to construct group-specific probes directed towards more conserved regions of the genomes. However, in routine testing molecular hybridization has no real advantage over ELISA with respect to sensitivity (De Bokx and Cuperus, 1987). When later on the polymerase chain reaction (PCR) (Saiki *et al.*, 1988) became more easily accessible, a new interest in the genome of plant viruses and viroids as a basis for designing detection methods became apparent. Many DNA amplification methods have been developed (Birkenmeyer and Mushahwar, 1991). Vunsh *et al.* (1990) were the first to combine reverse transcription and PCR for detection of an RNA virus. Especially during the last few years many applications of PCR-based detection methods have been described. PCR can be done directly in tubes where the virus particles have been trapped by immunocapture (Wetzel *et al.*, 1992; Candresse *et al.*, 1993). In a recent paper Nolasco *et al.* (1993) described a combined immunocapture and PCR method that can be performed in a microtiter plate. This test has the typical analytical sensitivity of assays based on the polymerase chain reaction, is not more laborious than ELISA, and allows an equivalent degree of automation. Lair *et al.* (1994) developed a method in which viral RNA is amplified using a single temperature procedure in order to detect citrus tristeza virus in plant nucleic acid extracts. Henson and French (1993) published a recent review on the polymerase chain reaction and plant disease diagnosis.

## IV. SENSITIVITY VERSUS SPECIFICITY

In designing a detection method one always has to compromise between sensitivity and specificity. Analytical sensitivity and analytical specificity can be differentiated from diagnostic sensitivity and diagnostic specificity as defined by Sheppard *et al.* (1986). The term analytical sensitivity is a measure of the precision of the test, or the minimum amount detectable within a given system. The term analytical specificity is frequently used to describe the degree of cross-reactivity in a test system. The diagnostic sensitivity of a test equals the percentage of infected samples that are found infected by the test. The diagnostic specificity of a test equals the percentage of non-infected samples that are found negative by the test. In the ideal situation both diagnostic sensitivity and diagnostic specificity are 100%. General experience, however, is that this situation is seldom reached and that attempts to increase the diagnostic sensitivity, result in a decrease in diagnostic specificity. It is also true that attempts to increase the diagnostic specificity of a test will in most cases result in lower diagnostic sensitivity. This dilemma is well illustrated as follows: A physician that tells every patient that consults him, that he or she is suffering from influenza, will reach the highest possible score with respect to the detection of influenza: 100%. However, the diagnostic specificity of his detection method is 0%: none of the non-influenza patients is indicated as negative. As soon as the physician tries to be more specific in his diagnosing of patients he will notice that his diagnostic sensitivity goes down. He will indicate false-negatives: he will miss a case of influenza every now and then. This story reminds us that we should never be satisfied with a test having a diagnostic sensitivity of 100%, without knowing its diagnostic specificity. Diagnostic sensitivity may be so low, that the test is not at all useful. What should be done when one has to choose between tests with different values for diagnostic sensitivity and diagnostic specificity? Sheppard *et al.* (1986) have already answered this question:

> High diagnostic sensitivity is required where false-negative results cannot be tolerated and where false-positive results are acceptable (e.g. situations where disease eradication is imperative irrespective of supplies). High diagnostic specificity is required for tests in which false-positive results cannot be tolerated, but false-negative results can be tolerated (situations where the disease can be controlled or managed without major economic losses or disruption of seed supply). Frequently in serial testing, an initial test of high diagnostic sensitivity and low diagnostic specificity is used to 'screen out' 'positive' samples. These 'positive' samples are then subjected to a second test having a low diagnostic sensitivity and a high diagnostic specificity.

In those cases where the prevalence of the pathogen is low, high diagnostic specificity is more important for a test than high diagnostic sensitivity, because this will prevent that too many false-positive results are obtained. However, the diagnostic sensitivity not being 100%, every now and then results in a small number of false-negative results. This may not be acceptable if one is dealing with quarantine organisms.

# V. DETECTION LIMITS FOR VIRUSES AND VIROIDS IN FIELD SAMPLES

A new detection method is almost always developed in a laboratory using test samples of relatively pure viruses or viroids and one can optimize the method to a point that it has high analytical sensitivity and a high specificity. It is then always a difficult but exciting part in detection research to implement a detection method that has been developed in the laboratory, for large-scale testing of field samples. A field sample, mostly a homogenate of plant parts in a buffer solution, usually is a complex mixture of many components. The pathogen to be detected is often only one amongst many components, and sometimes is present in only small amounts. In some instances the other components of such a sample interfere in the test in such a way that the final results become inconclusive. For example, extracts of *Arabis* mosaic virus-infected blackcurrant buds inhibited the ELISA-reaction (Clark and Adams, 1977). The signal which identifies the specific reaction of an antibody or molecular probe with the target pathogen may be too small in relation to the background noise caused by nonspecific reactions that occur with the other components in the sample.

It is obvious that a good detection method for large-scale testing of field samples should have a high signal to noise ratio (S:N). A high S:N ratio insures that the equivocal results in the grey interval between the unequivocal readings for healthy and diseased samples occurs relatively infrequently. In the grey interval one finds the false-positive and the false-negative results of a test. A false-positive reaction will cause a producer to be denied certification for his product; a false-negative will cause the buyer of a product to get a product that is inferior to that expected on the basis of the accompanying certificate. Both situations should be avoided. How can a S:N ratio be maximized? In short, by increasing the signal and/or by decreasing the noise.

## A. Decreasing Noise in Immunoassays

Decreasing the background noise in immunoassays is acceptable as long as this does not decrease the signal. A very efficient way to increase the analytical specificity of serological tests would be production of more specific antibodies and/or antibodies with a low tendency to adhere indiscriminately to different kinds of material present in the test sample. Antibodies of the IgG type in general are less 'sticky' than the IgM types.

Another way to prevent nonspecific reactions is to block the reactive groups with additives such as skimmed milk powder (Johnson *et al.*, 1984). However, experiences in our laboratory with the detection of shallot latent virus proved that in many cases the signal is also reduced and that the optimal amount of additive differs from sample to sample, making the method impractical for large-scale testing. However, Van den Heuvel and Peters (1989) were very successful in increasing the absorbance values of infected plant material in a cocktail ELISA for potato leafroll virus while the background signals were reduced, by adding sodium diethyldithiocarbamate or ethylenediaminetetraacetic acid to the reaction mixture.

Nonspecific reactions of components in field samples can also be prevented by the removal of these components by partial purification of the sample. Many times a rather simple sample preparation procedure such as a ten- to 200-fold dilution of the homogenate with buffer has a tremendous effect on ELISA-tests (Clark and Adams, 1977; Flegg and Clark, 1979). In other cases where more complicated steps have to be performed and large numbers of samples have to be handled, this will not be an acceptable solution. Many nucleic-acid based tests need rather intensive sample preparation and this is a serious disadvantage of these tests.

## B. Increasing Signal in Indexing Procedures

Increasing the signal for a serological or a nucleic-acid based test is only worthwhile if the noise is not increased at the same time at the same rate. The signal is often amplified by sophisticated systems that build 'Christmas tree'-like structures with markers on a specifically bound antibody (e.g. biotin-avidin systems). However, similar structures will be built with markers on an antibody bound nonspecifically. Thus not improving the signal-to-noise ratio.

Inducing the multiplication of the pathogen to be tested in physiologically active plant material as a means of increasing test sensitivity has already been mentioned before. It also is a very effective way to improve the signal-to-noise ratio. For example, the sprouting of dormant seed potato tubers before testing has been applied successfully for many years in the Netherlands, but the disadvantage is that this step takes time. For this reason more sensitive methods that can eliminate the requirement for multiplication are being actively sought. A shorter indexing time would enable seed-potato growers to obtain their certification earlier which would enhance sales to markets with early planting seasons.

The latest developments in attempts to increase the signal in detection methods is the use of PCR to amplify the number of specific nucleotide sequences of the pathogen. The introduction of the so-called reverse transcription-PCR (RT-PCR) provided a simple method to amplify parts of ssRNA genomes and has already been applied for viruses in host tissues (Robertson et al., 1991; Wetzel et al., 1992) in dormant corms (Vunsh et al., 1990), tubers (Barker et al., 1993; Spiegel and Martin, 1993), woody tissues (Korschinek et al., 1991; Rowhani et al., 1993), and even aphids (Hadidi et al., 1993). Hadidi and Yang (1990) used the method to detect viroids. In those cases where a simple PCR did not work the use of nested primers proved to be helpful (Van der Wilk et al., 1994). The PCR by itself is not a detection method, it has to be followed by polyacrylamide-gel electrophoresis and/or molecular hybridization with a specific probe. This makes the use of PCR-based tests relatively time-consuming. However, new methods in which two differently marked primers are used, one of which is used to bind the amplified product to a solid base, will enable an ELISA-type format for the evaluation of the PCR products in microtiter plates. For example, Rasmussen et al. (1994) used one primer that was covalently bound to the wall of a microwell and a free primer for the amplification reaction. The amplification product bound to the well was then detected by hybridizing with a biotinylated probe, followed by coupling to a streptavidin-alkaline phosphatase conjugate, incubation with a fluorochrome substrate, and reading of the fluorescence.

Similar systems are now also available commercially.

The main advantage of PCR, its sensitivity, is also its main disadvantage. Individual samples have to be handled very carefully in order to prevent cross-contamination of samples. The use of a Pollähne-press, for example, to prepare samples of plant sap with just a simple washing of the rollers between samples will not be possible because of virus carry-over which will lead to false positive test results. The implementation of PCR for routine large-scale testing will largely depend on the development of equipment that can be used to prepare and handle large numbers of samples without cross-contamination.

## C. Selection of Indexing Method

How should one choose a method to index for viruses and viroids? This is a question that cannot be answered simply, even though it is always tempting to use a method with high analytical sensitivity. Is an ELISA-method for potato virus $Y^N$ ($PVY^N$) that can detect 10 ng of the virus in a microplate well inferior to a molecular hybridization dot blot test that detects 30 pg of the virus per spot? The latter test has the best analytical sensitivity. As can be seen from Table 1 compiled from our own experiments, 200 µl of sap that represents 0.4 g of leaves is used in ELISA, whereas due to the characteristics of the hybridization test only 3 µl of sap representing 0.006 g of leaves can be applied. Consequently the concentration of virus, expressed as ng virus per g of leaves, that can be detected by ELISA and the hybridization test is in the same order of magnitude. The big advantage of the hybridization test with respect to sensitivity is lost due to other detrimental characteristics of the test. De Bokx and Cuperus (1987) when comparing cDNA hybridization and three modifications of ELISA for the detection of PVY in dormant potato tubers found that the hybridization technique yielded slightly better results than the standard ELISA procedure. However, the hybridization technique was time-consuming and involved hazardous reagents not acceptable for routine analyses. Similar results were found

*Table 1.* Sensitivity of cDNA hybridization versus that of ELISA to detect PVY, and dependence on sample size.

| Test | Analytical sensitivity (ng) | Lowest concentration detected |
|------|-----------------------------|-------------------------------|
| ELISA | 2.0 | 2 ng/0.4 g = 5 ng/g |
| cDNA hybridization | 0.03 | 0.03 ng/0.006 g = 5 ng/g |
| | Sample size | |
| ELISA | 200 µl | corresponding to c. 0.4 g of leaves |
| cDNA hybridization | 3 µl | corresponding to c. 0.006 g of leaves |

when a hybridization test for the detection of viroids was compared with a method using bi-directional electrophoresis (Huttinga *et al.*, 1986).

This analysis indicates that in designing an indexing method for viruses and viroids many factors have to be considered. Sensitivity is important, as is specificity, but they are not the only determining factors. Ease of sample preparation and simplicity of the test itself, time available for the test, and last but not least, the cost of a test is also a very important factor.

## VI. CONCLUSION

Indexing for viruses and viroids is an important aspect in the free trade of plant material all over the world. Many methods are available that can detect viruses and viroids in a quick and reliable way. They are almost always based on intrinsic properties of the pathogens and can detect molecular units at the pg level in individual tests. Whether or not these methods are useful for indexing depends to a large extend on many factors, like simplicity of the test, the amount of time that is needed to perform the test, whether or not specific and /or expensive chemicals (e.g. rapidly decaying radiochemicals) or equipment are needed, and whether or not highly trained personnel is needed to perform the test or to read the results.

It is now generally realized that indexing the quality of a product is not something that should be done only when the product is sold. Indexing by itself cannot improve the quality of a product, it only assesses the quality of it at a certain point in time. The whole production process for a crop, particularly propagation material, should be monitored and be directed toward providing a high-quality product. This means that pathogen-free plant material has to be carefully monitored throughout the growing season and from season to season. Propagation material should be indexed regularly and spread of viruses and viroids prevented by appropriate measures.

For good plant production procedures with regular checks on the health status of the material, the sensitivity of the test is not the most crucial factor. An early infection that is missed in one assay will be found in a subsequent test, because in the mean time it has multiplied and increased to a detectable level.

## REFERENCES

Bar-Joseph, M. and Malkinson, M. (1980). Hen egg yolk as a source of antiviral antibodies in the enzyme-linked immunosorbent assay (ELISA): a comparison of two plant viruses. *Journal of Virological Methods* **1**, 179-183.

Barker, H., Webster, K.D. and Reavy, B. (1993). Detection of potato virus Y in potato tubers: a comparison of polymerase chain reaction and enzyme-linked immunosorbent assay. *Potato Research* **36**, 13-20.

Birkenmeyer, L.G. and Mushahwar, I.K. (1991). DNA probe amplification methods. *Journal of Virological Methods* **35**, 117-126.

Bos, L. (1977). Seed-borne viruses. *In* 'Plant Health and Quarantine in International Transfer of Genetic Resources' (W.B. Hewitt and L. Chiarappa, eds), pp. 36-69.

CRC Press, Cleveland, Ohio.

Bos, L. (1978). 'Symptoms of Virus Diseases in Plants'. 3rd edition. Centre for Agricultural Publishing and Documentation, Wageningen.

Bos, L. (1992). New plant virus problems in developing countries: a corollary of agricultural modernization. *Advances in Virus Research* **41**, 349-407.

Candresse, T., Revers, F., Lanneau, M., Macquaire, G., Wetzel, T. and Dunez, J. (1993). Use of PCR to detect two fruit tree viruses and to analyze their molecular variability. Abstracts Book of IXth International Congress of Virology, p. 108. Glasgow, 8-13 August 1993.

Clark, M.F. and Adams, A.N. (1977). Characteristics of the microplate method of enzyme-linked immunosorbent assay for the detection of plant viruses. *Journal of General Virology* **34**, 475-483.

De Bokx, J.A. (1987). Characterization and identification of potato viruses and viroids. Biological properties. *In* 'Viruses of Potatoes and Seed-Potato Production' (J.A. de Bokx and J.P.H. van der Want, eds), pp. 58-82. Centre for Agricultural Publishing and Documentation, Wageningen.

De Bokx, J.A. and Cuperus, C. (1987). Detection of potato virus Y in early-harvested potato tubers by cDNA hybridization and three modifications of ELISA. *EPPO Bulletin* **17**, 73-79.

Derrick, K.S. (1973) Quantitative assay for plant viruses using serologically specific electron microscopy. *Virology* **56**, 652-653.

Diener, T.O. (1979). 'Viroids and Viroid Diseases'. John Wiley & Sons, New York.

Dodds, J.A., Morris, T.J. and Jordan, R.L. (1984). Plant viral double-stranded RNA. *Annual Review of Phytopathology* **22**, 151-168.

Flegg, C.L. and Clark, M.F. (1979). The detection of apple chlorotic leafspot virus by a modified procedure of enzyme-linked immunosorbent assay (ELISA). *Annals of Applied Biology* **91**, 61-65.

Francki, R.I.B., Fauquet, C.M., Knudson, D.L. and Brown, F. (Eds) (1991). 'Classification and Nomenclature of Viruses'. Archives of Virology, Supplement 2. Springer-Verlag, Wien and New York.

Frison, E.A., Bos, L., Hamilton, R.I., Mathur, S.B. and Taylor, J.W. (Eds) (1990). 'FAO/IBPGR Technical Guidelines for the Safe Movement of Legume Germplasm'. FAO/IBPGR, Rome.

Hadidi, A. and Yang, X. (1990). Detection of pome fruits viroids by enzymatic cDNA amplification. *Journal of Virological Methods* **30**, 261-269.

Hadidi, A., Montasser, M.S., Levy, L., Goth, R.W., Converse, R.H., Madkour, M.A. and Skrzeckowski, L.J. (1993). Detection of potato leafroll and strawberry mild yellow-edge luteoviruses by reverse transcription-polymerase chain reaction amplification. *Plant Disease* **77**, 595-601.

Henson, J.M. and French, R. (1993). The polymerase chain reaction and plant disease diagnosis. *Annual Review of Phytopathology* **31**, 81-109.

Huttinga, H., Mosch, W.H.M. and Treur, A. (1986). Detection of viroids by molecular hybridization and bi-directional electrophoresis in the Netherlands. *In* 'Viroids of plants and their detection'. Proceedings of an International Seminar, pp. 93-100. Warsaw, 12-20 August 1986.

Johnson, D.A., Gautsch, J.W., Sportsman, J.R. and Elder, J.H. (1984). Improved technique utilizing nonfat dry milk for analysis of proteins and nucleic acids transferred to nitrocellulose. *Gene Analysis Techniques* **1**, 3-8.

Jordan, R. and Hammond, J. (1991). Comparison and differentiation of potyvirus isolates and identification of strain-, virus, subgroup-specific and potyvirus group-common epitopes using monoclonal antibodies. *Journal of General Virology* **72**, 25-36.

Köhler, G. and Milstein, C. (1975). Continuous cultures of fused cells secreting antibody of predetermined specificity. *Nature* **256**, 495-497.

Korschineck, I., Himmler, G., Sagl, R., Steinkeller, H. and Katinger, H.W.D. (1991). A PCR membrane spot assay for the detection of plum pox virus RNA in bark of infected trees. *Journal of Virological Methods* **31**, 139-145.

Lair, S.V., Mirkov, T.E., Dodds, J.A. and Murphy, M.F. (1994). A single temperature amplification technique applied to the detection of citrus tristeza viral RNA in plant nucleic acid extracts. *Journal of Virological Methods* **47**, 141-151.

Maat, D.Z. and Huttinga, H. (1987). Serology. *In* 'Viruses of Potatoes and Seed-Potato Production' (J.A. de Bokx and J.P.H. van der Want, eds), pp. 45-57. Centre for Agricultural Publishing and Documentation, Wageningen.

Milne, R.G. and Luisoni, E. (1977). Rapid immune electron microscopy of virus preparations. *In* 'Methods in Virology VI' (K. Maramorosch and H. Koprowski, eds), pp. 270-274. Academic Press, New York.

Morris, T.J. and Smith, E.M. (1977). Potato spindle tuber disease: procedures for the detection of viroid RNA and certification of disease-free potato tubers. *Phytopathology* **67**, 145-150.

Nolasco, G., de Blas, C., Torres, V. and Ponz, F. (1993). A method combining immunocapture and PCR amplification in a microtiter plate for the detection of plant viruses and subviral pathogens. *Journal of Virological Methods* **45**, 201-218.

Ouchterlony, O. (1968). 'Handbook of Immunodiffusion and Immunoelectrophoresis'. Prog. Allergy, Ann. Arbor. Publish, Michigan.

Owens, R.A. and Diener, T.O. (1981). Sensitive and rapid diagnosis of potato spindle tuber viroid disease by nucleic acid hybridization. *Science* **213**, 670-672.

Rasmussen, S.R., Rasmussen, H.B., Larsen, M.R., Hoff-Jørgensen R. and Cano, R.J. (1994). Combined polymerase chain reaction-hybridization microplate assay used to detect bovine leukemia virus and *Salmonella*. *Clinical Chemistry* **40**, 200-205.

Roberts, I.M., Milne, R.G. and Van Regenmortel, M.H.V. (1982). Suggested terminology for virus/antibody interactions observed by electron microscopy. *Intervirology* **18**, 147-149.

Robertson, N.L., French, R. and Gray, S.M. (1991). Use of group-specific primers and the polymerase chain reaction for the detection and identification of luteoviruses. *Journal of General Virology* **72**, 1473-1477.

Rowhani, A., Chay, C., Golino, D.A. and Falk, B.W. (1993). Development of a polymerase chain reaction technique for the detection of grapevine fanleaf virus in grapevine tissue. *Phytopathology* **83**, 749-753.

Saiki, R.K., Gelfand, G.H., Stoffel, S., Scharf, S.J., Higuchi, R., Horn, G.T., Mullis, K.B. and Ehrlich, H.A. (1988). Primer directed enzymatic amplification of DNA with a thermostable DNA polymerase. *Science* **239**, 487-491.

Saldarelli, P., Minafra, A., Martelli, G.P. and Walter, B. (1994). Detection of grapevine leafroll-associated closterovirus III by molecular hybridization. *Plant Pathology* **43**,

91-96.

Sänger, H.L., Klotz, G., Riesner, D., Gross, H.J. and Kleinschmidt, A.K. (1976). Viroids are single-stranded covalently closed circular RNA molecules existing as highly base-paired rod-like structures. *Proceedings National Academy of Sciences* **73**, 3852-3856.

Schumacher, J., Meyer, N., Riesner, D. and Weidemann, H.L. (1986). Routine technique for diagnosis of viroids and viruses with circular RNAs by 'return'-gel electrophoresis. *Journal of Phytopathology* **115**, 332-343.

Schumacher, J., Randles, J.W. and Riesner, D. (1983). A two-dimensional technique for the detection of circular viroids and virusoids. *Analytical Biochemistry* **135**, 288-295.

Sela, I., Reichman, M. and Weissbach, A. (1984). Comparison of dot molecular hybridization and enzyme-linked immunosorbent assay for detecting tobacco mosaic virus in plant tissues and protoplasts. *Phytopathology* **74**, 385-389.

Sheppard, J.W., Wright, P.F. and Desavigny, D.H. (1986). Methods for the evaluation of EIA tests for use in the detection of seed-borne diseases. *Seed Science & Technology* **14**, 49-59.

Spiegel, S. and Martin, R.R. (1993). Improved detection of potato leafroll virus in dormant potato tubers and microtubers by the polymerase chain reaction and ELISA. *Annals of Applied Biology* **122**, 493-500.

Valverde, R.A., Nameth, S.T. and Jordan, R.L. (1990). Analysis of double-stranded RNA for plant virus diagnosis. *Plant Disease* **74**, 255-258.

Van den Heuvel, J.F.J.M. and Peters, D. (1989). Improved detection of potato leafroll virus in plant material and in aphids. *Phytopathology* **79**, 963-967.

Van der Wilk, F., Korsman, M.G. and Zoon, F. (1994) Detection of tobacco rattle virus in nematodes by reverse transcription and polymerase chain reaction. *European Journal of Plant Pathology* **100**, 109-122.

Van Slogteren, D.H.M. (1955). Serological micro-reactions with plant viruses under paraffin oil. Proceedings of the 2nd Conference on Potato Virus Diseases, pp. 51-54. Lisse-Wageningen 1954.

Vetten, H.J., Ehlers, U. and Paul, H.L. (1983). Detection of potato virus Y and A in tubers by enzyme-linked immunosorbent assay after natural and artificial break of dormancy. *Phytopathologische Zeitschrift* **108**, 41-53.

Vunsh, R., Rosner, A. and Stein, A. (1990). The use of the polymerase chain reaction (PCR) for the detection of bean yellow mosaic virus in gladiolus. *Annals of Applied Biology* **117**, 661-669.

Wetzel, T., Candresse, T., Macquaire, G., Ravelonandro, M. and Dunez, J. (1992). A highly sensitive immunocapture polymerase chain reaction method for plum pox potyvirus detection. *Journal of Virological Methods* **39**, 27-37.

# 4

# DETECTING PROPAGULES OF PLANT PATHOGENIC FUNGI

Sally A. Miller

Department of Plant Pathology, The Ohio State University, Ohio Agricultural Research and Development Center, Wooster, Ohio, USA

## I. INTRODUCTION

Detection of propagules of plant pathogenic fungi in seed and vegetative propagating materials is an important component of effective disease management. It is also necessary in cases of pathogens for which specific restrictions on importation have been made by governments to protect domestic crop production from introduced threats. However, comparatively few fungal pathogens are routinely included in indexing or certification programs for major crops. There are a number of historical reasons for this, including the availability of fungicides that allowed economic production of a crop in the presence of a particular fungal pathogen, as well as a lack of rapid, accurate tests that could be performed routinely without the need for expertise in morphological identification of fungi. As fungicides continue to be lost, however, due to governmental regulations, environmental concerns, reluctance of chemical manufacturers to retain or pursue crop use labels, particularly for "minor" crops, and development of fungicide resistance in fungal populations, it is becoming more and more critical that growers begin with clean planting materials. With the advent of biotechnology and its application to the detection and identification of fungi, specific, sensitive tools are becoming available to detect fungal propagules.

Advances in Botanical Research Vol. 23
Incorporating Advances in Plant Pathology
ISBN 0-12-005923-1

These tools are primarily derived from immunoassay and nucleic acid hybridization technologies. Immunoassays offer speed, ease-of-use, specificity, sensitivity and increasingly wider availability. Many immunoassays for detection of plant pathogenic fungi now employ monoclonal antibodies, which may allow the differentiation of closely related fungal taxa. Nucleic acid hybridization-based techniques, including dot-blot assays and the polymerase chain reaction (PCR), offer high levels of sensitivity and specificity. Advances in sample preparation and nucleic acid detection methods are improving the general usefulness of nucleic acid-based techniques as routine detection tools. The application of these techniques to the detection of plant pathogenic fungi, particularly in seeds and vegetative propagating materials, is the focus of this chapter.

## II. TRADITIONAL METHODS

Traditional methods for detection of plant pathogenic fungi in seeds and vegetative propagating materials may vary depending on the fungus and the host plant. Specific examples of some of the more commonly used techniques, including their major limitations, are presented here:

*Visual inspection of symptoms* - Visual inspection of plants for symptoms of disease in the field, or on samples brought into a testing facility, is a major component of plant health certification programs for fungal pathogens. Diagnoses are based on the presence of characteristic symptoms, and where sporulation is present, general confirmation can be made by light microscopy. However, while this approach is rapid and relatively inexpensive, the obvious disadvantages are lack of specificity and inability to detect asymptomatic infections.

*Blotter method* - For detection of *Leptosphaeria maculans* (anamorph: *Phoma lingam*) in crucifer seed, sterilized trays are covered with blotter paper soaked in a solution of the herbicide 2,4-Dichlorophenoxyacetic acid (2,4-D). Seeds may be pre-treated with a surface disinfestant to reduce contamination by other fungi. From each seedlot tested, 2500 seeds are placed on each of four trays, which are then covered with plastic and incubated for 10-11 days in the light at 20 C. Seeds are then evaluated visually for infection by *L. maculans*. The presence of the pathogen can be confirmed microscopically, and determination of virulence or avirulence can be assessed by inoculation of cabbage cotyledons and pigment production *in vitro* (Maguire and Gabrielson, 1983). In some cases, the use of 2,4-D is replaced by freezing the seed, as for detection of *Alternaria radicina* in carrot seed (Pryor *et al.*, 1994). This assay is clearly time-consuming, and the presence of the pathogen of interest may be obscured by faster-growing saprophytic fungi. If characteristic sporulation does not occur, accurate identification may be impossible. The assay also lacks specificity, unless labor-intensive and time-consuming plant inoculation tests are carried out.

*Direct plating on agar media* - Plating seeds on agar media has been used for detection of many seedborne fungi. Some assays utilize only water agar, e.g. detection of *Phoma betae* on beet seed (Mangan, 1983), although semi-selective media have also been developed (Mathur and Lee, 1978; Pryor *et al.*, 1994). These media do not completely control growth of saprophytic fungi, and other limitations mentioned for the blotter method also apply (Cunfer, 1983). Roebroeck *et al.* (1990)

developed a method for detecting *Fusarium oxysporum* in gladiolus by macerating corms in a semi-selective medium, incubating for one week, then plating the macerate on the same medium plus agar. Colonies of *Fusarium oxysporum* could be identified, but the test did not differentiate between pathogenic and nonpathogenic isolates.

*Growing-on test* - This method involves the planting of seed and evaluation of disease incidence in the seedlings. It is particularly appropriate for biotrophic fungi that cannot be cultured. Large numbers of seed may need to be sown to detect pathogens present at low levels, requiring a large amount of greenhouse space. This method is time-consuming and due to space limitations not appropriate for testing large numbers of seed lots if results are needed quickly. It also requires environmental conditions appropriate for disease development and the visible expression of characteristic disease symptoms.

*Culture indexing* - This is a method for detecting fungi and bacteria in vegetative propagating material, which involves simply plating portions of stems or other planting material, after surface disinfestation, on semi-selective or nonselective agar media or into tubes of liquid culture media. Cultures are incubated for 10-12 days; if growth is observed during that time, the cutting is rejected. This method is used for detection of *Fusarium*, *Verticillium* and other systemic fungal pathogens (Raju and Olson, 1985). This is a time-consuming and non-specific method, since the growth of any fungus (or bacterium) is cause for rejection.

*Bait tests* - A root tip bait test has been developed for detection of *Phytophthora fragariae* var. *fragariae* in strawberries (Duncan, 1980). Approximately 20-50 mm of root tips are cut from strawberry runners and mixed with soilless planting mix in a ratio of 1 part roots: 3 parts mix. The mixture is placed in a pot into which bait plants, the highly susceptible *Fragaria semperflorens* var. *alpina* ('Baron Solemacher') are then transplanted. Bait plants collapse 3-6 weeks after infection and the presence of *P. f. fragariae* is confirmed by microscopic analysis of roots for the presence of typical oospores and by red coloration of the stele. This test is considered to be quite sensitive but is also very time-consuming. A self-bait test has been developed for detection of *P. f. rubi* in raspberry plants, but it suffers from the same limitations as the strawberry assay (Duncan, 1990).

## III. SEROLOGICAL TECHNIQUES

The development and application of immunoassays for detection of fungal plant pathogens has expanded significantly in the last 10 -15 years. This has been possible in large part by the introduction of monoclonal antibodies and sensitive immunoassay formats, especially the enzyme-linked immunosorbent assay (ELISA), and to some extent by the development of more highly specific polyclonal antibodies. Immunoassays have been developed for a wide array of fungal plant pathogens, including numerous fungi of interest in indexing programs (Table I) (Miller and Martin, 1988; Dewey *et al.*, 1991; Schots *et al.*, 1994). Some assays have been commercialized, including kits for the detection of fungi in the genera *Phytophthora*, *Pythium* and *Rhizoctonia*, and *Septoria tritici*, *S. nodorum* and *Pseudocercosporella herpotrichoides* (Petersen *et al.*, 1990; Smith *et al.*, 1990; Mittermeier *et al.*, 1990; Miller

Table I. Some examples of immunoassays for detection of fungal propagules in planting material.

| Fungus | Assay Type[a] | Antibody Type | Target Plant Tissue | Reference |
|---|---|---|---|---|
| Verticillium dahliae | Immuno-fluorescence | Polyclonal | Potato tubers | Nachmias and Krikun, 1984 |
| Tilletia controversa | DAS-ELISA | Monoclonal | Wheat seed | Banowetz et al., 1984 |
| Sirococcus strobilinus | Indirect ELISA | Monoclonal | Seed | Mitchell and Sutherland, 1986 |
| Sirococcus strobilinus | Immuno-blot | Monoclonal | Spruce seed | Mitchell, 1988 |
| Phytophthora fragariae | DAS-ELISA | Polyclonal | Strawberry roots | Amouzou-Alladaye et al., 1988 |
| Phytophthora spp. | Indirect ELISA | Polyclonal | Strawberry roots | Mohan, 1988, 1989 |
| Colletotrichum sp. | Indirect ELISA | Polyclonal | Anemone corms | Barker and Pitt, 1988 |
| Rhizoctonia spp. | DAS-ELISA | Polyclonal | Poinsettia stem cuttings | Benson, 1992 |
| Fusarium oxysporum f. sp. narcissi | Indirect ELISA | Polyclonal | Narcissus bulbs | Linfield, 1993 |
| Spongospora subterranea | Indirect ELISA | Polyclonal | Potato tubers | Harrison et al., 1993 |
| Leptosphaeria maculans V strain | Indirect ELISA | Monoclonal | Seed, stem tissue | Stace-Smith et al., 1993 |
| Colletotrichum acutatum | Indirect ELISA | Monoclonal | Strawberry petioles | Barker et al., 1994 |

[a]Indirect ELISA = Antigen capture, indirect enzyme-linked immunosorbent immunoassay (ELISA); DAS-ELISA = Double antibody sandwich ELISA.

and Joaquim, 1993). However, these kits have been designed primarily for rapid field diagnostics (Benson, 1991; Ellis and Miller, 1993), as an aid in making fungicide application decisions (Mittermeier et al., 1990; Timmer et al., 1993), and for use in diagnostic clinics (Pscheidt et al., 1992), and are not being used routinely in indexing programs. For many of the pathogens of concern in certification programs and especially those of quarantine significance, distinctions between closely related taxa, i.e. forma speciales or physiologic race, may be required. This necessitates the use of highly specific assays capable of making these distinctions, as well as detecting propagules at low concentrations in the planting material.

## A. Antibodies

### 1. Polyclonal antibodies

Although numerous immunoassays have been developed for fungal pathogens by utilizing polyclonal antisera (see reviews by Dewey, 1988 and Dewey *et al.*, 1991), a high degree of specificity, without sacrificing sensitivity, has been difficult to achieve. Many types of immunogen preparations have been used, including whole cells (Guy and Rath, 1990; Kraft and Boge, 1994), crude mycelial or spore extracts (Dewey and Brasier, 1988; Harrison *et al.*, 1990), extracellular culture filtrates (Smith *et al.*, 1990; Kim *et al.*, 1991; Brill *et al.*, 1994), excreted toxins (Bhatnagar *et al.*, 1989; Tseng, 1989; Ward *et al.*, 1990), crude or partially purified soluble proteins (Unger and Wolf, 1988; Höxter *et al.*, 1991; Velicheti *et al.*, 1993), and cell wall extracts (Linfield, 1993), resulting in varying degrees of specificity for the target fungus. There also appears to be no consistent advantage in using any particular life stage of the target fungus, i.e. spores, mycelia, resting structures, etc.. The specificity of a polyclonal antiserum can be improved to some extent by absorbing it with heterologous antigens to remove cross-reacting antibodies, or in some cases by diluting the antiserum to a level that eliminates low levels of cross-reactivity. However, these approaches have seldom been successful for production of antisera specific at the level of species or lower. In addition, cross-absorption by any means usually results in a significant reduction in antibody titer. For example, Mohan (1989) used affinity chromatography in an attempt to eliminate non-specific antibodies that cross-reacted with antigens from *Phytophthora cactorum* in an antiserum raised against *P. fragariae*. The resulting antiserum was low in titer and still not specific to *P. fragariae*.

The difficulty in obtaining highly specific polyclonal antisera against crude or partially purified extracts may be due, at least in part, to the presence of immuno-dominant high molecular weight polysaccharides. As shown for *Penicillium, Aspergillus, Botrytis* and *Monascus* spp. (Notermans *et al.*, 1988; Cousin *et al.*, 1990) these molecules may mask the presence of more unique antigens in the immunogen preparation. Researchers have approached this problem by using as immunogens various purified fungal components, such as ribosomes (Takenaka, 1992), mycelial proteins excised from electrophoretic gels (Lind, 1990), and lectins (Kellens and Peumans, 1991). In the latter case, antisera were produced that distinguished between several anastomosis groups of *Rhizoctonia solani*. However, in general, the practical limit of specificity for polyclonal antisera produced using purified fungal components appears to be at the level of species.

Another approach to the problem of specificity is to take advantage of variations in the test animal's response to immunization or to manipulate the animal's immune response. Early-bled, relatively low-titer antisera have been shown to contain more specific antibodies to *Botrytis cinerea* than later-bled, high-titer antisera (Ricker *et al.*, 1991). These antibodies reacted with low molecular weight glycolipids present in the target fungus but not in other fungi and were used in an assay to detect low levels of *B. cinerea* in juice from infected grapes. Presumably, this approach might be used to prepare antisera against fungi of interest in indexing programs, although reactivity of antisera with low molecular weight compounds may limit the types of assays that can be developed (Ricker *et al.*, 1991). Another approach is to induce immunological tolerance in the test animal in order to produce a very specific antiserum. Del Sorbo *et al.* (1993) introduced mycelial antigens of *Fusarium oxysporum* f. sp. *dianthi* or *F. oxysporum* f. sp. *lycopersici* into neonatal mice through lactation, then immunized the mice with the heterologous antigen. Antisera that could be used to differentiate the two *formae speciales* by Western blotting and radioimmunoassay were produced in both combinations. This technique would not be likely to yield large enough amounts of antiserum for routine use in an ELISA or other commonly used assays, however, and may be most useful as a means of preparing animals for the production of very specific monoclonal antibodies.

Immunoassays utilizing polyclonal antisera raised against fungal pathogens have been developed and intended for use in indexing planting material (Table I). However, such assays face (or have faced) problems in acceptance in such programs, primarily due to requirements for specificity and/or assay sensitivity that polyclonal antisera-based assays may not be able to meet. In addition, batch-to-batch variation in antisera, either from different animals or from different bleeds of the same animal, may result in inconsistencies in sensitivity and reactivity with target as well as non-target fungi. Unless the assays are produced commercially, i. e. the assay for detection of *Rhizoctonia* in poinsettia cuttings (Table I), or by groups dedicated to, and funded for, long-term antiserum production, it may not be practical to carry out the extensive testing that new batches of antisera usually require.

## 2. Monoclonal antibodies

Many of the problems associated with polyclonal antisera can be eliminated by the use of monoclonal antibodies. The production of monoclonal antibodies was first reported in the United Kingdom by Kohler and Milstein (1975), and since then monoclonal antibodies have been developed for all classes of plant pathogens. Several review articles include information on the development and/or use of monoclonal antibodies for detection of plant pathogenic fungi (Dewey *et al.*, 1991; Dewey, 1992; Miller and Joaquim, 1993), and experimental protocols are well documented (Goding, 1987; Harlow and Lane, 1988). Briefly, monoclonal antibodies are produced by the fusion of ß-lymphocytes isolated from the spleen of an immunized animal with myeloma cells originally derived from an animal of the same species and capable of continuous growth in culture. Mice have been, by far, the preferred experimental animal for hybridoma production. After the fusion, cells are incubated in a medium that is selective for hybrid products. Surviving hybrid cells, or hybridomas, are screened for the production of antibodies of interest, isolated by limiting dilution or other method, and cloned. Since each lymphocyte produces only one type of antibody, the resultant hybridomas also produce monospecific, or monoclonal, antibodies. Clones may be stored frozen in liquid nitrogen for an indefinite period of time. Because hybridomas are theoretically immortal, a constant, uniform supply of antibodies can be made available.

Many different types of immunogens have been used in the production of monoclonal antibodies, including crude mycelial extracts (Petersen *et al.*, 1990; Yuen *et al.*, 1993; Stace-Smith *et al.*, 1993), cell-free washings from the surface of mycelial cultures (Dewey *et al.*, 1989b; Bossi and Dewey, 1992; Thornton at al., 1993), glutaraldehyde-fixed whole cells (Hardham *et al.*, 1986; Estrada-Garcia *et al.*, 1989), partially purified secreted proteins (Matthew and Brooker, 1991; Salinas and Schots, 1994), fractionated hyphal walls (Wong *et al.*, 1988), and extracts of culture supernatants (Xia *et al.*, 1992). While in theory it should be fairly straightforward to develop hybridomas secreting highly specific monoclonal antibodies from any type of fungal immunogen, in practice this is not always the case. As shown for the production of polyclonal antisera, non-specific immunodominant antigens may mask the presence of more specific epitopes, resulting in the development of few ß-lymphocytes producing antibodies against these epitopes. Finding the hybridomas produced from these rare lymphocytes requires extensive screening, often involving thousands of hybridoma cultures. For example, Stace-Smith *et al.* (1993) reported that it was necessary to screen over 12,000 culture wells, each containing multiple hybridomas, to eventually identify two hybridomas secreting monoclonal antibodies that could be used to differentiate the weakly virulent (WV) and highly virulent (V) strains of *Leptosphaeria maculans*, causal agent of blackleg of canola. This problem has been approached primarily by manipulating the immunogen, although as seen for polyclonal antibodies, it is sometimes still difficult to eliminate immunodominant, non-specific epitopes (Dewey, 1992). Induction of immunological tolerance through the use of immunosuppressive drugs (Hamilton *et al.*, 1990), or by immunization of neonatal mice with non-target immunogens (tolerogens) directly (Burns *et al.*, 1994) or through lactation, as described above, may be appropriate in such cases.

The problems noted above notwithstanding, there have been numerous examples of successful production of specific monoclonal antibodies against fungal pathogens during the past decade. Some of these are listed in Table II. Early work by Hardham *et al.* (1986) demonstrated that monoclonal antibodies could be developed with varying degrees of specificity, from the level of genus to that of a single isolate. These monoclonal antibodies were also shown to label different structures on zoospores and cysts of *Phytophthora cinnamomi*. In subsequent studies with other oomycetes, especially *Pythium* spp., species and/or genus specificity has been achieved (Estrada-Garcia *et al.*, 1989; Yuen *et al.*, 1993). Among the true fungi, a few studies have resulted in the development of monoclonal antibodies that differentiate isolates at subspecific levels. Wong *et al.* (1988) reported production of a monoclonal antibody capable of differentiating *Fusarium oxysporum* f. sp. *cubense* race 4 from other races of this pathogen. As noted above, Stace-Smith *et al.* (1993) produced two monoclonal antibodies capable of differentiating weakly and highly virulent strains of *Leptosphaeria maculans*. These highly specific monoclonal antibodies are more the exception than the rule, however, and most studies have resulted in production of monoclonal antibodies specific at the species or genus level (Table II). In fact, specific attempts to raise monoclonal antibodies capable of differentiating races of *Phytophthora megasperma* f. sp. *glycinea* (*P. sojae*) and *Colletotrichum lindemuthianum* were not successful (Wycoff and Ayers, 1990; Pain *et al.*, 1992).

Only a few monoclonal antibodies have been produced for use specifically in detecting fungi on or in seed or vegetative propagating material (Table I). In one of the first applications of this technology to plant pathogenic fungi, Banowetz *et al.* (1984) attempted to develop monoclonal antibodies for differentiation of *Tilletia controversa*, causal agent of dwarf bunt disease of wheat, from the morphologically similar *T. caries*, which causes common bunt disease. The presence of spores that resemble *T. controversa* on seed has resulted in rejection of shipments of wheat by China, where dwarf bunt disease has not been reported and a zero tolerance has been established. The authors were not able to find a monoclonal antibody that could be used to make a qualitative distinction between the two species, although quantitative differences in the binding of some monoclonal antibodies to teliospore surface antigens were observed.

S.A. Miller

*Table II.* Specificity of some monoclonal antibodies (Mabs) developed against plant pathogenic fungi.

| Mabs Produced Against | Highest Level of Specificity Achieved | Reference |
|---|---|---|
| *Tilletia controversa* | Genus | Banowetz *et al.,* 1984 |
| *Phytophthora cinnamomi* | Isolate | Hardham *et al.,* 1986 |
| *Sirococcus strobilinus* | Genus[1] | Mitchell, 1986 |
| *Fusarium oxysporum* f. sp. cubense | Race | Wong *et al.,* 1988 |
| *Ophiostoma ulmi* | Subspecies | Dewey *et al.,* 1989b |
| *Pythium aphanidermatum* | Species | Estrada-Garcia *et al.,* 1989 |
| *Phytophthora sojae* | Species | Wycoff and Ayers, 1990 |
| *Septoria nodorum, S. tritici* | Species | Petersen *et al.,* 1990 |
| *Rhizoctonia solani* | Anastomosis group | Matthew and Brooker, 1991 |
| *Botrytis* | *B. cinerea* and *B. fabae* but not *B. allii* | Bossi and Dewey, 1992 |
| *Colletotrichum lindemuthianum* | *C. lindemuthianum, C. orbiiculare* and *C. trigolii* but not other *Colletotrichum* spp. | Pain *et al.,* 1992 |
| *Pythium ultimum* | Species | Yuen *et al.,* 1993 |
| *Leptosphaeria maculans* | Strain (V and VV) | Stace-Smith *et al.,* 1993 |
| *Rhizoctonia solani* | Species | Thornton *et al.,* 1993 |
| *Mucor racemosus* | Order Mucorales | DeRuiter *et al.,* 1993 |
| *Pyrenophora* | Genus | Burns *et al.,* 1994 |
| *Botrytis* | Genus | Salinas and Schots, 1994 |
| *Armillaria* spp. | Species | Priestly *et al.,* 1994 |
| *Heterobasidium anosum* | Genus[1] | Galbraith and Palfreyman, 1994 |
| *Colletotrichum acutatum* | Species | Barker *et al.,* 1994 |

[1] Reaction of monoclonal antibodies with other species in the same genus as the target pathogen not reported.

Monoclonal antibodies were developed successfully to detect *Sirococcus strobilinus,* the causal agent of shoot blight in pine, fir and spruce seedlings, in seed of these species (Mitchell, 1986). Monoclonal antibodies to the highly virulent strain of *L. maculans* may prove valuable in testing seeds of cruciferous crops, including canola, for infestation by the pathogen (Stace-Smith *et al.,* 1993). They could be used for both detection of *L. maculans* in seed lots and differentiation of the weakly and highly virulent strains, possibly replacing laborious and time-consuming host inoculation. Monoclonal antibodies specific for *Colletotrichum acutatum,* causal agent of blackspot disease of strawberries, were developed and used in an indirect ELISA to detect the pathogen in infected strawberry petioles (Barker *et al.,* 1994). Unfortunately, the species-specific monoclonal antibodies reacted only with conidia and therefore the assay required sporulation of the fungus. Combination of ELISA with the standard paraquat test, which induces sporulation on petiole tissue, could still improve the efficiency and accuracy of detection of this pathogen. Preliminary reports have been published recently on attempts to develop monoclonal antibodies to *Pyrenophora graminea* (Burns *et al.*, 1994), *Spongospora subteranea* (Harrison *et al.*, 1994) and *Verticillium dahliae* and *V. albo-atrum* (van de Koppel and Schots, 1994). If development work is successfully completed, these monoclonal antibodies could form the basis for time-saving detection assays for barley seed, potato tubers and rose cuttings and other crops, respectively. Although not originally developed for seed analysis, monoclonal antibodies that react specifically with *Septoria nodorum* (Petersen *et al.*, 1990) might also be useful in testing cereal seed.

## B. Immunoassays

Antibodies and/or antisera can be employed in a number of different assay formats for detection of plant pathogenic fungi. The type of assay used ultimately depends on several factors, including i) the fungal tissue type to be detected, i.e. spores, mycelial extracts, etc.; ii) the type of plant and plant parts assayed; iii) the level of technological sophistication available to the laboratory or individuals carrying out the testing; iv) the volume of samples to be tested at a time and over an extended period; v) the level of specificity and sensitivity required; and vi) if monoclonal antibodies are used, specific characteristics of the antibodies. For example, immunofluorescence may be the assay of choice if the antibody used detects spore surface antigens, the appropriate equipment and trained personnel are available, and the number of samples is not extremely large. On the other hand, ELISA would be more appropriate for detection of soluble extracts of target fungi, in systems where quantitation is needed and a large throughput of samples is required.

### 1. *Enzyme-linked immunosorbent assay (ELISA)*
While there are many variations of ELISA, the types most commonly used to detect fungi are the double antibody sandwich assay (DAS-ELISA) and indirect ELISA (Table I). In a typical DAS-ELISA, a specific capture antibody is immobilized onto a solid surface such as the wells of a microtiter plate. The sample is added, and unbound material is washed away. Bound antigen is detected by the addition of a detecting antibody that has been conjugated with an enzyme, and unbound material is again washed away. The presence of the detecting antibody is determined through the addition of a substrate for the enzyme. The amount of color that develops is proportional to the amount of antigen present in the sample. The intensity of the color is recorded numerically through the use of automated equipment, although for qualitative uses, color change can be determined by eye. Indirect ELISA is a similar assay, except that the enzyme is conjugated to a general detecting antibody, e.g. goat anti-mouse or sheep anti-rabbit antibody, that binds to the specific antibody that has already bound to the target antigen. In indirect ELISA, the antigen may be bound to the solid substrate (antigen-capture) or a triple antibody sandwich may be used. In the latter, specific antibody produced in one animal species is bound to the solid substrate, while a second specific antibody produced in another animal species binds to the bound antigen. The "sandwich" is detected by an antibody/enzyme conjugate

that binds to the second antibody. In another type of sandwich assay, $F(ab')_2$ fragments from target-specific antibodies are used as the capture reagent, and a specific second antibody produced in the same animal species binds to antigen bound to the $F(ab')_2$ fragments. The "sandwich" in turn is detected by the addition of a general detecting antibody-enzyme conjugate that reacts specifically with the Fc portion of the second antibody (Barbara and Clark, 1982). Indirect, triple antibody and $F(ab')_2$-based ELISAs require more steps than DAS-ELISA, but do not require conjugation of antibody and enzyme, since general detecting antibody-enzyme conjugates are available commercially. There may also be differences in specificity and sensitivity between these assays (Bar-Joseph and Salomon, 1980). ELISAs can also be carried out on membranes made of nitrocellulose, nylon or other materials (Sherwood, 1987; Mitchell, 1988; Dewey et al., 1989a). Some of the commonly used configurations of dot immunobinding assays include dot blot, slot blot and dipstick assays. Dot immunobinding assays are very similar to microtiter plate ELISAs, using many of the same reagents and protocols, except that a precipitating substrate is required for the former, while soluble substrates are used for the latter. Membrane-based assays offer speed, convenience and often increased sensitivity over microtiter plate ELISAs. For example, Mitchell (1988) developed a monoclonal antibody-based dot immunobinding assay to detect S. strobilinus in extracts of spruce seed. The assay was 5-25 times more sensitive than antigen-capture indirect ELISA, detecting 1-5 ng S. strobilinus antigen. It could be carried out in seed testing laboratories where relatively sophisticated equipment used in microplate ELISA might not be available. A possible drawback of dot-immunobinding assays is that results may be difficult to quantify, although densitometers can be used to estimate the relative amount of color development on the membranes.

## 2. Immunofluorescence

While immunofluorescence has been used in indexing systems primarily for the detection of bacterial plant pathogens (see Chapter 2), applications of immuno-fluorescence for detection of fungi have been reported as well (see Miller and Martin, 1988). Both indirect and direct immunofluorescence methods are used; in the former, specific antibodies bound to their target antigens are detected by using second antibodies conjugated with fluorescent dyes such as fluorescein isothiocyanate (FITC) or rhodamine isothiocyanate (RITC). Fluorescence, indicating the presence of the target antigen, is visualized microscopically. In direct immunofluorescence, the fluor-escent dye is conjugated directly to the specific, detecting antibody. The latter technique is advantageous because the number of steps are reduced, but some sensitivity may be lost, particularly when monoclonal antibodies are used (Salinas et al., 1994).

Immunofluorescence may be particularly useful for the detection of fungal spores, which are often too large to bind to microtiter plates in ELISA without prior disruption. Since the spores or other fungal cells are visualized by immunofluor-escence microscopy, fungal morphology and subtle variations in antibody binding patterns may be used to reduce the frequency of false positive results. Studies utilizing monoclonal antibodies to zoosporic fungi have clearly shown differences in antibody binding sites on zoospores and cysts between different taxa (Hardham et al., 1985; Estrada-Garcia et al., 1989), as have similar studies with spores of higher fungi (Wong *et al.*, 1988; Salinas and Schots, 1994). Immunofluorescence can also be applied *in situ*, allowing visualization of fungi in plant tissue. Nachmias and Krikun (1984) developed an immunofluorescence assay to detect *Verticillium dahliae* in potato tubers, and suggested that such an assay could be used in seed potato certification. Improvements in immunofluorescence protocols, such as carrying out all reaction steps in 96-well filtration plates (Salinas and Schots, 1994), may simplify the technique significantly and allow higher, faster throughput of samples.

The disadvantages of immunofluorescence compared to ELISA are primarily the requirements of immunofluorescence for specialized, expensive equipment and well-trained personnel, lack of quantitation (except where propagules can be counted), and technical problems such as autofluorescence of plant tissue or soil particles. However, these problems can be overcome for appropriate systems and immunofluorescence has the potential for routine use in testing laboratories for detection of fungi as well as bacteria (Salinas and Schots, 1994).

## IV. NUCLEIC ACID-BASED TECHNIQUES

Nucleic acid-based techniques have been applied to the detection and identification of plant pathogenic fungi for only about ten years, and during that time several major innovations in the technology have been made and successfully applied to fungi. Some of the fungi that have been studied are of interest in seed or plant health indexing programs, and these techniques have the potential to be used to detect and identify them in a rapid, accurate, cost-effective way. The techniques and their applications to detection or identification of plant pathogenic fungi that will be reviewed in this chapter are 1) "traditional" dot-blot and related DNA hybridization assays, and 2) the polymerase chain reaction (PCR) assay, including random amplified polymorphic DNA (RAPD) analysis. "Traditional" DNA hybridization assays include dot-blot, slot blot and tissue blot assays, utilizing cloned DNA probes or probes developed by newer techniques such as PCR or RAPD, or through direct sequencing of specific fragments. These are either truly taxon (usually spe-cies)-specific probes, which can be used directly in a dot blot or related assay with pure fungal cultures or with extracts of diseased plant material, or probes that produce specific identifiable patterns in restriction fragment length polymorphism (RFLP) analyses. It is generally agreed (Schots *et al.*, 1994) that while RFLP analysis is very useful in research laboratories for genetic, population and taxonomic studies, it is unlikely to be used routinely in diagnostic laboratories. Therefore, uses of RFLP techniques will only be mentioned briefly, where they have been used specifically for pathogen detection or identification (usually in Southern hybridizations) or where they provide an informative adjunct to PCR.

## A. Nucleic Acid Probes

### 1. Probe development

Nucleic acid probes are sequences of nucleic acids that are labelled with a marker and used to detect complementary nucleic acid sequences in a sample. Principles of, and methodology for, development of nucleic acid probes are available in laboratory manuals (e. g. Sambrook *et al.* 1989) and will not be presented in detail here. DNA probes have been developed since the late-1980s for detection or identification of several fungal plant pathogens (Table III), although few of them are of significance in seed or vegetative propagating materials. Both specific probes, which detect complementary DNA sequences present only in the target pathogen in an application such as dot blot hybridization, and RFLP probes, which detect differences in banding patterns between target and non-target species in Southern hybridizations or hybridizations on dried gels (Table III), have been reported. Cloned probes are produced by digesting chromosomal or mitochondrial DNA with restriction enzymes, ligating the fragments into plasmids or cosmids digested with the same enzymes, transforming competent bacteria (*Escherichia coli*) with the recombinant plasmids, selecting recombinant clones on antibiotic-containing media and screening the clones for appropriate specificity by colony hybridization (Sambrook *et al.* 1989). The highest level of sensitivity is obtained by selecting recombinant clones that hybridize intensely with DNA of the target fungus, indicating that the cloned DNA fragment is present in high copy number. In one of the earliest studies on plant pathogenic fungi, Rollo *et al.* (1987) developed cloned DNA probes to detect *Phoma tracheiphila*, the causal agent of mal secco disease of lemon, in a dot-blot assay. This work was followed soon after by several papers describing the development of cloned DNA probes for the detection of *Phytophthora parasitica* (Goodwin *et al.*, 1989) and *P. citrophthora* (Goodwin *et al.*, 1990b). In these cases the probes were produced using chromosomal DNA digested with the restriction enzymes *Eco*RI and *Hind*III. Two *P. citrophthora* probes were highly specific, reacting with isolates of the pathogen from citrus, walnut, cherry and kiwi, but not from cacao or with other *Phytophthora* species. The *P. parasitica* probes were also highly specific at the species level. All of the cloned probes reported were shown by Southern hybridizations to be medium to high copy number DNA sequences. Recently, Möller *et al.* (1993) reported using *P. infestans* chromosomal DNA as the source of a cloned 430 base pairs (bp) multicopy probe. Although absolute species specificity was not achieved and the probe reacted with isolates of closely related *P. mirabilis* and *P. phaseoli*, it was suggested that the probe could be of diagnostic value since the latter two fungi are not found in potato tissue.

Cloned DNA probes have been developed for two pathogens of turfgrass, *Leptosphaeria korrae* (Tisserat *et al.*, 1991) and *Ophiosphaerella herpotricha* (Sauer *et al.*, 1993), which are difficult to diagnose by traditional means. Both were cloned from genomic DNA and represent multicopy sequences. However, the two *L. korrae* probes were produced by cloning discrete electrophoretically separated bands of *Eco*RI-digested total *L. korrae* DNA. One of the probes was highly specific to *L. korrae*, and did not hybridize to any other fungal DNA tested. This level of species specificity is better than that obtained with monoclonal antibodies to this pathogen (Nameth *et al.*, 1990). The method of cloning electrophoretically separated DNA fragments was also used by Stammler *et al.* (1993) to develop DNA probes for the detection of *Phytophthora fragariae* var. *rubi*. The pathogen causes root rot of raspberries that is spread by planting infected nursery stock, and a rapid, practical means of detecting the pathogen in planting stocks is needed. Attempts to clearly differentiate *P. fragariae* from other *Phytophthora* species by serological techniques

*Table III.* Some examples of the use of DNA probes for the detection and/or identification of plant pathogenic fungi.

| Pathogen | Probe Source[1] | Label | Assay | Reference |
|---|---|---|---|---|
| *Phoma tracheiphila* | Cloned DNA | [32]P | Dot-blot | Rollo *et al.,* 1987 |
| *Gaeumannomyces graminis* | Cloned mtDNA | [32]P | Dried gel[2] | Henson, 1989 |
| *Phytophthora parasitica* | Cloned chromosomal DNA | [32]P; sulfonated | Dot-blot; tissue blot | Goodwin *et al.,* 1989, 1990a |
| *Phytophthora citrophthora* | Cloned chromosomal DNA | [32]P | Dot-blot; tissue blot | Goodwin *et al.,* 1990b |
| *Leptosphaeria korrae* | Cloned genomic DNA | [32]P | Slot-blot | Tisserat *et al.,* 1991 |
| *Gaeumannomyces graminis* vars. | Cloned mtDNA | [32]P; DIG[3] | Southern | Bateman *et al.,* 1992 |
| *Ophiosphaerella herpotricha* | Cloned genomic DNA | [32]P | Slot-blot | Sauer *et al.,* 1993 |
| *Phytophthora capsici, megakarya, cinnamomi, palmivora, Phytophthora* spp. | rDNA ITS I | [32]P | PCR[4] & dot-blot | Lee *et al.,* 1993 |
| *Phytophthora infestans* | Cloned chromosomal DNA | Biotin | Southern | Möller *et al.,* 1993 |
| *Phytophthora fragariae* var. *rubi,* var. *fragariae* | Cloned total DNA | DIG | Dot-blot; Southern | Stammler *et al.,* 1993 |
| *Fusarium culmorum* | Cloned genomic DNA | DIG | Southern | Koopmann *et al.,* 1994 |
| *Leptosphaeria maculans* | Amplified chromosomal DNA | DIG | Southern | Schäfer and Wöstemeyer, 1994 |
| *Pyrenophora teres, graminea* | Cloned total DNA | [32]P | Dot-blot | Husted, 1994 |
| *Pseudocercosporella herpotrichoides* R-type | Cloned genomic DNA | [32]P | Slot-blot | Nicholson and Rezanoor, 1994 |
| *Pythium ultimum* | rDNA ITS | DIG | PCR & dot-blot | Levesque *et al.,* 1994 |

[1] mtDNA = mitochondrial DNA; rDNA = ribosomal DNA; ITS = internal transcribed spacer region (of rDNA).
[2] RFLP gels dried and hybridized directly with labelled probe.
[3] DIG = digoxigenin
[4] PCR = Polymerase chain reaction

(Table I) have not been entirely successful. Several cloned probes were capable of detecting *P. fragariae* in a dot blot assay, but could not differentiate it from *P. fragariae* var. *fragariae* or *P. cambivora*. However, the probes did not hybridize with DNA from *Pythium ultimum* or the other *Phytophthora* species tested. Some of the probes gave restriction patterns on Southern blots that could be used to differentiate and identify these groups of fungi.

Recently Husted (1994) reported on the development of cloned DNA probes to differentiate *Pyrenophora graminea* and *P. teres*, seedborne pathogens of barley. Although both are serious pathogens of barley, *P. graminea* does not cause epidemics and can be controlled by the use of clean or fungicide-treated seed. Cloned probes were developed that could differentiate the two pathogens, although weak cross-hybridization to DNA from some other fungi was noted in all cases. However, the target species could easily be distinguished from non-target species by Southern hybridization.

The production of cloned DNA probes as described above is a time-consuming process, and recently other methods have been employed to develop highly specific DNA probes without the necessity of cloning. PCR is being used to amplify sequences of fungal DNA, which can be labelled and used as probes in dot blot or RFLP analyses. Levesque *et al.* (1994) produced DNA probes highly specific to *Pythium ultimum* through PCR amplification of the internal transcribed spacer (ITS1) region of nuclear ribosomal DNA (rDNA), followed by restriction mapping and selection of species-specific fragments. Both the restriction fragment probes and a probe made from the entire ITS1 were specific for *P. ultimum* in a dot blot hybridization assay. A similar approach was used by Ward and Gray (1992) to develop a probe for identification of fungi in the *Gaeumannomyces-Phialophora* complex. The amplified mitochondrial rDNA fragment produced was broadly specific for fungi within this complex, which could be differentiated by RFLP analysis using the probe.

Where DNA sequence information is available, oligonucleotide probes can be synthesized and used to identify various fungal taxa. Lee *et al.* (1993) developed 20 bp oligonucleotide probes complementary to species-specific regions of ITS1 which could differentiate *Phytophthora capsici*, *P. cinnamomi*, *P. megakarya* and *P. palmivora* (Table III). A *Phytophthora* genus-specific probe was also generated in this way. Randomly generated DNA fragments produced by RAPD PCR have also been used as a source of highly specific probes. DNA fragments of different sizes generated by this method were selected, labelled and used as specific probes in the detection and differentiation of aggressive strains of *Leptosphaeria maculans*, causal agent of blackleg disease of crucifers (Schäfer and Wöstmeyer, 1994). Another method of specific probe production involves the use of RAPD primers with embedded restriction sites (Fani *et al.*, 1993). Bands of amplified DNA can be selected, cloned, sequenced, then used as specific DNA probes for microbial detection or identification.

## 2. Probe labelling

One of the principle limitations to widespread practical use of nucleic acid hybridization techniques utilizing DNA probes has been the lack of stable, "user-friendly" probe labels. Most probes have been labelled with the radioactive marker [32]P (Table III), which provides great sensitivity but is relatively unstable and is not practical for most diagnostic laboratories. Other markers, such as biotin or digoxigenin offer greater convenience, shelf-life and safety than radioactive labels. In early studies, non-specific reactions and reduced sensitivity were problematic (Zwadyk *et al.*, 1986; Audy *et al.*, 1991). However, commercial kits for non-radioactive labelling of nucleic acid probes are now available and significantly streamline the process of tagging probes. There are also a number of choices for detection of the probe labels, including colorimetric or chemiluminescent reactions (Tullis, 1994).

Where alternative probe labels have been compared directly with [32]P for detection or identification of fungi, sensitivity of the alternate labels has been as good or better than the radioactive label. Goodwin *et al.* (1989) found that sulfonated DNA probes were as sensitive as [32]P labelled probes in the detection of *Phytophthora parasitica*, with a detection limit between 1 and 10 ng purified DNA. Digoxigenin labelled probes, which are used more commonly, were shown by Bateman *et al.* (1992) to be quicker and more sensitive than [32]P-labelled probes when used with a chemiluminescent substrate in the detection of *Gaeumannomyces graminis*.

## B. Dot-Blot or Squash Hybridization

There are numerous nucleic acid hybridization procedures that are carried out on a membrane surface, the most common of which are dot- or slot-blot and squash- or tissue-blot hybridization, each utilizing specific nucleic acid probes. In these assays, small amounts of denatured nucleic acids are immobilized on a membrane, usually nitrocellulose or nylon, blocked and probed with a specific nucleic acid probe. Hybridization between the sample and probe is detected using one of the marker systems described above. Plant samples can be extracted and applied to the membrane (dot- or slot-blot), or squashed directly onto the membrane (squash- or tissue-blot hybridization), then dried and stored before analysis. In this way, samples can be prepared in less than optimal conditions and mailed to laboratories for testing. Since 1985 there have been relatively few assays of this type developed for the direct detection of plant pathogenic fungi in plant tissue samples, and fewer still that might be useful for detection of fungi in seeds and vegetative propagating materials (Table III). Sensitivity of these assays for direct detection of fungi in infected plant tissue has not always been sufficient for routine use, although detection limits of $\leq 1$ ng purified DNA have been routinely reported for dot-blot assays (Goodwin *et al.*, 1989, 1990b; Tisserat *et al.*, 1991; Sauer *et al.*, 1993). Thus, while the root-infecting fungi *Leptosphaeria korrae* and *Ophiosphaerella herpotricha* could be detected in turfgrass tissue samples by using slot blot assays with [32]P-labelled cloned DNA probes (Tisserat *et al.*, 1991; Sauer *et al.*, 1993), it was not possible to detect *Phytophthora citrophthora* (Goodwin *et al.*, 1990b) or *Phoma tracheiphila* (Rollo *et al.*, 1987) in citrus roots by this method. The dot blot and Southern hybridization methods developed for *Phytophthora fragariae* were also considered questionable for routine pathogen detection in raspberry or strawberry tissue due to lack of sensitivity (Stammler and Seemüller, 1994). The failure to detect *P. citrophthora* and *P. fragariae* var. *rubi*, should not be a result of very low fungal concentrations in citrus or raspberry roots, since these pathogens can be detected routinely in root tissue by means of a monoclonal antibody-based immunoassay (Timmer at al., 1993; Ellis and Miller, 1993).

## C. Polymerase Chain Reaction (PCR)

The polymerase chain reaction is a relatively new technique (Saiki *et al.*, 1985, 1988) that has gained broad acceptance very quickly in many areas of science (see Bej *et al.*,

1991; Mullis *et al.*, 1994). Its use in fungal identification, genetics, taxonomy and other areas has increased rapidly in the last four years, and it is bound to become more widely applied in laboratories throughout the world. In-depth reviews and protocols for PCR and related techniques are available in books and laboratory manuals (Wöstmeyer *et al.*, 1992; Foster *et al.*, 1993; Henson and French, 1993; White, 1993; Mullis *et al.*, 1994). Conceptually the method is quite simple. Sequences of DNA are amplified exponentially through repetitive cycles of DNA synthesis. In each cycle, double stranded DNA is converted to single strands at high temperatures, followed by annealing of oligonucleotide primers to target DNA at reduced temperatures and finally extension of the targeted DNA sequences through the action of a heat stable DNA polymerase, such as *Taq* polymerase. The primers are short, single stranded sequences of DNA that anneal under the appropriate conditions with complementary sequences of template DNA. The amplified product, which can represent a million-fold or higher amplification of the target sequence, can be detected on an agarose gel, by dot-blot hybridization, or used in RFLP analysis. More specific and sensitive methods for detection of PCR products have also been developed (Tullis, 1991).

### 1. Specific PCR primers

For plant pathogenic fungi, specific primers have most often been based on sequences of genes encoding ribosomal RNA (rDNA; Table IV). These genes have been relatively well studied in fungi (Appels and Honeycutt, 1986), and contain both highly conserved coding regions and variable ITS regions. They are also highly repetitive and easy to isolate by PCR. Primers homologous to sequences of the conserved small and large rDNA genes flanking one or both of the ITS regions (ITS1 and ITS2) are used to amplify these regions, which can be sequenced and compared for nucleotide differences between taxa. These differences can then be used to develop highly specific primers. The "universal" rDNA primers developed by White *et al.* (1990) have been used in numerous studies (Lee and Taylor, 1992; Mills *et al.*, 1992; Levesque *et al.*, 1994; Tisserat *et al.*, 1994), although other primers of this type have also been used (Xue *et al.*, 1992; Johanson and Jeger, 1993; Poupard *et al.*, 1993). For example, Tisserat *et al.* (1994) used the universal primers ITS4 and ITS5 to amplify the ITS1, ITS2 and 5.8s rDNA sequences between the large and small rRNA genes of *Ophiosphaerella korrae* and *O. herpotricha*. The resultant 590 bp fragment from *O. herpotricha* and the 590 and 1,019 bp fragments from *O. korrae* were sequenced, variable regions were located, and specific primer pairs were designed for each species. Both sets of primers specifically amplified a 454 bp product, from fungal DNA extracts as well as from extracts of *L. korrae*- or *L. herpotricha*-infected turfgrass plants.

A similar approach was utilized by Xue *et al.* (1992) to develop primers to identify and differentiate highly (HV) and weakly virulent (WV) isolates of *Leptosphaeria maculans*. They used primers homologous to the ends of the 17s and 5.8s rRNA genes flanking the ITS1 region. The amplified products from both HV and WV isolates contained the entire ITS1 region, and were sequenced and compared. DNA sequence data indicated 66.5 - 87.8% similarity between the HV and WV isolates in the ITS1 region. The sequence differences were used to design pathotype-specific primers that amplified a 220 bp product. The fragment was readily amplified from

*Table IV.* Examples of the use of specific polymerase chain reaction (PCR) assays for the detection and/or identification of plant pathogenic fungi.

| Pathogen | Primer Source[1] | Product Size (bp) | Reference |
|---|---|---|---|
| *Colletotrichum gloeosporioides* | rDNA ITS | 450 | Mills *et al.*, 1992 |
| *Gaeumannomyces graminis* | mtDNA | 188 | Schesser *et al.*, 1991; Henson *et al.*, 1993 |
| *Leptosphaeria maculans* | rDNA ITS1 | 220 | Xue *et al.*, 1992 |
| *Mycosphaerella fijiensis, M. musicola* | rDNA ITS 1 | 1000 | Johanson and Jeger, 1993 |
| *Ophiosphaerella korrae, O. herpotricha* | rDNA ITS | 454 | Tisserat *et al.*, 1994 |
| *Phytophthora fragariae* var. *rubi, P. f. fragariae* | Cloned nuclear DNA | 3006 | Stammler and Seemüller, 1993 |
| *P. parasitica* | Cloned chromosomal | 1000 | Érsek *et al.*, 1994 |
| *P. citrophthora* | DNA | 650 | Érsek *et al.*, 1994 |
| *Peronospora tabacina* | Cloned RAPD fragment | 232 | Wiglesworth *et al.*, 1994 |
| *Pseudocercosporella herpotrichoides* | rDNA ITS 1 and ITS 2 | 615 | Poupard *et al.*, 1993 |
| *Verticillium dahliae, V. albo-atrum* | rDNA ITS 1 and ITS 2 | 334 | Nazar *et al.*, 1991 |
| *V. dahliae* | mitochondrial ssu RNA | 140 | Li *et al.*, 1994 |
| *V. dahliae, V. albo-atrum,* NL group | Cloned repetitive DNA | 580 300 | Carder *et al.*, 1994 |

[1] Abbreviations: rDNA = ribosomal DNA; ITS = internal transcribed spacer region of rDNA; mtdna = mitochondrial DNA; ssu = small subunit.

fungal extracts and infected canola tissue.

When taxon-specific DNA sequences, such as cloned species-specific probes are available, all or part (terminal regions) of the fragment can be sequenced, and primers designed to amplify the sequence by PCR. Stammler and Seemüller (1993), sequenced a 3035 bp DNA fragment (Table III) from *Phytophthora fragariae* var. *rubi* that contained the small subunit rRNA gene (1784 bp), as well as flanking variable regions. Primers were selected from the terminal regions of the fragment, and amplified a 3006 bp product from *P. f. rubi* and *P. f. fragariae* DNA but not from DNA of other *Phytophthora* or *Pythium* species. The product was amplified from the roots of 90% of the *P. f. rubi*-infected raspberry plants tested, compared to 45% detection by Southern hybridization with the original probe. This approach has also been taken to develop primers for the detection of other *Phytophthora* species (Érsek *et al.*, 1994) and *Gaeumannomyces graminis* and related fungi (see Table IV). Alternate sources of fungal DNA that have been used for primer development include a gene for toxin production (Jones and Dunkle, 1993) and cloned RAPD fragments (Wiglesworth *et al.*, 1994).

## 2. Characteristics of specific PCR

In addition to the specificity that can be designed into PCR, this technique offers a level of sensitivity previously unattainable. Detection of femtogram amounts of purified fungal DNA by PCR have been reported (e.g. Mills *et al.*, 1992; Stammler and Seemüller, 1993), and where comparisons have been made, PCR has been far more sensitive than DNA probes. In several studies, the use of PCR allowed consistent detection of fungal pathogens in plant tissue, which was not always possible with DNA probes (Stammler and Seemüller, 1993; Henson *et al.*, 1993; Érsek *et al.*, 1994). The sensitivity and specificity of PCR require strict attention to a number of parameters during amplification, including especially annealing temperature and concentration of primers, DNA template, nucleotides and salts. The type of DNA polymerase used may also affect sensitivity; we have found that addition of *Pfu* polymerase to *Taq* polymerase in the reaction mixture results in significant increases in sensitivity for some systems (Miller, *unpublished*). Inhibitors in plant tissue, possibly phenolic compounds, can interfere with amplification and must be carefully controlled, by dilution of the sample or by other methods (see Henson and French, 1993). Hu *et al.* (1993) used an internal control template to determine the presence of inhibitors in sunflower tissue infected with *Verticillium dahliae*. This type of control would be very helpful in routine use of PCR for detection of fungi in a wide range of plant samples, which are likely to be encountered in a diagnostic testing facility. Since very little template DNA is required for PCR, rather crude sample extraction techniques, such as boiling (Henson *et al.*, 1993) or grinding in sodium hydroxide followed by dilution in buffer (Wang *et al.*, 1993) are often effective for sample preparation, significantly reducing the time required for PCR analysis. One of the greatest advantages of PCR, its exquisite sensitivity, can also be the cause of false positives due to the presence of even minute amounts of contaminant DNA. This can be controlled by segregating work areas and equipment, especially pipets, using specialized aerosol free pipet tips, autoclaving buffers and other solutions in small amounts, and maintaining a clear separation between samples throughout DNA extraction and amplification. A negative control containing all reaction components except the template DNA must be included in every set of reactions to assess the probability of contamination.

## 3. Applications of specific PCR

PCR assays have been developed for several fungal plant pathogens in seed or vegetative propagating materials (Table IV). Of particular interest are assays for detection of *Phytophthora fragariae* var. *rubi* and *P. f. fragariae* (Stammler and Seemüller, 1993), other *Phytophthora* species (Lee *et al.*, 1993; Érsek *et al.*, 1994), *Verticillium dahliae* and *V. albo-atrum* (Nazar *et al.*, 1991; Li *et al.*, 1994; Carder *et al.*, 1994) and *Leptosphaeria maculans* (Xue *et al.*, 1992). A preliminary report has been published on development of PCR primers for detection of *Colletotrichum acutatum*, although cross-hybridization with some isolates of other *Colletotrichum* species was observed (Mills *et al.*, 1994). PCR will undoubtedly be applied to other fungi for use in indexing systems. However, at the moment these systems are mainly restricted to research laboratories, where equipment and trained personnel are available. As it stands, PCR is a technique that requires a fairly sophisticated laboratory, with an initial moderately costly investment in equipment, including at a minimum an automated thermal cycler, a spectrophotometer, equipment for electrophoresis and photography, and repeating pipets. While most research laboratories have this equipment available, diagnostic laboratories may not. The cost of supplies alone for PCR is at least $1 US per sample (per reaction tube, if at least 30 samples are tested at a time), about one-half of which is the cost of commercial DNA polymerase. Labor costs may also be significant, depending on the sample DNA extraction protocol and the method used to visualize the amplified products. Most laboratories would be limited at best to 30-60 samples (reaction tubes) per day per technician. However, it is possible that costs will decrease as PCR protocols are simplified and the technique is used more widely.

*4. Randomly primed PCR (RAPD technique)*
Randomly primed PCR is a very new technique (Williams *et al.*, 1990) that has nonetheless been applied widely to plant pathogenic fungi. In this technique, known as RAPD (random amplified polymorphic DNA) analysis, single arbitrary nona- or decamer oligonucleotide primers complementary to unknown sequences throughout the genome are used to direct DNA amplification. The primers, which are available commercially, should have a G + C content of 50 - 80% and must not contain palindromic sequences (Williams *et al.*, 1990). Primer annealing takes place under conditions of low stringency, usually resulting in the amplification of one or more fragments less than 3000 bp in size. Band polymorphisms between individuals visualized on gels are known as RAPDs. RAPD analysis provides much, if not all, of the same genetic information as RFLP analysis, but is much simpler and faster to carry out, and does not require large amounts of highly purified DNA or the use of radioisotopes (Wöstmeyer *et al.*, 1992; Black, 1993). Since sequence data is not required for primer design, RAPD analysis can be used even for fungi that are poorly characterized genetically. It is a relatively easy means of identifying fungi at the sub-specific level, and several systems have been developed for potential use in pathogen indexing or quarantine programs. An important disadvantage of RAPD analysis over specific PCR is that pure cultures of the fungi of interest are usually required. Contamination of the sample with DNA from other microorganisms or plant tissue might hopelessly obscure the results of a RAPD analysis.

RAPD analyses also may serve as a source of DNA fragments for the production of specific probes or primers (Lee *et al.*, 1993; Levesque *et al.*, 1994; Manulis *et al.*, 1994). Amplified DNA fragments unique to a group of interest can be recovered, cloned and labelled for use in a dot blot or similar assay. They may also be partially or completely sequenced, and the sequences used to design primers for specific PCR. Genetic markers produced in this way have been called SCARs (sequence character-ized amplified regions) (Paran and Michelmore, 1993). Use of such probes or primers would allay concerns about possible lack of reproducibility of RAPD analyses between laboratories (Black, 1993).

Manulis *et al.* (1994) screened 30 different arbitrary decamer primers for the ability to distinguish between pathogenic races of *Fusarium oxysporum* f. sp. *dianthi* and nonpathogenic isolates from carnation cuttings. Of the 22 primers that gave reproducible patterns, all produced identical RAPD patterns that differed from the patterns of nonpathogenic isolates and other *F. oxysporum formae speciales*. There was also evidence of differentiation of race 2 and race 4 isolates by some primers. The authors noted that the RAPD assay could be used for large scale screening of *Fusarium* cultures from carnation in Israel, where cuttings are routinely indexed for this pathogen. Preliminary results have been published on attempts to develop RAPD markers for pathogenic isolates of *F. oxysporum* f. sp. *gladioli*, which causes yellows and corm rot of gladiolus and related crops (Mes *et al.*, 1994). Polymorphisms between race 1 and race 2 isolates were found for 23 of 80 primers tested, and one primer resulted in amplification of a potentially diagnostic fragment. After further work, this system could be used as an adjunct to selective isolation of the pathogen from corms to detect latent infections, or to develop a specific probe. RAPD assays have also been developed for the detection of seedborne *Leptosphaeria maculans* (Goodwin and Annis, 1991). Like the DNA probe (Schäfer and Wöstemeyer, 1994), specific PCR (Xue *et al.*, 1992) and serological (Stace-Smith *et al.*, 1993) assays mentioned in previous sections of this chapter, the RAPD assay also clearly differentiated the aggressive or virulent types from the nonaggressive, or avirulent types. As a final example, Hamelin *et al.* (1993) identified seven RAPD markers that could be used to distinguish between the North American (NA) and European (EU) races of *Gremmeniella abietina*, which causes Scleroderris canker of conifers. Quarantines have been established in North America to restrict the spread of the EU race. The RAPD assay was used to identify *G. abietina* directly from single fruiting bodies of the fungus, eliminating the need for culturing (about one month). Clearly the routine use of this assay could significantly streamline the identification process for this pathogen.

## V. CONCLUSION

Development of new biotechnological techniques for detection of plant pathogenic fungi is currently in an intermediate stage of growth. Considerable effort has been expended on the development of serological assays, but even with the advent of monoclonal antibodies, it has not always been possible to achieve distinction between taxa at the sub-specific level. However, immunoassay formats·such as ELISA are generally quick and easy to use, from sample preparation to assay interpretation. Many standard microtiter plate formats can be completed in 1 - 3 hrs, and 10 minute assays that do not require specialized equipment and are suitable for field use have been developed (Miller and Joaquim, 1993). These assays have an important role in crop production, but for certification or quarantine programs, greater specificity may be needed. On the other hand, nucleic acid hybridization-based assays (including PCR) do not appear to be limited in the level of specificity that can be achieved, but are cumbersome, more time-consuming than immunoassays and require more specialized equipment, facilities and worker training. Problems with sensitivity of dot-blot and related DNA-DNA hybridization assays have been overcome by the use of PCR, in which extremely low amounts of DNA can be detected, even in complex samples. Assays combining the ease of use of an immunoassay with the exquisite specificity and sensitivity of PCR would probably be quickly accepted in indexing and other diagnostic programs. It is likely that the flaws of both serological and DNA-based assays will be addressed in future research, culminating in improved systems for fungal detection.

There are several pathogens of interest in indexing programs that have been studied thoroughly, especially *Leptosphaeria maculans* and *Phytophthora fragariae*. Serological and DNA-based assays have been developed for these fungi, and illustrate the advantages and disadvantages of each approach. Although there may be something lacking in each type of assay, they are superior to the traditional assays in the amount of information that can be obtained and the time required to obtain it. What may not be advantageous in these and other assays is cost. Cost is a critical consideration for diagnostic testing in agriculture, and total costs, including equipment, supplies and labor must be reasonable if such tests are to be used widely. In agricultural diagnostics, there is always a question about how much the end user is willing to pay for the information provided. This of course means that it is especially important that costs be reduced as much as possible in order to enable routine use of these tools. Large-scale production of standardized reagents and assays by private or public concerns should help reduce assay cost and would also improve reproducibility between laboratories. Questions of patentability of antibodies, probes and primers should also be considered. While the ability to patent encourages invention, it may increase costs to the ultimate end user.

The scientific advances of the last five to ten years in biotechnology have been very encouraging from the viewpoint of solving problems in crop production. Tools that make it easier to provide pathogen-free seed and vegetative propagating material will become increasingly important as consumers continue to express concern over the use of fungicides in agricultural commodities. Further research in the development and application of new technology, and eventually its routine use, can only benefit agricultural producers and consumers.

## ACKNOWLEDGMENTS

I thank Ravindra G. Bhat for critically reading the manuscript and for helpful comments and suggestions, and Marcella E. Grebus for providing the cost analysis for PCR assays.

## REFERENCES

Amouzou-Alladaye, E., Dunez, J. and Clerjeau, M. (1988). Immunoenzymatic detection of *Phytophthora fragariae* in infected strawberry plants. *Phytopathology* 78, 1022-1026.

Appels, N. and Honeycutt, R. L. (1986). rDNA: Evolution over a billion years. *In* 'DNA Systematics' (Dutta, S. K., ed.), Vol. 2, pp. 81-135. CRC Press, Boca Raton, FL.

Audy, P., Parent, J.-G. and Asselin, A. (1991). A note on four nonradioactive labelling systems for dot hybridization detection of potato viruses. *Phytoprotection* 72, 81-86.

Banowetz, G. M., Trione, E. J., and Krygier, B. B. (1984). Immunological comparisons of teliospores of two wheat bunt fungi, *Tilletia* species, using monoclonal antibodies and antisera. *Mycologia* **76**, 51-62.

Barbara, D. J. and Clark, M. F. (1982). A simple indirect ELISA using F(ab')₂ fragments of immunoglobulin. *Journal of General Virology* **58**, 315-322.

Bar-Joseph, M. and Saloman, R. (1980). Heterologous reactivity of tobacco mosaic virus strains in enzyme-linked immunosorbent assays. *Journal of General Virology* **47**, 509-412.

Barker, I. and Pitt, D. (1988). Detection of the leaf curl pathogen of anemones in corms by enzyme-linked immunosorbent assay (ELISA). *Plant Pathology* **37**, 417-422.

Barker, I., Brewer, G., Cook, R. T. A., Crossley, S. and Freeman, S. (1994). Strawberry blackspot disease. *In* 'Modern Assays for Plant Pathogenic Fungi: Identification, Detection and Quantification' (Schots, A., Dewey, F. M. and Oliver, R., eds.), pp. 179-182. CAB International, Oxford.

Bateman, G. L., Ward, E. and Antoniw J. F. (1992). Identification of *Gaeumannomyces graminis* var. *tritici* and *G. graminis* var. *avenae* using a DNA probe and non-molecular methods. *Mycological Research* **96**, 737-742.

Bej, A. K., Mahbubani, M. II. and Atlas, R. M. (1991). Amplification of nucleic acids by polymerase chain reaction (PCR) and other methods and their application. *Critical Reviews in Biochemistry and Molecular Biology* **26**, 301-334.

Benson, D. M. (1991). Detection of *Phytophthora cinnamomi* in azalea with commercial serological assay kits. *Plant Disease* **75**, 478-482.

Benson, D. M. (1992). Detection by enzyme-linked immunosorbent assay of *Rhizoctonia* species on poinsettia cuttings. *Plant Disease* **76**, 578-581.

Bhatnagar, D., Neucere, J. N. and Cleveland, T. E. (1989). Immunochemical detection and aflatoxigenic potential of *Aspergillus* species and antisera prepared against enzymes specific to aflatoxin biosynthesis. *Food and Agricultural Immunology* **1**, 225-234.

Black, W. C., IV (1993). PCR with arbitrary primers: approach with care. *Insect Molecular Biology* **2**, 1-6.

Bossi, R. and Dewey, M. F. (1992). Development of a monoclonal antibody-based immunodetection assay for *Botrytis cinerea*. *Plant Pathology* **41**, 472-482.

Brill, L. M., McClary, R. D. and Sinclair, J. D. (1994). Analysis of two ELISA formats and antigen preparations using polyclonal antibodies to *Phomopsis longiicolla*. *Phytopathology* **84**: 173-179.

Burns, R., Vernon, M. L. and George, E. L. (1994). Monoclonal antibodies for the detection of *Pyrenophora graminea*. *In* 'Modern Assays for Plant Pathogenic Fungi: Identification, Detection and Quantification' (Schots, A., Dewey, F. M. and Oliver, R., eds.), pp. 199-203. CAB International, Oxford.

Carder, J. H., Morton, A., Tabrett, A. M and Barbara, D. J. (1994). Detection and differentiation by PCR of subspecific groups within two *Verticillium* species causing vascular wilts in herbaceous hosts. *In* 'Modern Assays for Plant Pathogenic Fungi: Identification, Detection and Quantification' (Schots, A., Dewey, F. M. and Oliver, R., eds.), pp. 91-97. CAB International, Oxford.

Cousin, M. A., Dufrenne, J., Rombouts, F. M. and Notermans, S. (1990). Immunological detection of *Botrytis* and *Monascus* species in food. *Food Microbiology* **7**, 227-235.

Cunfer, B. M. (1983). Epidemiology and control of seed-borne *Septoria nodorum* on wheat. *Seed Science and Technology* **11**, 707-718.

Del Sorbo, G., Scala, F., Capparelli, R., Iannelli, D. and Noviello, C. (1993). Differentiation between two formae speciales of *Fusarium oxysporum* by antisera produced in mice immunologically tolerized at birth through lactation. *Phytopathology* **83**, 1178-1182.

De Ruiter, G. A., Van Bruggen-van der Lugt, A. W., Bos, W., Notermans, S. H., Rombouts, F. M. and Hofstra, H. (1993). The production and partial characterization of a monoclonal IgG antibody specific for molds belonging to the order Mucorales. *Journal of General Microbiology* **139**, 1557-1564.

Dewey, F. M. (1988). Development of immunological diagnostic assays for fungal plant pathogens. *Brighton Crop Protection Conference-Pests and Diseases,* British Crop Protection Council, pp. 777-786.

Dewey, F. M. (1992). Detection of plant-invading fungi by monoclonal antibodies. *In* 'Techniques for the Rapid Detection of Plant Pathogens' (J. M. Duncan and L. Torrance, eds.). pp. 47-65. Blackwell Scientific, Oxford.

Dewey, F. M. and Brasier, C. M. (1988). Development of ELISA for *Ophiostoma ulmi* using antigen coated wells. *Plant Pathology* **37**, 28-35.

Dewey, F. M., MacDonald, M. M. and Phillips, S. I. (1989). Development of mono-clonal-antibody-ELISA, -DOT-BLOT, and -DIP-STICK immunoassays for *Humicola lanuginosa* in rice. *Journal of General Microbiology* **135**, 361-374.

Dewey, F. M., Munday, C. J. and Brasier, C. M. (1989). Monoclonal antibodies to specific components of the Dutch elm disease pathogen *Ophiostoma ulmi*. *Plant Pathology* **38**, 9-20.

Dewey, M., Evans, D., Coleman, J., Priestly, R., Hull, R., Horsley, D. and Hawes, C. (1991). Antibodies in plant science. *Acta Botanica Neerlandica*, **40**, 1-27.

Duncan, J. M. (1980). A technique for detecting red stele (*Phytophthora fragariae*) infection of strawberry stocks before planting. *Plant Disease* **64**, 1023-1025.

Duncan, J. M. (1990). *Phytophthora* species attacking strawberry and raspberry. *Bulletin OEPP/EPPO* **20**, 107-115.

Ellis, M. A. and Miller, S. A. (1993). Using a *Phytophthora*-specific immunoassay kit to diagnose raspberry Phytophthora root rot. *HortScience* **28**, 642-644.

Érsek, T., Schoelz, J. E. and English, J. T. (1994). PCR amplification of species-specific DNA sequences can distinguish among *Phytophthora* species. *Applied and Environmental Microbiology* **60**, 2616-2621.

Estrada-Garcia, M., Green, J. R., Booth, J. M., White, J. G. and Callow, J. A. (1989). Monoclonal antibodies to cell surface components of zoospores and cysts of the fungus *Pythium aphanidermatum* reveal species-specific antigens. *Experimental Mycology* **13**, 348-355.

Fani, R., Damiani, G., Di Serio, C., Gallori, E., Griffoni, A. and Bazzicalupo, M. (1993). Use of random amplified polymorphic DNA (RAPD) for generating specific DNA probes for microorganisms. *Molecular Ecology* **2**, 243-250.

Foster, L. M., Kozak, K. R., Loftus, M. G., Stevens, J. J. and Ross, I. K. (1993). The polymerase chain reaction and its application to filamentous fungi. *Mycological Research* **97**, 769-781.

Galbraith, D. N. and Palfreyman, J. W. (1994). Detection of *Heterobasidium annosum* using monoclonal antibodies. *In* 'Modern Assays for Plant Pathogenic Fungi: Identification, Detection and Quantification' (Schots, A., Dewey, F. M. and Oliver,

R., eds.), pp. 105-110. CAB International, Oxford.

Goding, J. W. (1987). 'Monoclonal Antibodies: Principles and Practice.' 2nd. ed. Academic Press, London.

Goodwin, P. H. and Annis, S. L. (1991). Rapid identification of genetic variation and pathotype of *Leptosphaeria maculans* by random amplified polymorphic DNA assay. *Applied and Environmental Microbiology* **57**, 2482-2486.

Goodwin, P. H., Kirkpatrick, B. C. and Duniway, J. M. (1989). Cloned DNA probes for identification of *Phytophthora parasitica*. *Phytopathology* **79**, 716-721.

Goodwin, P. H., English, J. T., Neher, D. A., Duniway, J. M. and Kirkpatrick, B. C. (1990a). Detection of *Phytophthora parasitica* from soil and host tissue with a species-specific DNA probe. *Phytopathology* **80**, 277-281.

Goodwin, P. H., Kirkpatrick, B. C. and Duniway, J. M. (1990b). Identification of *Phytophthora citrophthora* with cloned DNA probes. *Applied and Environmental Microbiology* **56**, 669-674.

Guy, P. L. and Rath, A. C. (1990). Enzyme-linked immunosorbent assays (ELISA) to detect spore surface antigens of *Metarhizium anisopliae*. *Journal of Invertebrate Pathology* **55**, 435-436.

Hamelin, R. C., Ouellette, G. B. and Bernier, L. (1993). Identification of *Gremmeniella abietina* races with random amplified polymorphic DNA markers. *Applied and Environmental Microbiology* **59**, 1752-1755.

Hamilton, A. J., Batholomew, M. A., Fenelon, L. E., Figueroa, J. and Hay, R. J. (1990). A murine monoclonal antibody exhibiting high species specificity for *Histoplasma capsulatum* var. *capsulatum*. *Journal of General Microbiology* **136**, 331-335.

Hardham, A. R., Suzaki, E. and Perkin, L. (1985). The detection of monoclonal antibodies specific for surface components on zoospores and cysts of *Phytophthora cinnamomi*. *Experimental Mycology* **9**, 264-268.

Hardham, A. R., Suzaki, E. and Perkin, L. (1986). Monoclonal antibodies to isolate-, species-, and genus-specific components on the surface of zoospores and cysts of the fungus *Phytophthora cinnamomi*. *Can. J. Bot.* **64**, 311-321.

Harlow, E. and Lane, D. (1988). 'Antibodies, A Laboratory Manual', Cold Spring Harbor Laboratory, Cold Spring Harbor, New York, 726 pp. .

Harrison, J. G., Barker, H., Lowe, R. and Rees, E. A. (1990). Estimation of amounts of *Phytophthora infestans* mycelium in leaf tissue by enzyme-linked immunosorbent assay. *Plant Pathology* **39**, 274-277.

Harrison, J. G., Rees, E. A., Barker, H. and Lowe, R. (1993). Detection of spore balls of *Spongospora subterranea* on potato tubers by enzyme-linked immunosorbent assay. *Plant Pathology* **42**, 181-186.

Harrison, J. G., Lowe, R., Wallace, A. and Williams, N. A. (1994). Detection of *Spongospora subterranea* by ELISA using monoclonal antibodies. *In* 'Modern Assays for Plant Pathogenic Fungi: Identification, Detection and Quantification' (Schots, A., Dewey, F. M. and Oliver, R., eds.), pp. 23-27. CAB International, Oxford.

Henson, J. M. (1989). DNA probe for identification of the take-all fungus, *Gaeumann-omyces graminis*. *Applied and Environmental Microbiology* **55**, 284-288.

Henson, J. M and French, R. (1993). The polymerase chain reaction and plant disease diagnosis. *Annual Review of Phytopathology* **31**, 81-109.

Henson, J. M., Goins, T., Grey, W., Mathre, D. E. and Elliott, M. L. (1993). Use of polymerase chain reaction to detect *Gaeumannomyces graminis* DNA in plants grown in artificially and naturally infested soil. *Phytopathology* **83**, 283-287.

Höxter, H., Miedaner, Th., Sander, E. and Geiger, H. H. (1991). Quantitative assessment of *Microdochium nivale* in rye with ELISA. *Journal of Plant Diseases and Plant Protection* **98**, 13-17.

Hu, X., Nazar, R. N. and Robb, J. (1993). Quantification of *Verticillium* biomass in wilt disease development. *Physiological and Molecular Plant Pathology* **42**, 23-36.

Husted, K. (1994). Development of species-specific probes for identification of *Pyrenophora graminea* and *P. teres* by dot-blot or RFLP. *In* 'Modern Assays for Plant Pathogenic Fungi: Identification, Detection and Quantification' (Schots, A., Dewey, F. M. and Oliver, R., eds.), pp. 191-197. CAB International, Oxford.

Johanson, A. and Jeger, M. J. (1993). Use of PCR for detection of *Mycosphaerella fijiensis* and *M. musicola*, the causal agents of Sigatoka leaf spots on banana and plantain. *Mycological Research* **97**, 670-674.

Jones, M. J. and Dunkle, L. D. (1993). Analysis of *Cochliobolus carbonum* races by PCR amplification with arbitrary and gene-specific primers. *Phytopathology* **83**, 366-370.

Kellens, J. T. C. and Peumans, W. J. (1991). Biochemical and serological comparison of lectins from different anastomosis groups of *Rhizoctonia solani*. *Mycological Research* **95**, 1235-1241.

Kim, Y. S., Jellison, J., Goodell, B., Tracy, V., and Chandhoke, V. (1991). The use of ELISA for the detection of white- and brown-rot fungi. *Holzforschung* **45**, 403-406.

Kohler, G. and Milstein, C. 1975. Continuous cultures of fused cells secreting antibody of predefined specificity. *Nature* **45**, 495-497.

Koopmann, B., Karlovsky, P., and Wolf, G. (1994). Differentiation between *Fusarium culmorum* and *Fusarium graminearum* by RFLP and with species-specific DNA probes. *In* 'Modern Assays for Plant Pathogenic Fungi: Identification, Detection and Quantification' (Schots, A., Dewey, F. M. and Oliver, R., eds.), pp. 37-46. CAB International, Oxford.

Kraft, J. M. and Boge, W. L. (1994). Development of an antiserum to quantify *Aphanomyces euteiches* in resistant pea lines. *Plant Disease* **78**, 179-183.

Lee, S. B. and Taylor, J. W. (1992). Phylogenetic relationships of five fungus-like *Phytophthora* species inferred from the internal transcribed spacers of ribosomal DNA. *Molecular Biology and Evolution* **9**, 636-653.

Lee, S. B., White, T. J. and Taylor, J. W. (1993). Detection of *Phytophthora* species by oligonucleotide hybridization to amplified ribosomal DNA spacers. *Phytopathology* **83**, 177-181.

Levesque, C. A., Vrain, T. C. and De Boer, S. H. (1994). Development of a species-specific probe for *Pythium ultimum* using amplified ribosomal DNA. *Phytopathology* **84**, 474-478.

Li, K.-N., Rouse, D. I. and German, T. L. (1994). PCR primers that allow intergeneric differentiation of ascomycetes and their application to *Verticillium* species. *Applied and Environmental Microbiology* **60**, 4324-4331.

Lind, V. (1990). Isolation of antigens for serological identification of *Pseudocercosporella herpotrichoides* (Fron) Deighton. *Journal of Plant Diseases and Plant Protection* **97**, 490-501.

Linfield, C. A. (1993). A rapid serological test for detecting *Fusarium oxysporum* f. sp. *narcissi* in *Narcissus*. *Annals of Applied Biology* **123**, 685-693.

Maguire, J. D. and Gabrielson, R. L. (1983). Testing techniques for *Phoma lingam. Seed Science and Technology* **11**, 599-605.

Mathur, S. B. and Lee, S. L. N, (1978). A quick method for screening wheat samples for *Septoria nodorum. Seed Science and Technology* **6**, 925-926. `

Mangan, .A. (1983). The use of plain water agar for the detection of *Phoma betae* on beet seed. *Seed Science and Technology* **11**, 607-614.

Manulis, S., Kogan, N., Reuven, M. and Ben-Yephet, Y. (1994). Use of the RAPD technique for identification of *Fusarium oxysporum* f. sp. *dianthi* from carnation. *Phytopathology* **84**, 98-101.

Matthew, J. S. and Brooker, J. D. (1991). The isolation and characterization of polyclonal and monoclonal antibodies to anastomosis group 8 of *Rhizoctonia solani. Plant Pathology* **40**, 67-77.

Mes, J. J., van Doorn, J., Roebroeck, E. J. A. and Boonekamp, P. M. (1994). Detection and identification of *Fusarium oxysporum* f. sp. *gladioli* by RFLP and RAPD analysis. *In* 'Modern Assays for Plant Pathogenic Fungi: Identification, Detection and Quantification' (Schots, A., Dewey, F. M. and Oliver, R., eds.), pp. 63-68. CAB International, Oxford.

Miller, S. A. and Joaquim, T. R. (1993). Diagnostic techniques for plant pathogens. *In* 'Biotechnology in Plant Disease Control' (I. Chet, ed.). pp. 321-339. Wiley-Liss, New York.

Miller, S. A. and Martin, R. R. (1988). Molecular diagnosis of plant disease. *Annual Review of Phytopathology* **26**, 409-432.

Mills, P. R., Sreenivasaprasad, S. and Brown, A. E. (1992). Detection and differentiation of *Colletotrichum gloeosporioides* isolates using PCR. *FEMS Microbiology Letters* **98**, 137-144.

Mills, P. R., Sreenivasaprasad, S. and Brown, A. E. (1994). Detection of the anthracnose pathogen *Colletotrichum. In* 'Modern Assays for Plant Pathogenic Fungi: Identification, Detection and Quantification' (Schots, A., Dewey, F. M. and Oliver, R., eds.), pp. 183-189. CAB International, Oxford.

Mitchell, L. A. (1986). Derivation of *Sirococcus strobilinus* specific monoclonal antibodies. *Canadian Journal of Forestry Research* **16**, 939-944.

Mitchell, L. A. (1988). A sensitive dot immunoassay employing monoclonal antibodies for detection of *Sirococcus strobilinus. Plant Disease* **72**, 664-667.

Mitchell, L. A. and Sutherland, J. R. (1986). Detection of seed-borne *Sirococcus strobilinus* with monoclonal antibodies in an enzyme-linked immunosorbent assay. *Canadian Journal of Forestry Research* **16**, 945-948.

Mittermeier, L., West, S., Dercks, W. and Miller, S. A., (1990). Field results with a diagnostic system for the identification of *Septoria nodorum* and *Septoria tritici. In* 'Brighton Crop Protection Conference-Pests and Diseases', pp. 757-762, British Crop Protection Council, Surrey, UK.

Mohan, S. B., 1988. Evaluation of antisera raised against *Phytophthora fragariae* for detecting the red core disease of strawberries by enzyme-linked immunosorbent assay. *Plant Pathology* **37**, 206-216.

Mohan, S. B., 1989. Cross-reactivity of antiserum raised against *Phytophthora fragariae* with other *Phytophthora* species and its evaluation as a genus-detecting antiserum. *Plant Pathology* **38**, 352-263.

Möller, E. M., de Cock, A. W. A. M. and Prell, H. H. (1993). Mitochondrial and nuclear DNA restriction enzyme analysis of the closely related *Phytophthora* species *P. infestans, P. mirabilis,* and *P. phaseoli. J. Phytopathology* 139, 309-321.

Mullis, K, B., Ferré, F. Gibbs, R. A. (eds.) (1994). 'The Polymerase Chain Reaction.' Birkhauser, Boston. 458 pp.

Nachmias, A. and Krikun, J. (1984). Diagnosis of *Verticillium dahliae* in potato seed tubers with immunoassay techniques. *Potato Research* 27, 423-426.

Nameth, S. T., Shane, W. W. and Stier, J. C. (1990). Development of a monoclonal antibody for detection of *Leptosphaeria korrae,* the causal agent of necrotic ringspot of turfgrass. *Phytopathology* 80:1208-1211.

Nazar, R. N., Hu., X., Schmidt, J., Culham, D., and Robb, J. (1991). Potential use of PCR-amplified ribosomal intergenic sequences in the detection and differentiation of verticillium wilt pathogens. *Physiological and Molecular Plant Pathology* 39, 1-11.

Nicholson, P. and Rezanoor, H. N. (1994). DNA probe for the R-type of eyespot disease of cereals *Pseudocercosporella herpotrichoides. In* 'Modern Assays for Plant Pathogenic Fungi: Identification, Detection and Quantification' (Schots, A., Dewey, F. M. and Oliver, R., eds.), pp. 17-22. CAB International, Oxford.

Notermans, S. Veeneman, G. H., van Zuylen, C. W. E. M. Hoogerhout, P. and van Boom, J. (1988). (1→5)-linked ß-D-galactofuranosides are immunodominant in extracellular polysaccharides of *Penicillium* and *Aspergillus* species. *Molecular Immunology* 25, 975-978.

Pain, N. A., O'Connell, R. J., Bailey, J. A. and Green, J. R. (1992). Monoclonal antibodies which show restricted binding to four *Colletotrichum* species: *C. lindemuthianum, C. malvarum, C. orbiculares* and *C. trifolii. Physiological and Molecular Plant Pathology* 40, 111-126.

Paran, I. and Michelmore, R. W. (1993). Development of reliable PCR-based markers linked to downy mildew resistance genes in lettuce. *Theoretical and Applied Genetics* 85, 985-993.

Petersen, F. P., Rittenburg, J. H., Miller, S. A. and Grothaus, G. D. (1990). Development of monoclonal antibody-based immunoassays for detection and differentiation of *Septoria nodorum* and *S. tritici* in wheat. *In* 'Brighton Crop Protection Conference-Pests and Diseases', pp. 751-756, British Crop Protection Council, Surrey, UK.

Poupard, P., Simonet, P., Cavelier, N. and Bardin, R. (1993). Molecular characterization of *Pseudocercosporella herpotrichoides* isolates by amplification of ribosomal DNA internal transcribed spacers. *Plant Pathology* 42, 873-881.

Priestly, R., Mohammed, C. and Dewey, F. M. (1994). The development of monoclonal antibody-based ELISA and dipstick assays for the detection and identification of *Armillaria* species in infected wood. *In* 'Modern Assays for Plant Pathogenic Fungi: Identification, Detection and Quantification' (Schots, A., Dewey, F. M. and Oliver, R., eds.), pp. 149-156. CAB International, Oxford.

Pryor, B. M., Davis, R. M. and Gilbertson, R. L. (1994). Detection and eradication of *Alternaria radicina* on carrot seed. *Plant Disease* 78, 452-456.

Pscheidt, J. W., Burket, J. Z., Fischer, S. L. (1992). Sensitivity and clinical use of *Phytophthora*-specific immunoassay kits. *Plant Disease* 76, 928-932.

Raju, B. C. and Olson, C. J. (1985). Indexing systems for producing clean stock for disease control in commercial floriculture. *Plant Disease* 69, 189-192.

Ricker, R. W., Marois, J. J., Dlott, J. W., Bostock, R. M. and Morrison, J. C. (1991).

Immunodetection and quantification of *Botrytis cinerea* on harvested wine grapes. *Phytopathology* **81**:404-411.

Roebroeck, E. J. A., Groen, N. P. A. and Mes, J. J. (1990). Detection of latent *Fusarium oxysporum* in gladiolus corms. *Acta Horticulturae* **266**, 468-476.

Rollo, F., Amica, A., Francesco, F. and Silvestro, I. (1987). Construction and characterization of a cloned probe for the detection of *Phoma tracheiphila* in plant tissues. *Applied Microbiology and Biotechnology* **26**, 352-257.

Saiki, R. K., Scharf, S. J., Faloona, S., Mullis, K. B., Horn, G. T., Erlich, M. A. and Arnheim, N. (1985). Enzymatic amplification of ß-globin genomic sequences and restriction site analysis for diagnosis of sickle cell anaemia. *Science* **230**, 1350-1354.

Saiki, R., Gelfand, D. H., Stoffel, S., Scharf, S. J., Higuchi, R., Horn, G.T., Mullis, K. B. and Erlich, H. A. (1988). Primer directed enzymatic amplification of DNA with a thermostable DNA polymerase. *Science* **239**, 487-494.

Salinas, J. and Schots, A. (1994). Monoclonal antibodies-based immunofluorescence test for detection of conidia of *Botrytis cinerea* on cut flowers. *Phytopathology* **84**, 351-356.

Salinas, J., Schober, G. and Schots, A. (1994). Monoclonal antibodies for detection of conidia of *Botrytis cinerea* on cut flowers using an immunofluorescence assay. *In* 'Modern Assays for Plant Pathogenic Fungi: Identification, Detection and Quantification' (Schots, A., Dewey, F. M. and Oliver, R., eds.), pp. 173-178. CAB International, Oxford.

Sambrook, J., Fritsch, E. F. and Maniatis, T. (1989). *Molecular Cloning; A Laboratory Manual*, 2nd ed. Cold Spring Harbor Laboratory Press, Cold Spring Harbor, N. Y.

Sauer, L. M., Hulbert, S. H. and Tisserat, N. A. (1993). Identification of *Ophiosphaerella herpotricha* by cloned DNA probes. *Phytopathology* **83**, 97-102.

Schäfer, C. and Wöstemeyer, J. (1994). Molecular diagnosis of the rapeseed pathogen *Leptosphaeria maculans* based on RAPD-PCR. *In* 'Modern Assays for Plant Pathogenic Fungi: Identification, Detection and Quantification' (Schots, A., Dewey, F. M. and Oliver, R., eds.), pp. 1-8. CAB International, Oxford.

Schesser, K., Luder, A. and Henson, J. M. (1991). Use of polymerase chain reaction to detect the take-all fungus, *Gaeumannomyces graminis*, in infected wheat plants. *Applied and Environmental Microbiology* **57**, 553-556.

Schots, A., Dewey, F. M. and Oliver, R. (eds.). (1994). 'Modern Assays for Plant Pathogenic Fungi: Identification, Detection and Quantification'. CAB International, Oxford.

Sherwood, J. L. (1987). Comparison of a filter paper immunobinding assay, western blotting and an enzyme-linked immunosorbent assay for the detection of wheat streak mosaic virus. *Journal of Phytopathology* **118**, 68-75.

Smith, C. M., Saunders, D. W., Allison, D. A., Johnson, L. E. B., Labit, B., Kendall, S. J. and Hollomon, D. W. (1990). Immunodiagnostic assay for cereal eyespot: novel technology for disease detection. *Brighton Crop Protection Conference-Pests and Diseases*, British Crop Protection Council, pp. 763-770.

Stace-Smith, R., Bowler, G., MacKenzie, D. J. and Ellis, P. (1993). Monoclonal antibodies differentiate the weakly virulent from the highly virulent strain of *Leptosphaeria maculans*, the organism causing blackleg of canola. *Canadian Journal of Plant Pathology* **15**, 127-133.

Stammler, G. and Seemüller, E. (1993). Specific and sensitive detection of *Phyto-*

*phthora fragariae* var. *rubi* in raspberry roots by PCR amplification. *Journal of Plant Diseases and Protection* **100**, 394-400.

Stammler, G. and Seemüller, E. (1994). Detection of *Phytophthora fragariae* var. *rubi* in infected raspberry roots by PCR. *In* 'Modern Assays for Plant Pathogenic Fungi: Identification, Detection and Quantification' (Schots, A., Dewey, F. M. and Oliver, R., eds.), pp. 135-139. CAB International, Oxford.

Stammler, G., Seemüller, E. and Duncan, J. M. (1993). Analysis of RFLPs in nuclear and mitochondrial DNA and the taxonomy of *Phytophthora fragariae*. *Mycological Research* 97, 150-156.

Takenaka, S. (1992). Use of immunological methods with antiribosome serums to detect snow mold fungi in wheat plants. *Phytopathology* **82**, 896-901.

Thornton, C. R., Dewey, F. M. and Gilligan, C. A. (1993). Development of monoclonal antibody-based immunological assays for the detection of live propagules of *Rhizoctonia solani* in soil. *Plant Pathology* **42**, 763-773.

Timmer, L. W., Menge, J. A., Zitco, S. E., Pond, E. Miller, S. A., and Johnson, E. L. V. (1993). Comparison of ELISA techniques and standard isolation methods for *Phytophthora* detection in citrus orchards in Florida and California. *Plant Disease* 77, 791-796.

Tisserat, N. A., Hulbert, S. H. and Nus, A. (1991). Identification of *Leptosphaeria korrae* by cloned DNA probes. *Phytopathology* **81**, 917-921.

Tisserat, N. A., Hulbert, S. H. and Sauer, K. M. (1994). Selective amplification of rDNA internal transcribed spacer regions to detect *Ophiosphaerella korrae* and *O. herpotricha*. *Phytopathology* **84**, 478-482.

Tseng, T.-C. (1989). Immunoassay in mycotoxin research. *In* 'Phytochemical Ecology: Allelochemicals, Mycotoxins and Insect Pheromones and Allomones' (C. H. Chou and G. R. Waller, eds.). pp. 363-370. Institute of Botany, Academica Sinica Monograph Series No. 9, Taipei, ROC.

Tullis, R. H. (1994). Ultrasensitive nonradioactive detection of PCR reactions: and overview. *In* 'The Polymerase Chain Reaction' (K. B. Mullis, F. Ferre and R. A. Gibbs, eds.). pp. 123-133. Birkhauser, Boston.

Unger, J. -G. and Wolf, G. (1988). Detection of *Pseudocercosporella herpotrichoides* (Fron) Deighton in wheat by indirect ELISA. *Journal of Phytopathology* **122**, 281-286.

van de Koppel, M. M and Schots, A. (1994). A double (monoclonal) antibody sandwich ELISA for the detection of *Verticillium* species in roses. *In* 'Modern Assays for Plant Pathogenic Fungi: Identification, Detection and Quantification' (Schots, A., Dewey, F. M. and Oliver, R., eds.), pp. 99-104. CAB International, Oxford.

Velicheti, R. K., Lamison, C., Brill, L. M., and Sinclair, J. B. 1993. Immunodetection of *Phomopsis* species in asymptomatic soybean plants. *Plant Disease* 77, 70-73.

Wang, H., Qi, M. and Cutler, A. J. (1993). A simple method of preparing plant samples for PCR. *Nucleic Acids Research* 21, 4153-4154.

Ward., C. M., Wilkinson, A. P., Bramham, S., Lee, H. A., Chan, H. W.-S., Butcher, G. W., Hutchings, A., and Morgan, M. R. A. (1990). Production and characterization of polyclonal and monoclonal antibodies against Aflatoxin B1 oxime-BSA in an enzyme-linked immunosorbent assay. *Mycotoxin Research* **6**, 73-83.

Ward, E. and Gray, R. M. (1992). Generation of a ribosomal DNA probe by PCR and its use in identification of fungi within the *Gaeumannomyces-Phialophora* complex.

*Plant Pathology* **41**, 730-736.

White, B. A., (ed.) (1993). 'PCR Protocols: Current Methods and Applications', 392 pp. Humana Press, Totowa, New Jersey.

White, T. J., Bruns, T., Lee, S. and Taylor, J. (1990). Amplification and direct sequencing of fungal ribosomal RNA genes for phylogenetics. *In* 'PCR Protocols: A Guide to Methods and Applications' (Innis, M. A., Gelfand, D. H., Sninsky, J. J. and White, T. J., eds.), pp. 315-322. Academic Press, San Diego, CA.

Wiglesworth, M. D., Nesmith, W. C., Schardl, C. L., Li, D. and Siegel, M. R. (1994). Use of specific repetitive sequences in *Peronospora tabacina* for the early detection of the tobacco blue mold pathogen. *Phytopathology* **84**, 425-430.

Williams, J. G. K., Kubelik, A. R., Livak, K. J., Rafolski, J. A. and Tingey, S. V. (1990). DNA polymorphisms amplified by arbitrary primers are useful as genetic markers. *Nucleic Acids Research* **18**, 6531-6535.

Wong, W. C., White, M., and Wright, I. G. (1988). Production of monoclonal antibodies to *Fusarium oxysporum* f. sp. *cubense* race 4. *Letters in Applied Microbiology* **6**, 39-42.

Wöstemeyer, J., Schäfer, Kellner, M. and Weisfeld, M. (1992). DNA polymorphisms detected by random primer dependent PCR as a powerful tool for molecular diagnostics of plant pathogenic fungi. *Advances in Molecular Genetics* **5**, 227-240.

Wycoff, K. L. and Ayers, A. R. (1990). Monoclonal antibodies to surface and extracellular antigens of a fungal plant pathogen, *Phytophthora megasperma* f. sp. *glycinea*, recognize specific carbohydrate epitopes. *Physiological and Molecular Plant Pathology* **37**, 55-79.

Xia, J. Q., Lee, F. N. and Kim, K. S. (1992). Monoclonal antibodies to an extracellular component of *Pyricularia grisea*. *Canadian Journal of Botany* **70**, 1790-1797.

Xue, B., Goodwin, P. H. and Annis, S. L. (1992). Pathotype identification of *Leptosphaeria maculans* with PCR and oligonucleotide primers from ribosomal internal transcribed spacer sequences. *Physiological and Molecular Plant Pathology* **41**, 179-188.

Yuen, G. Y., Craig, M. L. and Avila, F. (1993). Detection of *Pythium ultimum* with a species-specific monoclonal antibody. *Plant Disease* **77**, 692-698.

Zwadyk, P., Cooksey, R. C. and Thornsberry, C. (1986). Commercial detection methods for biotinylated gene probes: comparison with $^{32}$P-labelled DNA probes. *Current Microbiology* **14**, 95-100.

# 5

# ASSESSING PLANT-NEMATODE INFESTATIONS AND INFECTIONS

K. R. Barker and E. L. Davis

*Plant Pathology Department, North Carolina State University
Raleigh, North Carolina, United States*

## I. INTRODUCTION

The diagnosis, quantification, and monitoring of nematode populations provide an information base for integrated pest/plant-health management, research and regulatory programs. The application of the inverse relationships of initial nematode population densities to plant growth and yield, a keystone for integrated management programs, has stimulated interest and activities in population-density assessments and identifications over the last three decades (Barker and Imbriani, 1984; Nickle, 1991; Oostenbrink, 1966; Seinhorst, 1965). Nematode infestations and infections are generally assessed by established soil and plant sampling methods, extraction of nematodes from samples or tissue-staining procedures, and subsequent nematode identification and quantification by light microscopy (Barker, 1985; Cobb,

Advances in Botanical Research Vol. 23
Incorporating Advances in Plant Pathology
ISBN 0-12-005923-1

1918; Southey, 1986). The identifications of most nematode taxa are based primarily on their detailed morphology (Nickle, 1991). Differential plant-host responses, biochemical, serological, and molecular characteristics, however, are becoming increasingly important for identification of plant-parasitic nematodes at both the species and sub-specific levels (Curran, 1992; Eisenback et al., 1981; Nickle, 1991).

The potential of estimating plant-parasitic nematode infestations in soil for crop management purposes was first considered at length by Cobb (1918). He also characterized the aggregated dispersal patterns of plant-parasitic nematodes in agricultural fields, a problem which still must be considered when assessing nematode infestations and plant infections. Furthermore, he developed some of the primary assay procedures still in use today. In the 1920s, Gerald Thorne used sampling procedures as a basis for assessing potential nematode damage caused by Heterodera schachtii on sugar beets in Utah (Barker and Imbriani, 1984). These and other early research efforts preceded more recent work that has focused on the basic relationships of crop yield to nematode numbers and host-parasite relationships.

Today, a number of states in the USA and programs in other countries offer a range of nematode assay and diagnostic services (Barker and Imbriani, 1984). Providers include university, agricultural extension, state, federal, and private diagnostic laboratories. Communication between growers and these diagnostic specialists is critical to establish a common knowledge base and guidelines for when and how to sample in order to assess nematode infestations and infections in agroecosystems (Dunn, 1987). Still, the irregular dispersal and infection patterns associated with plant-parasitic nematodes limit precise assessments of nematode infestations as well as infections. For example, the aggregated spatial pattern of the root-knot nematode, Meloidogyne incognita, may result in overestimation of tobacco yield-loss when based on infested-field assays (Noe and Barker, 1985). The widely divergent feeding habits of these plant-parasitic nematodes, the large number of species, and variations in host preference and virulence within some nematode species compound these problems. Because of nematode interactions with various plant pathogenic fungi and bacteria, as well as certain nematode species serving as vectors of viruses, assessment of nematode infections often must consider associated disease complexes.

Effective nematode management, including regulatory or quarantine programs, relies on accurate identification and quantification of nematode species and potential parasitic variants within a given field or crop. An "ideal" method of nematode identification would be accurate, rapid, simple, sensitive, and quantitative at both the species and sub-specific levels. In contrast to nematodes, the indexing of plant-pathogenic viruses has been particularly successful by exploiting serological and molecular differences among virus isolates (Halk and De Boer, 1986). This advanced technology is due, in part, to the limits of visual identification of viruses, except at the electron microscope level. In contrast, the staples of nematode diagnosis have centered on sampling and extraction equipment and microscopes. These methods have been extremely useful, but they are labor-intensive, time-consuming and require specialized training. Variability in nematode morphology and the frequent presence of intraspecific races or pathotypes pose major problems in identification and related disease diagnosis. The extent of the integration of biochemical and molecular techniques with established morphological and host-range data for assessment of nematode species and parasitic variants remains an important consideration.

## A.  Synopsis of Nematode Systematics

More than 15,000 nominal species of nematodes have been placed in some 200 families (Maggenti, 1991).  Many thousands of species remain to be studied and identified, especially in non-agricultural ecosystems.  Except for new introductions and subsequent spread of a given nematode species, the majority of crop damage caused directly by nematodes has been detected in temperate countries (Luc *et al.,* 1990).  The intensive agriculture and emphasis on nematode disease diagnosis and management in temperate crops may account, in part,  for this observation.  In contrast, new harmful species as well as genera of plant-parasitic nematodes continue to be discovered in the sub-tropical and tropical areas (Luc *et al.,* 1990).

The range of habitats that are colonized by nematodes exceeds that of any other group of multicellular organisms (Luc *et al.,* 1990).   Trophic groups include microbivorous and predacious nematodes that feed upon microorganisms in soil and aquatic habitats, and nematodes adapted to parasitism of animals, fungi, and plants.  The bacterial-feeding nematode, *Caenorhabditis elegans,* has emerged as an important biological model for animal systems, and our increasing understanding of this species will be useful in genetic, molecular, and basic biological studies of plant-parasitic nematodes (Riddle and Georgi, 1990; Wood, 1988).  The diversity found in nematode taxa, including size, shape, morphology, reproduction, survival mechanisms, and feeding habits, reflect their adaptive capacity.

Higher classification of nematodes has a history of only slightly more than 100 years (Maggenti, 1991) and has been subject to considerable discussion.  Most nematologists place these organisms in a separate phylum, Nematoda (or Nemata), with two classes, Adenophorea and  Secernentea, that contain a total of 16 orders.  Plant-parasitic nematodes are placed within only two of these orders, and the majority of plant parasites belong in the order Tylenchida.  Some plant-parasitic nematodes are classified in the order Dorylaimida, including all known species that vector plant viruses.  The recently published "Manual of Agricultural Nematology" (Nickle, 1991) provides a comprehensive treatment of the morphology, taxonomy and systematics of plant-parasitic as well as insect-parasitic nematodes.    All plant-parasitic species are obligate parasites possessing a common adaption for plant parasitism -- the presence of a protrusible, buccal spear (stylet) used for feeding.

## B.  Nature of Nematode Parasitism and Host Response

The responses of plants to infection by nematodes are closely related to the feeding habits of these obligate parasites (Fig. 1).  The ectoparasites, considered to be the most primitive group, feed from outside the root by inserting their stylets into the cells of root tips and small roots.  Ectoparasitic nematodes may be grouped as migratory browsers that move relatively frequently about the root surfaces as they feed, and the more specialized sedentary ectoparasitic feeders that feed from particular root tissues for extended periods of time.  Thus, assessment of plant "infections" by migratory browsers is limited to soil assays.  This often is true for

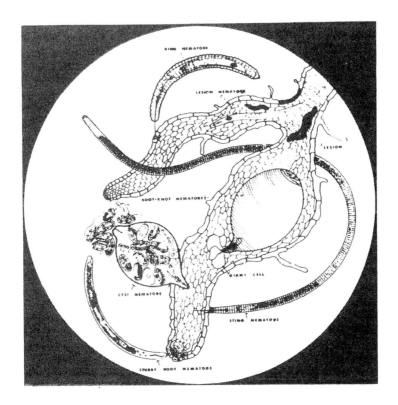

*Fig. 1.* A diagrammatic and hypothetical drawing illustrating several important types of plant-parasitic nematodes feeding on a single rootlet. Note the different sizes and shapes of the nematodes, the length of their stylets, feeding sites, and types of root damage caused by the various genera. (Courtesy of R. P. Esser).

sedentary ectoparasitic nematodes, as well, even though they modify plant root cells into specialized feeding sites (Hussey *et al.*, 1992; Wyss, 1981).

Nematodes that invade plant tissues, the endoparasites, are also grouped as either migratory or sedentary. Migratory endoparasites enter plants through roots, stems, or leaves and feed throughout cortical and/or other tissues, often causing extensive damage and necrosis of affected tissues. The sedentary endoparasites are considered to be the most advanced plant-parasitic nematodes. They enter plant tissue and modify plant cells into elaborate feeding sites that are essential to completion of their subsequent sedentary, parasitic life cycles (Hussey, 1989a; Jones, 1981). The complex morphological, physiological, and biochemical host and nematode responses associated with these interactions, when better characterized, may provide specific diagnostic markers or mechanisms to inhibit host infection by sedentary endoparasitic nematode species.

Responses to parasitism by nematodes in susceptible plant cultivars have been placed in five general categories:   1) stimulation - enhanced growth response sometimes resulting from a light attack from nematodes, including *Heterodera, Meloidogyne,* and *Xiphinema* species; 2) no visible reaction - this response occurs when parasitism is nondestructive, but may involve some tissue modification as affected by species of *Tylenchorhynchus* and *Criconemella*; 3) necrosis - tissue death and browning that typically involves root cortical and other parenchyma cells typified by parasitism by species of the genus *Pratylenchus*; 4) hyperplasia - this localized stimulation of cell division gives rise to gall tissues induced by species of *Meloidogyne, Ditylenchus,* and *Xiphinema*; and 5) hypertrophy - this striking increase in cell size gives rise to the specialized feeding cells such as, "nurse cells", "giant cells", or "syncytia" induced by a number of nematodes, including *Meloidogyne, Globodera, Heterodera, Nacobbus, Rotylenchulus* and *Tylenchulus* species (Pitcher, 1978). Plant responses to nematode infection also are also impacted by the associated abiotic and biotic environments, including interactions with other pathogens as well as symbionts (Barker *et al.*, 1993; Norton, 1978). Visual assessments of plant infections by endoparasitic nematodes may be accomplished by staining the suspected tissues with acid fuchsin or cotton blue and observing stained roots under a dissecting microscope (Barker, 1985; Southey, 1986).

. Plant resistance to nematode infection is generally related to the suppression of nematode reproduction and with an associated increase in crop yield (Barker, 1993). Tolerance is measured as a plant response, primarily yield, in the presence of nematode infection (Boerma and Hussey, 1992). Plant resistance mechanisms include morphological or preformed biochemical barriers, poor feeding-site establishment, and active plant-defense responses (Kaplan and Davis, 1987; Veech, 1981). Resistance has been evaluated primarily for sedentary endoparasites, since these nematodes establish an intimate host-parasite relationship. Accurate assessment of nematode infestations and infections, however, is critical for evaluation of plant resistance and tolerance to both ectoparasitic and endoparasitic nematodes, especially in field situations.

## C.  Nematode Dispersal Patterns

The patterns of nematode distribution over space and time must be considered when assessing infestations and plant infections for diagnostic or advisory purposes. The typical irregular, aggregated dispersal patterns of plant-parasitic nematodes in field soils poses problems for sampling soil and plant tissues, designing experiments involving these parasites, as well as selecting integrated management practices.

The vertical dispersal of plant-parasitic nematodes in the soil profile typically parallels that of host-root growth and distribution. Most plant-parasitic nematodes occur in the upper 30 cm of soil, but some are active at depths of 100 cm where deep-rooted plants are grown (Barker and Imbriani, 1984). For example, *Paratrichodorus minor* which vectors the corky ringspot virus, occurs at considerable soil depths, and this must be considered for the assessment of these pathogens and their management (Weingartner *et al.*, 1993). The horizontal migration of most plant-parasitic nematodes in soil is generally limited to a radius of less than 1 meter per year. The biological characteristics of certain plant-parasitic nematode species, however, as well as specific hosts, impact on their spatial dispersal patterns. Infections of susceptible plants by highly damaging species such as *Belonolaimus longicaudatus* or *Meloidogyne arenaria* typically occur in highly clustered patterns compared to less damaging species of nematodes. This is due to the severe restriction of root growth, associated tissue decay, and often plant death. *M. arenaria* and other species that deposit eggs in masses tend to have more clustered infections and dispersal patterns than those such as *Tylenchorhynchus claytoni* that deposit their eggs singly. Soil texture, topography, and management practices also affect nematode infection and dispersal patterns (Norton, 1978). Soil and water management practices, in particular, are often responsible for accelerated dispersal of nematodes within a field or geographic area (Barker, 1984).

## D. Nematodes and Disease Etiology

Definitive establishment of nematodes as a causal agent in plant disease via fulfillment of Koch's postulates is difficult because of their obligate parasitism (Mountain, 1954; Wallace, 1978). Symptoms of plants infected by root-parasitic nematodes, especially above-ground symptoms, are generally indicative of poor root function. Early nematicide experiments clearly demonstrated , however, the benefits to plant growth and yield incurred by reducing plant-parasitic nematode populations (Carter, 1943). Mountain (1954) conclusively demonstrated nematodes as plant pathogens in monoxenic experiments with *Pratylenchus* and plant roots. The root lesions formed in these tests, however, were never as severe as when nematode parasitism occurred in soil.

Wallace (1978; 1983) offered the thesis that the phenomenon of "one cause - one disease" is rare in nature. In reference to disease complexes involving nematode infections, Powell (1971) used the phrase "secondary pathogen" to refer to fungi and/or bacteria. Wallace (1978) expanded this concept with the term "determinant" as a substitute for causal agent. Determinants of disease complexes constitute the biotic and abiotic components that influence plant health. Thus, assessment of nematode infections often must consider the presence and role of multiple pathogens. Nematodes involved in synergistic interactions may serve as vectors of other pathogens, provide necrotic infection courts for pathogenic fungi or bacteria, induce physiological changes that enhance host susceptibility to other pathogens, or negate the normal host resistance to other pathogens (Hussey and McGuire, 1987; Khan, 1993; Powell, 1971). Infections by certain nematodes also interfere with the establishment of symbionts such as *Bradyrhizobium japonicum* on soybean (in the case of *Heterodera glycines*) and mycorrhiza on plants, especially on certain pine trees (Barham *et al.*, 1974; Hussey and McGuire, 1987).

Nematode infections enhance the development of a wide range of diseases caused by phytopathogenic fungi (Powell, 1971; Hussey and McGuire, 1987). The sedentary endoparasitic nematodes, especially *Meloidogyne* species, which alter host physiology and root morphology, are the primary pathogens involved in these disease complexes. Still, other endoparasitic nematodes, including *Pratylenchus* spp., *Hoplolaimus* spp., *Heterodera* spp., and *Globodera* spp. are important determinants in a number of fungal disease complexes. Synergistic effects of nematode infection and loss of disease resistance to fungal wilt diseases induced by species of *Fusarium* and *Verticillium* have been reported (Powell, 1971; Hussey and McGuire, 1987). Infections by a number of endoparasitic and ectoparasitic nematode genera also enhance the development of root diseases associated with soilborne fungi alone. Root rots and seedling diseases induced by species of *Pythium*, *Phytophthora*, *Rhizoctonia*, and *Cylindrocladium* are enhanced by the feeding of plant-parasitic nematodes (Powell, 1971; Hussey and McGuire, 1987). Powell (1971) suggests that this type of disease complex involving nematodes may be more important than those with the wilt fungi.

Disease complexes between nematodes and bacterial plant pathogens have been reported less frequently than fungal disease complexes (Norton, 1978). Bacterial wilt of tobacco and potato caused by *Pseudomonas solanacearum* is increased by concomitant infection with *Meloidogyne* species (Lucas, *et al.*, 1955; Weingartner and Shumaker, 1990). Root infection by *Criconemella xenoplax* has been associated with the increased susceptibility of peach trees to *Pseudomonas syringae* (Norton, 1978). A unique bacteria-nematode association often results in the important annual ryegrass toxicity animal disease in Australia. Concomitant infections of the seedheads of a number of grasses, including annual ryegrass, with *Clavibacter toxicus* and seed-gall forming nematodes, *Anguina* spp., bring about corynetoxin poisoning of sheep, horses, pigs, and cattle (McKay and Ophel, 1993). These highly lethal corynetoxins are the product of an association between the plant-pathogenic bacterium *C. toxius* and a bacteriophage (McKay and Ophel, 1993). *Anguina funesta* and other *Anguina* spp. serve as the carriers of the bacterium (McKay and Ophel, 1993). Diagnosis of the presence of the bacterium as a cause of livestock losses often is difficult. Two promising tests, however, have been developed. One test involves assessment of the nematode concentration and bacterial galls in the seed. The second test involves the use of ELISA (enzyme-linked immunosorbent assay) to detect the bacterium in emerging seedheads (McKay and Ophel, 1993).

Infestations by several ectoparasitic nematode species in the order Dorylaimida are important for a number of plant virus diseases. *Xiphinema*, *Longidorus*, and *Paralongidorus* spp. transmit the nepoviruses, and *Trichodorus* and *Paratrichodorus* spp. vector the tobraviruses (Hussey and McGuire, 1987). As these nematodes feed on virus-infected plants, virus particles become attached to the cuticular lining of the esophageal lumen or stylets. Although molting of nematode juveniles results in the removal of these viruses, adult stages remain viruliferous and may transmit these pathogens for many months. Thus, assessment of infestations of these nematodes has important implications in both nematode and virus infections. This is particularly true for the important virus diseases of woody fruits, brambles, strawberry, hops and potato (Hussey and McGuire, 1987). Initial acquisition of these viruses by the nematodes may occur as they feed on infected transplants, weeds, plants grown from virus-infected seeds, or plants inoculated by other means. Intensive population assessment for virus-vectoring nematodes may also provide reliable detection of associated virus(es) (Brown *et al.*, 1990). Recently, a biotin-avidin ELISA procedure was shown to be effective for assaying for the grape-fan-leaf virus (GFLV) in the vector *Xiphinema index* (Esmenjaud *et al.*, 1993). While the ELISA detected viruliferous *X. index*, the relative response varied with different field populations of the nematode (Fig. 2). Recently, reverse transcriptase-polymerase chain reaction (RT-

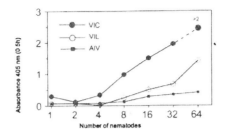

*Fig. 2.* Comparative ELISA response (absorbance) of viruliferous field populations of *Xiphinema index* from Champagne (ViC) and Languedoc (ViL). AiV is the virus-free greenhouse control population reared on grapevine (AXR1) (after Esmenjaud *et al.*, 1993).

PCR) has been used to detect GFLV in individual *X. index* (D. Esmenjaud, personal communication). The tylenchid nematode, *Criconemella xenoplax*, has been associated with the prunus necrotic ringspot virus in peach tree short-life disease, but no transmission of the virus by the nematode has been observed (Yuan *et al.*, 1990).

The association of nematodes and insects with specific plant diseases presents particular problems for assessment and disease management. The pine wilt disease caused primarily by *Bursaphelenchus xylophilus* is widespread in Japan, especially in stands of Japanese red pine and Japanese black pine (Mamiya, 1984). In nature this nematode is vectored by cerambycid beetles, *Monochamus* sp., but inoculations of pine trees with *B. xylophilus* alone results in tree death (Mamiya, 1984). The blue stain fungus, *Ceratocystis* spp., is associated with pinewood nematode disease (Mamiya, 1984). However, the rapid tree death is attributed to the accumulation of phytotoxins produced by the tree upon nematode infection (Oku, 1988). In contrast to the problem with pine wilt in Japan, *B. xylophilus* has been of limited local consequence in the United States. Nevertheless, the capacity of this nematode to survive in a dormant survival stage in pinewood chips, including products from North America, is posing major problems in the international commerce of wood products (Harmey and Harmey, 1993). Another insect-vectored nematode, *Radinaphelenchus cocopholus* induces the red ring disease of coconut (Griffith and Koshy, 1990). Although the insect-vector, the palm weevil (*Rhynchophorus palmerum*) is closely associated with this wilt-like disease, nematode infections introduced through the root system can induce the red ring malady.

## II. TRADITIONAL PROCEDURES FOR NEMATODE ASSESSMENTS

Indications of nematode disease are often difficult to assess because the majority of plant-parasitic nematodes are soil-inhabiting, microscopic, and induce relatively non-specific plant symptoms. Assessment of stem and foliar nematode pathogens is somewhat easier, but also requires a careful eye and trained personnel for diagnosis.

Hence, grower awareness of the potential for nematode damage is a critical component of the assessment process. Soil and plant samples often are taken in a uniform pattern in areas of suspected nematode infestation despite the typical irregular, aggregated distribution of nematodes in field soils (Barker and Campbell, 1981). The reduced efficiency of sampling is further confounded by the efficiency of the chosen nematode extraction procedure (Ferris, 1987; McSorley, 1987). Fluctuations in nematode populations within and between growing seasons, and the life stages required for nematode species identifications, also must be considered. Thus, in research, regulatory, and advisory programs, knowing when, where, what, and how to sample and accounting for inherent inefficiencies is paramount to accurate assessment of nematode infestations and infections (Barker and Campbell, 1981; Ferris, 1987). Integrating this information with nematode damage functions is especially critical for establishment of action thresholds for given nematode populations in agroecosystems (Barker and Imbriani, 1984; Barker and Noe, 1987).

## A. Signs and Symptoms

Associated symptoms and signs of nematode infections are invaluable in the initial phases of diagnosing plant-parasitic nematode problems. Signs of nematode infection include nematode structures such as cysts, egg masses, and survival structures that can be visualized without the aid of a microscope. The highly irregular growth patterns of annual crops within a field is a key symptom of most moderate-to-severe nematode infections. A general decline or dieback often is a characteristic symptom of late stages of nematode infections in perennial plants. These symptoms often occur in banana plantations when affected by *Radopholus similis* or *Helicotylenchus multicinctus* (Wallace, 1963). Similar declines in citrus are caused by *Radopholus citrophilus* and *Tylenchulus semipentrans* (Tarjan and O'Bannon, 1984). Specific symptoms vary, depending on stage of development of the plants when infected, numbers and kinds of nematodes, soil type and other environmental parameters. Unfortunately, the overall above-ground symptoms of nematode infections generally resemble those associated with any root injury, including stunting, incipient wilt, nutrient deficiencies, and unexplained yield loss. A few nematodes such as *Heterodera glycines* induce severe foliage chlorosis (nitrogen deficiency) resulting from the suppression of nodulation by nitrogen-fixing bacteria (Barker *et al.*, 1993; Hussey and McGuire, 1987). Some foliage-infecting nematodes such as *Aphelenchoides besseyi* produce even more specific above-ground symptoms. This nematode induces a pale yellow or white tip to the leaves of rice (Wallace, 1963). Rather clearly defined interveinal discoloration of leaves of strawberry and chrysanthemum is associated with *Aphelenchoides ritzembosa* (Wallace, 1963). This nematode, as well as *Anquina tritici* and *Ditylenchus dipsaci* also may induce extensive distortions of infected foliage tissues.

Below-ground symptoms of nematode infections vary from the striking and usually diagnostic root galls induced by *Meloidogyne* spp. to the slight root surface necrosis caused by *Xiphinema* spp. on strawberry or *Belonolaimus longicaudatus* on peanuts. Localized or severe root lesions, restricted root mass, and development of adventitious roots may also be indicative of nematode infection. Highly damaging ectoparasitic nematodes may also induce drastic modifications in root architecture. For example, large numbers of *Paratrichodorus minor* induce a severe stubby root effect on corn, and *Belonolaimus longicaudatus* causes a similar "coarse root" malady on a number of crop plants, including cowpea, peanut, and corn.

Plant symptoms alone, however, are not reliable for assessing most nematode infections. The presence of the nematode with the associated plant disease must be established for proper diagnosis. This assay may involve the observation of signs such as cysts of *Heterodera* and *Globodera*, egg masses of *Meloidogyne* or *Rotylenchulus* spp. attached to plant roots, or the identification of nematodes extracted from plant and soil samples.

## B.  Sampling

Representative soil and/or plant tissue samples are essential for qualitative and quantitative assessment of nematode infections and infestations. For soil inhabiting nematodes, a cylindrical sampling tube is often employed, but trowels and shovels can also be substituted (Barker and Campbell, 1981). Often a predetermined number of samples are taken and bulked per unit area prior to nematode extraction (Barker and Campbell, 1981). Samples taken in the root zone of plants with suspected infections, and multiple samples taken per unit area, can increase the likelihood of obtaining target organisms. Assessments within 50% of the true mean population levels can be achieved for various nematode species (Schmitt *et al.*, 1990). The number of samples that can be taken has practical and economic considerations, and the efficiency of intensive sampling for nematodes diminishes above a critical number of samples (McSorley, 1987). A highly precise, computer-based intensive soil sampling system combined with an ELISA test have provided highly reliable assessments of the potato cyst nematodes in the Netherlands (Schomaker and Been, 1992). This approach apparently allows remission of government required soil fumigation for farmers who grow  susceptible potatoes.

## C.  Soil Extraction and Bioassay

Available soil extraction procedures for nematode assessment have been described elsewhere in detail (Barker, 1985; Barker  *et al.*, 1986; Southey, 1986). In general, nematodes are separated from the soil phase by flotation and sieving, and/or encouraged to emerge from plant tissue by incubation in a wet environment. The relative efficiency of nematode extraction procedures and the time required may vary greatly (Table 1). The biology and population dynamics of target nematodes, the sampling time, and the type and condition of the host plant samples should be considered for selecting specific extraction procedures. For example, Baermann funnel extractions, while convenient, are dependent upon nematode motility. The life stages of nematodes that may occur in soil and/or in plant tissues range from relatively motile to quiescent forms, and include juveniles, adults, cysts, individual

Table 1. Relative efficiency of various methods of extracting nematodes (after Barker, 1985).

| Extraction | Percent efficiency[a] | Time required |
|---|---|---|
| *Meloidogyne* spp. | | |
| Juveniles from soil | | |
| Baermann trays/funnels (BF)[b] | 30-45 | 2-5 days |
| Centrifugal flotation (CF)[b] | 17-20 | 5-10 min |
| Elutriation + CF | 10-35 | 5-15 min |
| Sieving + BF | 30-54 | 2-5 days |
| Juveniles and egg fractions | | |
| Baermann trays/funnels | 20 | 2-5 days |
| Centrifugal flotation | 10 | 5-10 min |
| Elutriation + NaOCl | | |
| Egg extraction | 20-80 | 10-20 min |
| Sieving + BF-Seinhorst mist | 13-31 | 2-5days |
| *Criconemella* spp. | | |
| Centrifugal flotation | 25-62 | 5-10 min |
| Elutriation + CF | 70-79 | 5-15 min |
| Sieving + BF | 1-15 | 2-5 days |
| *Pratylenchus* spp. | | |
| Soil fraction | | |
| Baermann trays/funnels | 35-45 | 2-5 days |
| Centrifugal flotation | 17-29 | 5-10 min |
| Sieving + BF | 22-36 | 2-5 days |
| Soil and root fractions | | |
| Centrifugal flotation (soil only) | 10-30 | 5-10 min |
| Elutriation + Seinhorst mist | 20-80 | 2-14 days |
| Sieving + mist | 32-40 | 2-5 days |
| *Xiphinema* spp. | | |
| Baermann trays/funnels | 25-30 | 2-5 days |
| Elutriation + CF | 2-60 | 5-10 min |
| Sieving + BF | 27-35 | 2-5 days |

[a]Data do not reflect actual variation which may be 100% or more (Viglierchio, 1983). Rates vary with investigator and soil types.
[b]Combined with sieving.

eggs, egg masses, and various combinations thereof. When well-established plants are present, root as well as soil samples, may be used for assessments of endoparasitic and semi-endoparasitic nematode species. These endoparasites occur and reproduce within the roots or other plant parts on annual crops during the growing season and throughout the year on perennial plants. In contrast, ectoparasitic nematodes occur largely in the soil. Extraction methods are selected to target the nematode species and life stages present, their relative motility, and whether they are expected to be in the soil and/or in plant parts. When working with "unknown" samples, it's best to assay both root and soil samples. Specialized procedures also are available for assessing relative egg populations of *Meloidogyne*, *Globodera*, *Rotylenchulus*, and *Heterodera* spp. (Barker *et al.*, 1986; Brown, 1987; Southey; 1986).

Bioassays are particularly useful for assessing low-level infestations, host resistance, as well as nematicide efficacy for *Meloidogyne*, *Heterodera*, and *Globodera* spp. (Barker and Imbriani, 1984; Brown, 1987; Southey, 1986). When only juveniles or eggs of an identified nematode genus are present in samples, they can be grown on a suitable host to obtain the adult characteristics required for species identification. Root-gall indices from a tomato-bioassay for *Meloidogyne incognita* closely parallel temporal soil population shifts of juveniles of this species (Barker and Imbriani, 1984). The relationship between the number of infective units of root-knot nematodes and gall development in host bioassays can be used for assessing relative infestation levels in given fields on established plants, estimating expected yield losses, as well as projecting expected root-knot problems on a subsequent crop. Bioassays based on numbers of egg masses, cysts, and/or eggs are often used for assessments of plant resistance to *Meloidogyne*, *Heterodera*, and *Globodera* spp (Brown, 1987; Daykin and Hussey, 1985; Boerma and Hussey, 1992).

## D. Tissue Extraction and Staining

Nematode emigration during exposure of infected plant tissue samples to intermittent mist, as initially developed by or modified from Seinhorst (Barker, 1985; Southey, 1986), often provides reliable assessment of a range of endoparasitic nematode species. Diced plant tissue placed on a Baermann funnel or pan is a practical substitute for mist extractions (Dunn, 1993). Other techniques such as the flask-shaker procedure are suitable for some purposes but do not approach the efficiency of the Seinhorst mist (Barker, 1985).

Staining nematodes within plant tissues with acid fuchsin-glycerin (or lactophenol) or cotton blue-glycerin (Byrd *et al.*, 1983; Southey, 1986) enhances visualization of endoparasitic nematode infection using microscopy. Sodium hypochlorite-glycerin can be substituted for lactophenol to avoid the health hazards associated with phenol (Byrd *et al.*, 1983; Barker, 1985). In addition, acid fuchsin staining of extracted nematode eggs allows easy discrimination of individual eggs in samples that contain debris (Barker, 1985). The gelatinous matrix of *Meloidogyne* spp. egg masses can be stained with phloxine B to quantify the number of egg masses attached to roots (Daykin and Hussey, 1985).

# E. Nematode Identification

The identification of nematodes, at present, is based primarily on morphology and to a lesser extent  on cytology, host response, and biochemistry (Curran, 1992; Eisenback *et al.*, 1981; Nickle, 1991; Thorne, 1961; Triantaphyllou, 1985). Nematode size and shape, cuticular markings, types of esophagi and stylets, reproductive systems, and the presence or absence of males are some of the key diagnostic characteristics for these organisms. Figure 3 illustrates many of the morphological characters used to distinguish nematode species. Nematode genera can usually be distinguished by morphological study of juveniles or adults, but many species designations are based upon the anatomy of adult nematodes. Even the higher groups of nematodes are separated on the basis of relatively small differences in morphology (Maggenti, 1991). For example, the two classes within Nematoda were established, in part, on the presence or absence of a phasmid (caudal papillae connected with lateral precaudal glands in the Secernentea, previously called Phasmidia). Fortunately, more detailed and extensive descriptions of the two classes, as well as other taxa, are now available (Maggenti, 1991; Luc *et al.*, 1990). Because of the large number of nematode forms and species and their great variability, identification of these organisms by morphological characteristics alone often is very difficult (Thorne, 1961).

Signs and symptoms on plants are often characteristic of infection by some nematode species, but identifications should be substantiated by other criteria. For example, blackened seed galls, or "cockles", are characteristic of infection of cereals by *Anguina* spp. (Nickle, 1984). The shapes of infected root galls are helpful in distinguishing some species of *Meloidogyne*. *Meloidogyne hapla* typically induces the formation of small root galls with extensive proliferation of secondary roots. In contrast, *M. arenaria* populations typically induce the formation of many bead-like galls and restrictions in lateral root growth. High numbers of *M. arenaria*, however, may induce the formation of very large root galls that cannot be readily distinguished from those induced by *M. incognita* or *M. javanica*. Also, many field populations consist of mixtures of *Meloidogyne* species.

Differential host tests are useful in the diagnosis of some plant-parasitic nematodes. For example, the North Carolina differential hosts for root-knot nematodes have been used widely for differentiating the common species and their host races (Hartman and Sasser, 1985). These data should always be confirmed by parallel morphological and/or biochemical analyses, however, since some species variants do not conform to this host test. Unfortunately, the plants used in this host test may not provide host-range information that is applicable to crops grown in other geographic regions (Roberts, 1993). Still, the North Carolina Differential Host Test provides a uniform, accepted standard for description of the major *Meloidogyne* species and races. Sixteen potential races of *Heterodera glycines* based on a set of four soybean differentials have been proposed (Riggs and Schmitt, 1988). This system is extremely useful for describing parasitic variants of this pathogen. This host differential test, however, is somewhat arbitrary. Since *H. glycines* reproduces sexually, a "race" often represents a mixture of genotypes that may vary in virulence (Niblack, 1992). The differentiation of host races and pathotypes of other important nematode species, including *Ditylenchus dipsaci*, *Globodera rostochiensis*, and *G. pallida* (Dropkin, 1988)

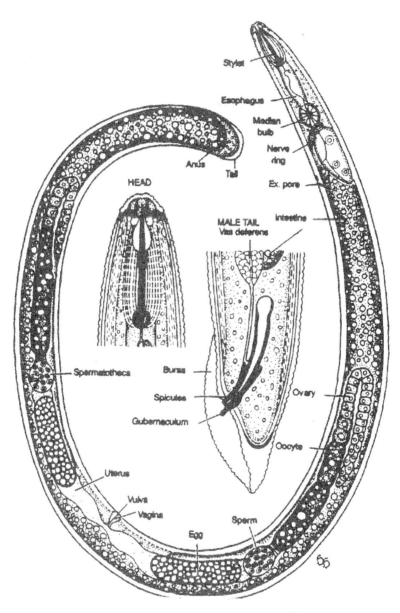

*Fig. 3.* Anatomy of a typical plant-parasitic nematode, *Hoplolaimus* sp. (Courtesy of Hedwig Triantaphyllou).

also are based on differential host tests. As in the following discussion, molecular techniques may become very useful for the identification of intraspecific variation.

## F. Defining Nematode Races, Pathotypes and Biotypes

The ability of populations within a given nematode species to parasitize specific hosts or cultivars of crop plants has often been used to classify variants of species into race, pathotypes, and biotypes (Triantaphyllou, 1987). In plant nematology, unfortunately, definitions of these terms and others including pathogenicity, virulence, and aggressiveness are not universally accepted (Barker, 1993). The term "race" has been used, primarily in the United States, to classify parasitic variants within nematode species based on differential host species, as with *Meloidogyne* spp. (Hartman and Sasser, 1985), and also based on selected genotypes within a host crop species, as with *H. glycines* (Schmitt and Shannon, 1992). The term 'pathotype' is used, particularly in Europe, to group intraspecific parasitic variants, primarily concerning potato cyst nematodes on differential potato genotypes (Cook and Evans, 1987). Numerous parasitic variants of *Ditylenchus dipsaci* have been termed as "races" based on their principal host species, even though variability in host range exists within these races (Nickle, 1991). One proposal includes "biotype" for parthenogenetic nematode populations with different host preferences, and "pathotype" for equivalent forms of amphimictic species (Sidhu and Webster, 1981). Dropkin (1988) suggested that the term 'pathotype' be used to designate any intraspecific variant of nematodes, based on variables such as differential host tests, in lieu of race, biotype, or strain. Still, Triantaphyllou (1987) recommended that "biotype" could be recognized as the basic unit for characterizing nematodes in reference to parasitic capabilities. In this context, biotype refers to a group of genetically closely related individuals sharing a common biological feature such as specified parasitic capability, preferably restricted to a single host with simple genetic basis for resistance.

Within the parthenogenetic *Meloidogyne* spp., both host and cytological races have been designated. Two cytological races, "A" and "B", have been described for both *M. hapla* and *M. incognita*, but little difference in host reaction is observed between cytological races (Triantaphyllou, 1985). In contrast, host races of *M. incognita* (races 1-4) and *M. arenaria* (races 1 and 2) are based on differential host responses (Hartman and Sasser, 1985). Because of a range of inherent problems with host-race determinations (Hartman and Sasser, 1985; Niblack, 1992; Triantaphyllou, 1987), the determination of biotypes or genotypes is becoming increasingly important for genetic studies, especially of amphimictic species. Communication among nematologists and other disciplines would certainly be improved if nematologists could come to agreement on definitions and use of these terms.

## G. Regulatory Issues

Successful quarantines of designated nematode species, as with other plant pathogens, demand rigorous assessment of infections and/or infestations. The potato cyst nematode, *Globodera rostochiensis*, is the most widely cited nematode, being on regulatory lists of 51 countries (Maas, 1987). The United States (US) quarantine for this pathogen has largely limited it to the Long Island area of New York, in contrast to an apparent failure to contain the spread of the soybean cyst nematode, *H. glycines,*. within the US (O'Bannon and Esser, 1987). Because it has a number of means to effectively spread over long distances, *H. glycines* now occurs in 29 states in the US and in Canada (Noel, 1992). These two species plus *Ditylenchus dipsaci*, *D. destructor*, *Heterodera schachtii*, *Meloidogyne chitwoodi*, and *Radopholus similis* are the primary nematode pests subject to regulatory measures. In total, more than 30 plant-parasitic nematode species are, or may be, targeted in regulatory programs in the US (O'Bannon and Esser, 1987). For the European and Mediterranean Plant Protection Organization (EPPO), seven nematode species -- *Aphelenchoides besseyi*, *D. destructor*, *G. rostochiensis*, *G. pallida*, *Nacobbus aberrans*, *R. similis*, *and Xiphinema americanum* are subject to regulatory measures (Maas, 1987). As international trade expands, regulatory assessments of nematode infections and infestations will likely become more important. This is exemplified by the European Community regulations on the import of coniferous products that have not been kiln-dried from regions where *B. xylophilus* is endemic (Harmey and Harmey, 1993).

## III. BIOTECHNOLOGY FOR ASSESSMENT OF NEMATODE INFECTIONS AND INFESTATIONS

Although morphology continues to be the mainstay in nematode identification, nucleic acid and protein techniques are becoming increasingly important for practical identifications and in nematode systematics (Burrows, 1990a; Caswell-Chen *et al.*, 1993; Curran, 1992; Curran and Robinson, 1993; Ferris and Ferris, 1992; Hyman, 1990; Hyman and Powers, 1991; Williamson, 1991). Molecular and biochemical data are less influenced by environmental parameters and more representative of nematode genotype (Williamson, 1991). The use of molecular methods for nematode identification has the capacity to be more objective, reliable, and sensitive than conventional diagnostic techniques, but must still be viewed as one component of a diagnostic strategy. Molecular identification may also be applicable to any nematode life-stage present in a sample. The choice of molecular or biochemical differences is not critical, but it must be accurate and provide information pertinent to its application. This is especially true for diagnoses at the sub-species levels that assess the parasitic ability of a given nematode genotype. The development of molecular tools for these types of identifications has great promise in plant nematology, but they should be related to and, if possible, integrated with morphological and host-range data to be most useful. The mode of nematode reproduction, ploidy, and genetic composition of nematode strains must be considered when developing molecular diagnostic tools. The Molecular Biology Committee of the Society of Nematologists has taken a proactive role in this regard by proposing a strain designation nomenclature based on that used for *C. elegans* and establishing an accessible database for this information (Bird and Riddle, 1994). As potential molecular means of nematode identification come to fruition, their reliability should be verified by extensive testing with nematode populations from different geographic regions.

## A. Protein and Isozyme Analyses

Total protein profiles determined by native and sodium dodecyl sulfate gel electrophoresis can distinguish species within *Meloidogyne, Heterodera, Globodera, Ditylenchus,* and *Radopholus* (Hussey, 1979; Williamson, 1991). Differences in glycoprotein profiles among species and populations of *Meloidogyne* and *Radopholus* have been obtained using lectin probes (Davis and Kaplan, 1992: Kaplan and Gottwald, 1992). Two-dimensional gel electrophoresis has been used to compare *Heterodera* isolates and *Meloidogyne* species (Ferris *et al.*, 1986), distinguish between *G. rostochiensis* and *G. pallida,* and to assess gene pool similarity among pathotypes of the potato cyst nematode (Bakker *et al.*, 1993). Isolines of *M. incognita* that vary in virulence on resistant tomato have been distinguished using 2-D electrophoresis (Dalmasso *et al.*, 1991). These findings are encouraging, but concerns about the labor involved, moderate sensitivity, complexity of protein patterns obtained, widespread applicability, and potential environmental or developmental influences may diminish their utility in diagnostics.

Electrophoretic enzyme phenotypes have proven particularly effective as important means of identifying *Meloidogyne, Heterodera, Globodera, Radopholus,* and *Anguina* at species and/or sub-species levels (Williamson, 1991). The reliable identification of the major species of *Meloidogyne* using isozyme phenotypes has been extensively verified through the International *Meloidogyne* Project (Esbenshade and Trianta-phyllou, 1985). The identification of *Meloidogyne incognita, M. arenaria,* and *M. javanica,* as well as *M. chitwoodi* and *M. naasi,* may be based on esterase phenotype alone (Esbenshade and Triantaphyllou, 1990). Since the esterase phenotypes for *M. hapla* and *M. incognita* are only slightly different, a second staining for malate dehydrogenase often is needed to differentiate these species (Fig. 4). Isozyme phenotypes determined from single *Meloidogyne* female proteins separated on an automated electrophoresis system (Esbenshade and Triantaphyllou, 1990) are now being used for *Meloidogyne* species diagnoses in a few states in the US and in several other countries. In addition to the assessment of *Meloidogyne* spp. infections, isozyme-based identifications facilitate the study of population dynamics of mixed species and confirm the identity of greenhouse cultures used in the evaluation of host resistance and other research purposes (Esbenshade and Triantaphyllou, 1990; Ibrahim and Perry, 1993). Esterase phenotype also is a reliable character for identifying race 'B' of *Meloidogyne* in pure cultures as well as in mixed-field populations (Fargette and Braaksma, 1990). These unusual root-knot populations, belonging to either *M. incognita* or *M. entrolobii,* often are a particular threat to agriculture due to their ready reproduction on normally resistant tomato, sweet-potato, and soybean. A potential new race of *M. chitwoodi* has recently been identified by isozyme phenotype (Van Meggelen *et al.*, 1994). Isozyme phenotpyes have recently been used to characterize populations of *G. pallida* in the United Kingdom (Phillips *et al.*, 1992) and *Meloidogyne* spp. populations from Spain (Cenis *et al.*, 1992). Differential esterase alleles have been used to demonstrate multiple matings among inbred lines of *H. glycines* (Triantaphyllou and Esbenshade, 1990).

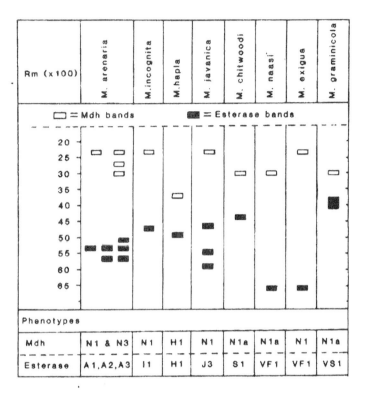

*Fig. 4.* Species-specific electrophoretic phenotypes of malate dehydrogenase (Mdh) and esterases helpful in differentiating eight species of *Meloidogyne*. On the same polyacrylamide gel (10-15% gradient), Mdh bands are bluish gray, those of esterases are black. (After Esbenshade and Triantaphyllou, 1990).

Three malate dehydrogenase phenotypes were observed in isoelectric focusing analyses of *Pratylenchus brachyurus* populations (Payan and Dickson, 1990).

## B. Serology

The specificity of antibody binding to target antigens has been used extensively in virology (Halk and De Boer, 1986), and the use of antibodies for nematode identification has recently been reviewed (Curran and Robinson, 1993; Schots, *et al.*, 1990; Williamson, 1991). Polyclonal antibodies are derived from animal blood sera and bind to multiple antigen epitopes, while monoclonal antibodies are derived from individual lymphocytes that produce one antibody that binds to one antigen epitope (Harlow and Lane, 1988). Experimental objectives, choice of immunogen, and screening procedures are critical considerations for antibody development. Two immunization strategies can be considered in the development of antibody probes for nematode diagnostics. A mixture of nematode antigens, ie. nematode homogenates, can be used to immunize animals, and antibodies that differentiate the target nematodes are subsequently assessed in a quantitative or qualitative manner. Alternatively, proteins that distinguish the target nematodes can be isolated and used to immunize animals followed by a primarily qualitative assessment of antibody specificity.

Polyclonal and monoclonal antibodies have been developed from nematode homogenates of *Meloidogyne, Heterodera, Globodera, Ditylenchus,* and *Bursaphelenchus* (Lawler and Harmey, 1993; Palmer *et al.*, 1992; Schots *et al.*,1990). Statistical analyses were used to partially discern differences in polyclonal and monoclonal antibody binding to several species of *Meloidogyne* (Davies and Lander, 1992). Robinson (1989) indicated that the ELISA approach may be used for quantitative detection of *Meloidogyne* spp. in soil and plant samples. This application was recently assessed for detection and quantification of root-knot nematodes from soil samples (Davies and Carter, 1994). Homogenates of *B. xylophilus* were used to produce polyclonal antibodies that can detect antigens of the pinewood nematode on the surface of wood (Lawler and Harmey, 1993).

Monoclonal antibodies have been developed from both nematode homogenates and isolated nematode proteins in efforts to characterize the esophageal gland secretions important in plant parasitism by *H. glycines* and *M. incognita* (Atkinson *et al.*, 1988; Davis *et al.*, 1992; Goverse *et al.* 1994; Hussey, 1989a and b). Monoclonal antibodies specific to esophageal gland antigens tend to bind to several species within a genus, but generally do not bind to other genera (Davis *et al.*, 1992; Goverse *et al.*, 1994; Hussey, 1989a). This approach may have potential for nematode species identification or assessing nematode infections, but the signals generated would have to be amplified for any practical considerations. The novel immunization techniques used in these investigations may facilitate the production of species-specific antibodies from isolated nematode antigens (Atkinson, 1988; Davis *et al.*, 1992; Goverse *et al.*, 1994).

Isolated, species-specific proteins were used to generate monoclonal antibodies that differentiate *Globodera rostochiensis* and *G. pallida* in immunoassays (Robinson *et al.*, 1993; Schots *et al.*, 1990). These monoclonal antibodies not only distinguish *G. pallida* and *G. rostochiensis,* but have been used in an ELISA (Fig. 5) for quantitative determination of potato cyst nematodes from soil samples (Schots *et al.*, 1992). The application of this technology for diagnosis and regulation of potato cyst nematode management in The Netherlands (Schomaker and Been, 1992) represents the best known example of the potential and use of serology for nematode identification.

## C. DNA Technologies

The increasing range of techniques to study DNA has stimulated much research on nematode mitochondrial, ribosomal, and genomic DNA and has provided potential tools for identification of nematodes as well as establishing and testing phylogenetic relationships for these organisms (Burrows 1990a; Curran, 1992; Ferris and Ferris, 1992; Hyman and Powers, 1991; Williamson, 1991). Caswell-Chen *et al.* (1993) offered

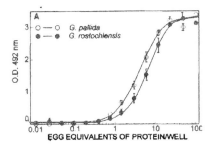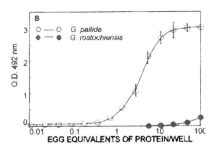

Fig. 5. Reactivities of WGP monoclonal antibodies with varying quantities of thermostable protein antigens isolated from either *Globodera rostochiensis* or *G. pallida*, and adsorbed to the wells of a microliter plate. On the X-axis the antigen quantity corresponding to the number of eggs from which it was isolated is given, on the Y-axis the optical density (ELISA) readings at 492 nm. (A) WGP 1 at $7.70 \times 10^{-9}$ M. (B) WGP 2 at $2.24 \times 10^{-10}$ M. (After Schots *et al.*, 1990).

an informative viewpoint on the application of biotechnology in nematology that includes a glossary of common molecular terminology. The objectives of molecular diagnostics should not be confused with the comparisons of molecular data used to assess nematode phylogeny and taxonomy (Curran 1992; Ferris and Ferris, 1992; Hyman and Powers, 1991). In classification, as stated by Curran (1992), "the choice of DNA sequences or genes to be studied is important and knowledge of the evolutionary relationship between sequences is crucial". In diagnosis, the objective is to obtain reliable molecular data to distinguish nematodes at the species or sub-species levels, not to imply relationships.

A full review of investigations of nematode DNA is beyond the scope of this treatment, so studies of plant-parasitic nematode DNA since those listed by Hyman (1990) and Curran (1992) have been summarized in Table 2. Restriction fragment length polymorphism (RFLP) analyses use restriction enzymes that cleave DNA at specific sequences to demonstrate differences in DNA sequence among nematodes as different (polymorphic) length bands (fragments) in an electrophoretic separation. RFLP analyses of genomic, mitochondrial, and ribosomal DNA are extremely useful to detect differences among both species and isolated populations within nematode species for research purposes (Table 2), and specific probes derived from polymorphisms may be used for practical diagnoses. The use of a specific DNA probe to detect purified mitochondrial DNA of *Meloidogyne* in various field soils (Hyman *et al.*, 1990) demonstrates the potential applicability of molecular probes to field samples

*Table II.* Summary of DNA analyses of plant-parasitic nematodes.[1]

| Genus | Level[2] | Strategy[3] | Reference |
|---|---|---|---|
| Meloidogyne | Spp | PCR, mtDNA primers, single juvenile/egg | Powers and Harris, 1993; Harris, *et al.* 1990 |
| | Spp | mtDNA sequence divergence | Powers, *et al.*, 1993 |
| | Spp | RFLP, PCR, mtDNA, 1-10 nematodes | Cenis, *et al.*, 1992 |
| | Spp | RAPD | Cenis, 1993 |
| | Spp | Species-specific oligonucleotide | Chacòn, *et al.*, 1991 |
| | Spp | Non-radioactive oligonucleotide | Chacòn, *et al.*, 1993 |
| | Spp | RFLP, genomic DNA | De Giorgi, *et al.*, 1991; Xue, *et al.*, 1992 |
| | Spp/Pop | RFLP, genomic DNA | Gàrate, *et al.*, 1991; Piotte, *et al.*, 1992 |
| | Spp/Pop | PCR, AFLP, mtDNA, rDNA | Xue, *et al.*, 1993 |
| | Pop | RFLP, genomic DNA | Carpenter, *et al.*, 1992 |
| | Pop | RFLP, mtDNA | Peloquin, *et al.*, 1993 |
| | Pop | RFLP, repeated DNA probe | Castagnone-Sereno, *et al.*, 1991 |
| Heterodera | Spp/Pop | RAPD, single cyst | Caswell-Chen, *et al.*, 1992 |
| | Spp/Pop | PCR, rDNA ITS sequence | Ferris, *et al.*, 1993 |
| Globodera | Spp | Species-specific DNA, non-radioactive | Burrows, 1990b |
| | Spp/Pop | RFLP, repetitive genomic DNA | Stratford, *et al.*, 1992 |
| | Pop | RFLP, mtDNA | Phillips, *et al.*, 1992 |
| Bursaphelenchus | Spp/Pop | RFLP, rDNA probe | Riga, *et al.*, 1992 |
| | Spp/Pop | PCR, repetitive DNA, fingerprinting | Harmey & Harmey, 1993 |
| | Spp/Pop | RFLP, *C. elegans* gene probe | Abad, *et al.*, 1991 |
| | Spp/Pop | PCR, *C. elegans*, DNA sequence | Beckenbach, *et al.*, 1992 |
| Xiphinema | Spp/Pop | RFLP, PCR, rDNA, ITS | Vrain, *et al.*, 1992; Vrain, 1993 |
| Ditylenchus | Spp/Pop | RFLP, PCR, rDNA ITS | Wendt, *et al.*, 1993 |
| Pratylenchus | Pop | RAPD | France & Brodie, 1994 |
| Radopholus | Spp | RAPD, rDNA ITS and IGS sequence | Kaplan, *et al.*, 1993 |

[1]Reports since those listed by Hyman (1990) and Curran (1992).
[2]Comparisons were made among nematode species (Spp) and/or populations (Pop) within species.
[3]Abbreviations: Polymerase chain reaction (PCR); mitochondrial deoxyribonucleic acid (mtDNA); restriction fragment length polymorphisms (RFLP); randomly amplified polymorphic DNA (RAPD); amplified fragment length polymorphisms (AFLP); ribosomal DNA (rDNA); intergenic (IGS) and internal transcribed (ITS) ribosomal spacer regions; *Caenorhabditis elegans*.

(Fig. 6). Non-radioactive DNA probes used to identify plant parasitic nematodes add an additional measure of practicality in routine diagnoses (Burrows, 1990b; Chacón *et al.*, 1993).

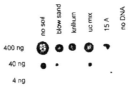

*Fig. 6.* Effects of soil samples on hybridization efficiency. Vertical columns designate soil samples, horizontal rows indicate amount of pBR322 target DNA added to reconstituted mixtures. Faint signals detectable by laser densitometry were observed with 40 ng of DNA in the krilium and 15 A soil samples, as well as in the 4 ng soil control preparation but do not appear in this reproduction. (After Hyman *et al.*, 1990).

Refinement of the polymerase chain reaction (PCR) by Saiki *et al.* (1988) has revolutionized molecular biology by providing the ability to use minute amounts of DNA template to amplify specific regions of DNA within the boundaries of chosen oligonucleotide primers that flank the region to be amplified. Amplifications of DNA from single juveniles of *M. incognita, M. arenaria, M. javanica, M. hapla,* and *M. chitwoodi* with specific primers that produce characteristic PCR products (Fig. 7) and are diagnostic of species after restriction enzyme digestion of similar size products (Powers and Harris, 1993). Random primers amplify DNA fragment length polymorphisms (RAPDs) that can discriminate among nematode species and populations (Table 2). Cloning and terminal end-sequencing of the relatively few RAPD polymorphisms between *R. similis* and *R. citrophilus* (Kaplan *et al.*, 1993) has been used to develop sequence-tagged sites (STS) and sequence characterized amplified regions (SCARs) that exclude or include, respectively, the original primer sequence (D. T. Kaplan, personal communication). This approach may be used to develop molecular markers applicable to genetic and diagnostic analyses of nematode populations. Specific amplifications using primers derived from conserved genomic, mitochondrial, and ribosomal DNA sequences, and *C. elegans* probes have generated polymorphisms among species and populations of *Meloidogyne,. Bursaphelenchus, Heterodera, Ditylenchus,* and *Xiphinema,* as well as DNA fingerprinting and sequence comparisons (Table 2). Phenograms derived from these types of analyses and sequence comparisons of isolated conserved DNA regions are particularly suited for analyses of nematode taxonomic relationships and phylogeny.

## D. Novel Approaches

Other compounds from nematodes may be useful in nematode identification. Pheromones have been used to distinguish species of *Radopholus, Globodera,* and *Bursaphelenchus* (Green *et al.*, 1970; Huettel, 1986; Riga and Webster, 1992). One

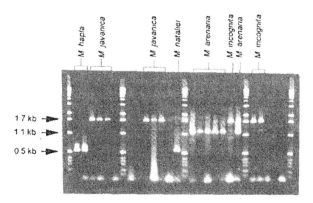

*Fig. 7.* Typical separation on a 1.0% agarose gel of products from PCR amplification of lysate from single *Meloidogyne* second-stage juveniles. The 1.7-kb product is characteristic of *Meloidogyne incognita* and *M. javanica*; the 1.1-kb product identifies *M. arenaria*, and the 0.52-kb product characterizes *M. chitwoodi, M. hapla, M. marylandi, M. naasi,* and *M. nataliei*. Empty lanes indicate failed reactions. Five lanes of 1 Kb DNA ladder (Gibco BRL, Gaithersburg, MD) are included as size standards. (After Powers and Harris, 1993).

plant-parasitic nematode sex pheromone has been isolated and identified from *H. glycines* (Huettel, 1986). Nematode sex pheromone bioassays usually measure attraction of males to females under controlled conditions, and thus, may not be practical or reliable for routine species diagnoses. Analysis of fatty acids has been used to identify plant pathogens (Johnk and Jones, 1993), but this analysis has not been applied for nematode identification.

The biochemical and molecular response of plants to nematode attack may offer a unique opportunity to assess nematode infections. Novel esterase bands were found in extracts from *Meloidogyne*-infected plant roots (Ibrahim and Perry, 1993). Physiological plant defense responses, production of pathogenesis-related proteins, and differential plant gene expression have been associated with nematode infection (Bowles *et al.*, 1991; Kaplan and Davis, 1987) but cannot be readily discerned from plant response to other pathogens. Altered expression of specific transcripts in syncytia induced by *H. schachtii* (Sijmons *et al.*, 1993) and giant-cells induced by *M. incognita* (Bird and Wilson, 1994; Opperman *et al.*, 1994; Wiggers *et al.*, 1991; Wilson *et al.*, 1994) have recently been reported. Nematode infection appears to modify quantitative, temporal, or spatial expression of "normal" plant genes. That qualitative signals can be isolated or quantitative shifts in gene expression be detected in plant tissue for the purposes of specific assessment of nematode infection is questionable.

## IV. CHALLENGES AND OPPORTUNITIES

'Accurate detection and assessment of nematode infections and infestations are becoming increasingly important as production agriculture is intensified to meet world food demands. Environmental, economic, and practical concerns have prompted restrictions in nematicide use and an increased emphasis in plant resistance and modified cultural practices for integrated pest management in sustainable agricultural systems. A greater understanding of the effects of agricultural practices on nematode biology and parasitism depends upon accurate assessment of nematode problems for the development of integrated management tactics. The urgent need for the incorporation of durable resistance to nematodes in many crop species is directly related to the challenges of monitoring highly variable populations of any given nematode species. Interdisciplinary research has increased at the university, regional, and national levels and ranges from evaluations of plant germplasm and the development of nematode-resistant cultivars to genome mapping and identification of molecular markers for plant resistance and nematode virulence genes.

Although morphological examinations of nematodes from soil and plant-tissue samples will continue to be an essential component for their identification, protein and nucleic acid analyses will undoubtedly be interfaced with traditional procedures used for nematode diagnostics. The automated isozyme analyses developed by Esbenshade and Triantaphyllou (1990) are now being used in a few states as primary diagnostic tools for *Meloidogyne* spp. determinations by research and extension programs. Serological detection of *G. rostochiensis* and *G. pallida* (Robinson *et al.*, 1993; Schots *et al.*, 1990; Shomaker and Been, 1992) is presently used in Europe. While these assays are useful for species' identifications, cropping history information and differential host tests currently are instructive for nematode diagnostic purposes at the sub-specific levels. The high frequency of occurrence of species mixtures in field populations still complicates the use of molecular techniques for advisory purposes. Molecular and biochemical techniques have also been used to detect viruliferous nematodes (Esmenjaud *et al.*, 1993) and have potential for analyses of nematode-disease complexes with plant pathogenic fungi and bacteria (McKay and Ophel, 1993).

The use of molecular markers to diagnose parasitic variability within plant-parasitic nematode populations has not yet been realized and may represent the most important potential contribution of molecular data to nematode identification. The development of these molecular probes will require genetically characterized nematode populations and plant genotypes, extensive testing, and may or may not reflect current host range data. Considerable progress has been made in developing diagnostic molecular probes for key species of major nematode taxa, including *Bursaphelenchus*, *Ditylenchus*, *Globodera*, *Heterodera*, *Meloidogyne*, *Pratylenchus*, *Radopholus* and *Xiphinema*. Still, the number of species studied in this regard represent only a small fraction of the total number of plant-parasitic nematodes. Also, the availability of the required expertise and equipment necessary for ready use of molecular diagnostic tools is limited primarily to the developed countries. The ultimate molecular tools for nematode diagnosis will need to be practical, or "user friendly", safe, and promote efficient, reliable, and economical nematode identification in a diversity of working environments.

Reliable assessments of nematode infestations and infections of materials subject to regulatory interstate and international quarantines continue to pose challenges (McSorley and Littell, 1993). Dependable molecular diagnostic probes suitable for assessing actual plant infections and plant-product infestations should prove invaluable in these programs. For example, DNA probes specific for *Bursaphelenchus xylophilus* and the immunological detection of *Bursaphelenchus* sp. antigens on wood surfaces may be used to monitor wood products subject to international quarantine measures (Lawler and Harmey, 1993; Harmey and Harmey, 1993). Establishing action thresholds will be another regulatory concern, particularly with the sensitivity of some molecular methods. Development of the technology for intraspecific differentiation of nematodes, based on parasitic ability, will require major decisions concerning quarantines that are based on identifications at the species level.

The technologies for information transfer on nematode diagnosis and management also are undergoing major change. In Florida, for example, an updated series of extension computer files are available in all county extension offices through a CD-ROM system (Dunn, 1993; Kucharek and Dunn, 1992). This is especially useful for coordination of activities by growers and advisory personnel as assessment methods and nematode problems evolve. Information transfer among research, advisory and regulatory programs will also be critical to develop and implement new methods of assessing nematode infestations and infections. Nematode-population assessment data are used as decision-making aids in approximately one-half of the states in the US. Unfortunately, the time required to collect soil and/or plant tissue samples and the sometimes questionable reliability of the results of available procedures limit the scope of these endeavors. The challenge of collecting and processing representative samples will continue even with improved methods for nematode identification and population assessments. New and improved molecular diagnostic procedures, including automated ELISA (M. P. Robinson, personal communication), should contribute toward more efficient and reliable assessments of nematode infections and infestations.

## REFERENCES

Abad, P., Tares, S., Brugier, N. and de Guiran, G. (1991). Characterization of the relationships in the pinewood nematode species complex (PWNSC) (*Bursaphelenchus* spp.) using a heterologous unc-22 DNA probe from *Caenorhabditis elegans*. *Parasitology* **102**, 303-308.

Atkinson, H. J., Harris, P. D., Halk, E. J., Novitski, Leighton-Sands, J., Nolan, P. and Fox, P. C. (1988). Monoclonal antibodies to the soya bean cyst nematode, *Heterodera glycines*. *Annals of Applied Biology* **112**, 459-469.

Bakker, J., Folkertsma, R. T., Rouppe van der Voort, J. N. A. M., de Boer, J M. and Gommers, F. J. (1993). Changing concepts and molelcular approaches in the management of virulence genes in potato cyst nematodes. *Annual Review of Phytopathology* **31**, 169-190.

Barham, R. O., Marx, D. H. and Ruehle, J. H. (1974). Infection of ectomycorrhizal and nonmycorrhizal roots of shortleaf pine by nematodes and *Phytophthora cinnamomi*. *Phytopathology* **64**, 1260-1264.

Barker, K. R. (1984). A history of the introduction and spread of nematodes. *In* 'The Movement and Dispersal of Agriculturally Important Biotic Agents' (D.K. MacKenzie, C.S. Barfield, G.G. Kennedy, R. D. Berger and D. J. Taranto, eds.). pp. 131-144. Claitors Books, Baton Rouge, LA.

Barker, K. R. (1985). Nematode extraction and bioassays. *In* 'An Advanced Treatise on *Meloidogyne*', Vol. II (K. R. Barker, C. C. Carter and J. N. Sasser [eds.]). pp. 19-35. North Carolina State University Graphics, Raleigh.

Barker, K. R. (1993). Resistance/tolerance and related concepts in plant nematology. *Plant Disease* **77**, 111-113.

Barker, K. R. and Campbell, C. L. (1981). Sampling nematode populations. *In* 'Plant-Parasitic Nematodes' (B. M. Zuckerman and R. A. Rohde, eds.). Vol. 3, pp. 451-474. Academic Press, NY.

Barker, K. R. and Imbriani, J. L. (1984). Nematode advisory programs. *Plant Disease* **68**, 735-741.

Barker, K. R. and Noe, J. P. (1987). Techniques in quantitative nematology. *In* 'Experimental Techniques in Plant Disease Epidemiology' (J. Kranz and J. Rotem, eds.). pp. 223-236. Springer-Verlag, Berlin, Heidelberg, New York.

Barker, K. R., Koenning, S. R., Huber, S. C. and Huang, J.-S. (1993). Physiological and structural responses of plants to nematode parasitism with *Glycine max-Heterodera glycines* as a model system. *In* 'International Crop Science I' (Buxton, D. R., R. Shibles, R. A. Forsberg, B. L. Blad, K. H. Asay, G. M. Paulsen and R. F. Wilson, eds.). pp. 761-777. Crop Science Society of America, Madison, WI.

Barker, K. R., Townshend, J. L., Bird, G. W., Thomason, I. J. and Dickson, D. W. (1986). Determining nematode population response to control agents. *In* 'Methods for Evaluating Pesticides for Controls of Plant Pathogens' (K. D. Hickey, ed.). pp. 283-296. American Phytopathological Society Press, St. Paul, MN.

Beckenbach, K., Smith, M. J. and Webster, J. M. (1992). Taxonomic affinities and intra- and interspecific variation in *Bursaphelenchus* spp. as determined by polymerase chain reaction. *Journal of Nematology* **24**, 140-147.

Bird, D. Mck. and Riddle, D.L. (1994). A genetic nomenclature for parasitic nematodes. *Journal of Nematology* **26**, 138-143.

Bird, D. Mck. and Wilson, M. A. (1994). DNA sequence and expression analysis of root-knot nematode-elicited giant cell transcripts. *Molecular Plant-Microbe Interactions* **7**, 419-424.

Boerma, H. R. and Hussey, R. S. (1992). Breeding plants for resistance to nematodes. *Journal of Nematology* **24**, 242-252.

Bowles, D. J., Gurr, S. J., Scollan, C. and Atkinson, H. J. (1991). Local and systemic changes in plant gene expression following root infection by cyst nematodes. *In* 'Biochemistry and Molecular Biology of Plant-Pathogen Interactions' (C. J. Smith, ed.). pp 225-236. Clarendon Press, Oxford.

Brown, D.J.F., Boag, B., Jones, A. T. and Topham, P. B. (1990). An assessment of soil-sampling density and spatial distribution required to detect viruliferous nematodes (Nematoda, Longidoridae and Trichodoridae) in fields. *Nematologia Mediterranea* **18**, 153-160.

Brown, R. H. (1987). Control strategies in low-value crops. *In* 'Principles and Practice of Nematode Control in Crops' (R.H. Brown and B.R. Kerry, eds.). pp. 351-387. Academic Press, Sidney.

Burrows, P. R. (1990a). The use of DNA to identify plant-parasitic nematodes. *Nematological Abstracts* **59**, 1-6.

Burrows, P. R. (1990b). The rapid and sensitive detection of the plant-parasitic nematode *Globodera pallida* using a non-radioactive biotinylated DNA probe. *Revue de Nematologie* **13**, 185-190.

Byrd, D. W., Kirkpatrick, T. and Barker, K. R. (1983). An improved technique for clearing and staining plant tissues for detection of nematodes. *Journal of Nematology* **15**, 142-143.

Carpenter, A. S., Hiatt, E. E., Lewis, S. A. and Abbott, A. G. (1992). Genomic RFLP analysis of *Meloidogyne arenaria* race 2 populations. *Journal of Nematology* **24**, 23-28.

Carter, W. (1943). A promising new soil amendment and disinfectant. *Science* **97**, 383-384.

Castagnone-Sereno, P., Piotte, C., Abad, P., Bongiovanni, M. and Dalmasso, A. (1991). Isolation of repeated DNA probe showing polymorphism among *Meloidogyne incognita* populations. *Journal of Nematology* **23**, 316-320.

Caswell-Chen, E. P., Williamson, V. M. and Wu, F. F. (1992). Random amplified polymorphic DNA analysis of *Heterodera cruciferae* and *H. schachtii* populations. *Journal of Nematology* **24**, 343-351.

Caswell-Chen, E. P., Williamson, V. M., and Westerdahl, B. B. (1993). Applied biotechnology in nematology. *Journal of Nematology Supplement* **25**, 719-730.

Cenis, J. L. (1993). Identification of four *Meloidogyne* spp. by random amplified polymorphic DNA (RAPD-PCR). *Phytopathology* **83**, 76-80.

Cenis, J. L., Opperman, C. H. and Triantaphyllou, A.C. (1992). Cytogenetic, enzymatic and restriction fragment length polymorphism variation of *Meloidogyne* spp. from Spain. *Phytopathology* **82**, 527-531.

Chacón, M. R., Parkhouse,M , Burrows, P. and Gàrate, T. (1993). The use of a digoxigenin-labelled synthetic DNA oligonucleotide for the rapid and sensitive identification of *Meloidogyne incognita*. *Fundamental and Applied Nematology* **16**, 495-499.

Chacón, M.R., Parkhouse, R. M. E., Robinson, M. P., Burrows, P. and Gàrate, T. (1991). A species specific oligonucleotide DNA probe for the identification of *Meloidogyne incognita*. *Parasitology* **103**, 315-319.

Cobb, N. A. 1918. Estimating the Nema Population in Soil. U.S. Department of Agriculture Technical Circular 1. 48 pp.

Cook, R. and Evans, R. (1987). Resistance and tolerance. *In* 'Principles and Practice of Nematode Control in Crops' (R.H. Brown and B. R. Kerry, eds.). pp. 179-231. Academic Press, Sidney.

Curran, J. (1992). Molecular taxonomy of nematodes. *In* 'Nematology from Molecule to Ecosystem.' (F. Gommers and P.W.Th. Maas, eds.). pp. 83-91. European Society of Nematologists. Wageningen, Netherlands.

Curran, J. and Robinson, M. P. (1993). Molecular aids to nematode diagnosis. *In* 'Plant-Parasitic Nematodes In Temperate Agriculture' (K. Evans, D. L. Trudgill and J. M. Webster, eds.). pp. 545-564. C.A.B. International, Wallingford, UK.

Dalmasso, A., Castagnone-Sereno, P., Bongiovanni, M and de Jong, A. (1991). Acquired virulence in the plant parasitic nematode *Meloidogyne incognita*. 2. Two-dimensional analysis of isogenic isolates. *Revue de Nematologie* **14**, 277-283.

Davies, K. G. and Carter, B. (1994). Optimization of an ELISA for the quantification of root-knot nematodes extracted from soil. *Journal of Nematology* **26**, 97-98. (Abstr.).

Davies, K. G. and Lander, E. B. (1992). Immunological differentiation of root-knot nematodes (*Meloidogyne* spp.) using monoclonal and polyclonal antibodies. *Nematologica* **38**, 353-366.

Davis, E. L. and Kaplan, D. T. (1992). Lectin binding to aqueous-soluble and body wall proteins from infective juveniles of *Meloidogyne* species. *Fundamental and Applied Nematology* **15**, 243-250.

Davis, E. L., Aron, L. M., Pratt, L. H. and Hussey, R. S. (1992). Novel immunization procedures used to develop monoclonal antibodies that bind to specific structures in *Meloidogyne* spp. *Phytopathology* **82**, 1244-1250.

Daykin, M. E. and Hussey, R. S. (1985). Staining and histopathological techniques in nematology. *In* 'An Advanced Treatise on *Meloidogyne*. Vol. II. Methodology' (K.R. Barker, C.C. Carter and J.N. Sasser, eds.). pp. 39-48. N.C. State University Graphics, Raleigh, NC.

De Giorgi, C., De Luca, F., Finetti-Sialer, M., Di Vito, M. and Lamberti, F. (1991). DNA fragments as molecular probes for the diagnosis of plant-parasitic nematodes. *Nematologia Mediterranea* **19**, 131-134.

Dropkin, V. H. (1988). The concept of race in phytonematology. *Annual Review of Phytopathology.* **26**, 145-161.

Dunn, R. A. (1987). Perspectives on extension nematology. *In* 'Vistas on Nemato-logy.' (J.A. Veech and D.W. Dickson, eds.). pp. 32-37. Society of Nematologists. Hyattsville, MD.

Dunn, R. A. (1993). 'Nematode Control Guide'. Florida Cooperative Extension Service, Institute of Food and Agricultural Sciences, University of Florida, Gainesville, FL.

Eisenback, J. D., Hirschmann, H., Sasser, J. N. and Triantaphyllou, A. C. (1981). 'A Guide to the Four Most Common Species of Root-knot Nematodes (*Meloidogyne* species) with pictorial key.' Cooperative Publishing, Departments of Plant Pathology and Genetics. North Carolina State University and United States Agency for International Development (USAID). Raleigh, NC. 48 pp.

Esbenshade, P. R. and Triantaphyllou, A. C. (1985). Identification of major *Meloidogyne* species employing enzyme phenotypes as differentiating characters. *In* 'An Advanced Treatise on *Meloidogyne*. Vol. I. Biology and Control' (J.N. Sasser and C.C. Carter, eds.). pp. 135-142.. N.C. State University Graphics, Raleigh, NC.

Esbenshade, P. R. and Triantaphyllou, A. C. (1990). Isozyme phenotypes for the identification of *Meloidogyne* species. *Journal of Nematology* **22**, 10-15.

Esmenjaud, D., Walter, B., Minot, J. C., Voisin, R. and Cornuet, P. (1993). Biotin-avidin ELISA detection of grapevine fanleaf virus in the vector nematode *Xiphinema index*. *Journal of Nematology* **25**, 401-405.

Fargette, M. and Braaksma, R. (1990). Use of the esterase phenotype in the taxonomy of the genus *Meloidogyne*. 3. A study of some 'B' race lines and their taxonomic position. *Revue de Nematologie* **13**, 375-386.

Ferris, H. (1987). Extraction efficiencies and population estimation. *In* 'Vistas on Nematology' (J.A. Veech and D.W. Dickson, eds.). pp. 59-63. Society of Nematologists. Hyattsville, MD.

Ferris, V. R. and Ferris, J. M. (1992). Integration of classical and molecular approaches in nematode systematics. *In* 'Nematology from Molecule to Ecosystem' (F. Gommers and P.W.Th. Maas, eds.). pp. 92-100. European Society of Nematologists. Wageningen, Netherlands.

Ferris, V. R., Ferris, J. M. and Faghihi, J. (1993). Variation in spacer ribsomal DNA in some cyst-forming species of plant-parasitic nematodes. *Fundamental and Applied Nematology* **16**, 177-184.

Ferris, V. R., Ferris, J. M., Murdock, L. L. and Faghihi, J. (1986). *Heterodera glycines* in Indiana, III. 2-D protein patterns of geographical isolates. *Journal of Nematology* **18**, 177-182.

France, R. A. and Brodie, B. B. (1994). Characterization of two populations of *Pratylenchus penetrans* based on differential reproduction on potato and DNA analysis. *Journal of Nematology* **26**, 100. (Abstr.).

Gàrate, T., Robinson, M. P., Chacón, M. R., Parkhouse, R. M. E. (1991). Characterization of species and races of the genus *Meloidogyne* by restriction enzyme analysis. *Journal of Nematology* **23**, 414-420.

Goverse, A., Davis, E. L. and Hussey, R. S. (1994). Monoclonal antibodies to the esophageal glands and stylet secretions of *Heterodera glycines*. *Journal of Nematology* **26**, 251-259.

Green, C. D., Greet, D. N. and Jones, F.G.W. (1970). The influence of multiple mating on the reproduction and genetics of *Heterodera rostochiensis* and *H. schachtii*. *Nematologica* **16**, 309-326.

Griffith, R. and Koshy, P. K. (1990). Nematode parasites of coconut and other palms. *In* 'Plant-parasitic Nematodes in Subtropical and Tropical Agriculture' (M. Luc, R.A. Sikora and J. Bridge, eds.). pp. 363-386. C.A.B. International, Wallingford, UK.

Halk, E. D. and De Boer, S. H. (1986). Monoclonal antibodies in plant disease research. *Annual Review of Phytopathology* **23**, 321-350.

Harlow, E. and Lane, D. (1988). 'Antibodies, A Laboratory Manual'. Cold Spring Harbor Laboratory, NY.

Harmey, J. H. and Harmey, M. A. (1993). Detection and identification of *Bursaphelenchus* species with DNA fingerprinting and polymerase chain reaction. *Journal of Nematology* **25**, 406-415.

Harris, T. S., Sandall, L. J. and Powers, T. O. (1990). Identification of single *Meloidogyne* juveniles by polymerase chain reaction amnplification of mitochondrial DNA. *Journal of Nematology* **22**, 518-524.

Hartman, K. M. and Sasser, J. N. (1985). Identification of *Meloidogyne* species on the basis of differential host test and perineal-pattern morphology. *In* 'An Advanced Treatise on *Meloidogyne*. Vol. II. Methodology' (K.R. Barker, C.C. Carter and J.N. Sasser, eds.). pp. 69-77. North Carolina State University Graphics, Raleigh, NC.

Huettel, R. N. (1986). Chemical communicators in nematodes. *Journal of Nematology* **18**, 3-8.

Hussey, R. S. (1979). Biochemical systematics of nematodes, A review. *Helminthological Abstracts (Ser B.)* **48**, 141-148.

Hussey, R. S. (1989a). Disease-inducing secretions of plant-parasitic nematodes. *Annual Review of Phytopathology* **27**, 123-141.

Hussey, R. S. (1989b). Monoclonal antibodies to secretory granules in esophageal glands of *Meloidogyne* species. *Journal of Nematology* **21**, 392-398.

Hussey, R. S. and McGuire, J. M. (1987). Interaction with other organisms. *In* 'Principles and Practice of Nematode Control in Crops' (R.H. Brown and B.R. Kerry, eds.). pp. 293-328. Academic Press. Sidney, NY.

Hussey, R. S., Mims, C. W. and Westcott III, S. W. (1992). Ultrastructure of root cortical cells parasitized by the ring nematode *Criconemella xenoplax*. *Protoplasma* **167**, 55-65.

Hyman, B. C. (1990). Molecular diagnosis of *Meloidogyne* species. *Journal of Nematology* **22**, 24-30.

Hyman, B. C. and Powers, T. O. (1991). Integration of molecular data with systematics of plant-parasitic nematodes. *Annual Review of Phytopathology* **29**, 89-107.

Hyman, B. C., Peloquin, J. J. and Platzer, E. G. (1990). Optimization of mitochondrial DNA-based hybridization assays to diagnostics in soil. *Journal of Nematology* **22**, 273-278.

Ibrahim, S. K. and Perry, R. N. (1993). Use of esterase patterns of females and galled roots for the identification of species of *Meloidogyne*. *Fundamental and Applied Nematology* **16**, 187-190.

Johnk, J. S. and Jones, R. K. (1993). Differentiation of populations of AG-2-2 of *Rhizoctonia solani* by analysis of cellular fatty acids. *Phytopathology* **83**, 278-283.

Jones, M. G. K. (1981). The development and function of plant cells modified by endoparasitic nematodes. *In* 'Plant-Parasitic Nematodes' (B. M. Zuckerman and R. A. Rohde, eds.). Vol. 3. pp. 255-280. Academic Press, NY.

Kaplan, D. T. and Davis, E. L. (1987). Mechanisms of plant incompatibility with nematodes. *In* 'Vistas on Nematology' (J.A. Veech and D.W. Dickson, eds.). pp. 267-276.. Society of Nematologists. Hyattsville, MD.

Kaplan, D. T. and Gottwald, T. R. (1992). Lectin binding to *Radopholus citrophilus* and *R. similis* proteins. *Journal of Nematology* **24**, 281-288.

Kaplan, D. T., Opperman, C. H. and Vanderspool, M. C. (1993). Characterization of genetic variation in *Radopholus* sibling species. *Phytopathology* **83**, 1360. (Abstr.).

Khan, M. W. (ed.). (1993). 'Nematode Interactions'. Chapman & Hall. London, UK.

Kucharek, T. and Dunn, R. A. (1992). 'Diagnosis and Control of Plant Diseases and Nematodes in a Home Garden'. *Circular 399-A*. Florida Cooperative Extension Service.

Lawler, C. and M. A. Harmey. (1993). Immunological detection of nematode antigens on the surface of a wood section. *Fundamental and Applied Nematology* **16**, 521-523.

Luc, M., Bridge, J. and Sikora, R. A. (1990). Introduction. *In* 'Plant-parasitic Nematodes in Subtropical and Tropical Agriculture' (M. Luc, R.A. Sikora and J. Bridge, eds.). pp. 11-27. C.A.B. International, Wallingford, UK.

Lucas, G. B., Sasser, J. N. and Kelman, A. (1955). The relationship of root-knot nematodes to Granville wilt resistance in tobacco. *Phytopathology* **45**, 537-540.

Maas, P.W.Th. (1987). Physical methods and quarantine. *In* 'Principles and Practice of Nematode Control in Crops' (R. H. Brown and B. R. Kerry, eds.). pp. 265-291. Academic Press, Sidney.

Maggenti, A. R. (1991). General nematode morphology. *In* 'Manual of Agricultural Nematology' (W.R. Nickle, ed.). pp. 3-46. Marcel Dekker, Inc., NY.

Mamiya, M. 1984. The pine wood nematode. *In* 'Plant and Insect Nematodes' (W.R. Nickle, ed.). pp. 589-621. Marcel Dekker, Inc., NY.

McKay, A. C. and K. M. Ophel. (1993). Toxigenic clavibacter/*Anguina* associations infecting grass seedheads. *Annual Review of Phytopathology* **31**, 151-167.

McSorley, R. (1987). Plot size and design for acquisition of field data in nematology. *In* 'Vistas on Nematology' (J.A. Veech and D.W. Dickson, eds.). pp. 52-58. Society of Nematologists. Hyattsville, MD.

McSorley, R. and R. L. Littell. (1993). Probability of detecting nematode infestations in quarantine samples. *Nematropica* **23**, 177-181.

Mountain, W. B. (1954). Studies of nematodes in relation to brown root rot of tobacco in Ontario. *Canadian Journal of Botany* **32**, 737-759.

Niblack, T. L. (1992). The race concept. *In* 'Biology and Management of the Soybean Cyst Nematode' (R.D. Riggs and J.A. Wrather, eds.). pp. 73-86. APS Press, St. Paul, MN.

Nickle, W. R. (ed.). (1984). 'Plant and Insect Nematodes'. Marcel Dekker, Inc., N.Y.

Nickle, W. R. (ed.). (1991). 'Manual of Agricultural Nematology'. Marcel Dekker, Inc., NY.

Noe, J. P. and Barker, K. R. (1985). Overestimation of yield loss of tobacco caused by aggregated spatial pattern of *Meloidogyne incognita*. *Journal of Nematology* **17**, 245-251.

Noel, G. R. (1992). History, distribution and economics. *In* 'Biology and Management of the Soybean Cyst Nematode' (R. D. Riggs and J. A. Wrather, eds.). pp. 1-13. APS Press, St. Paul, MN.

Norton, D. C. (1978). 'Ecology of Plant-Parasitic Nematodes'. John Wiley & Sons, NY.

O'Bannon, J. H. and Esser, R. P. (1987). Regulatory perspectives in nematology. *In* 'Vistas on Nematology' (J.A. Veech and D.W. Dickson, eds.). pp. 38-46. Society of Nematologists. Hyattsville, MD.

Oku, H. (1988). Role of phytotoxins in pine wilt diseases. *Journal of Nematology* **20**, 245-251.

Oostenbrink, M. (1966). Major characteristics of the relation between nematode and plants. *Mededlingen Landbouwhogeschool*. Wageningen **66-4**. 46 pp.

Opperman, C. H., Taylor, C. G. and Conkling, M. A. (1994). Root-knot nematode-directed expression of a plant root-specific gene. *Science* **263**, 221-223.

Palmer, H. M., Atkinson, H. J. and Perry, R. N. (1992). Monoclonal antibodies (MAbs) specific surface expressed antigens of *Ditylenchus dipsaci*. *Fundamental and Applied Nematology* **15**, 511-515.

Payan, L. A. and Dickson, D. W. (1992). Comparison of populations of *Pratylenchus brachyurus* based on isozyme phenotypes. *Journal of Nematology* **22**, 538-545.

Peloquin, J.J., Bird, D. Mck., Kaloshian, I. and Matthews, W. C. (1993). Isolates of *Meloidogyne hapla* with distinct mitochondrial genomes. *Journal of Nematology* **25**, 239-243.

Phillips, M. S., Harrower, B. E., Trudgill, D. L., Catley, M. A., Waugh, R. (1992). Genetic variation in British populations of *Globodera pallida* as revealed by isozyme and DNA analysis. *Nematologica* **38**, 304-319.

Piotte, C., Castagnone-Sereno, P., Uijthof, J., Abad, P., Bongiovanni, M. and Dalmasso, A. (1992). Molecular characterization of species and populations of *Meloidogyne* from various geographic origins with repeated-DNA homologous probes. *Fundamental and Applied Nematology* **15**, 271-276.

Pitcher, R. S. (1978). Interactions of nematodes with other pathogens. *In* 'Plant Nematology' (J.F. Southey, ed.). pp. 63-77. Her Majesty's Stationery Office, London.

Powell, N. T. (1971). Interactions between nematodes and fungi in disease complexes. *Annual Review of Phytopathology* **9**, 253-274.

Powers, T. O. and Harris, T. S. (1993). A polymerase chain reaction method for identification of five major *Meloidogyne* species. *Journal of Nematology* **25**,1-6.

Powers, T. O., Harris, T. S. and Hyman, B. C. (1993). Mitochondrial DNA sequence divergence among *Meloidogyne incognita, Romanomermis culicivorax, Ascaris suum* and *Caenorhabditis elegans*. *Journal of Nematology* **25**, 564-572.

Riddle, D. L. and Georgi, L. L. (1990). Advances in research on *Caenorhabditis elegans*, application to plant parasitic nematodes. *Annual Review of Phytopathology* **28**, 247-269.

Riga, E. and Webster, J. M. (1992). Use of sex pheromones in the taxonomic differentiation of *Bursaphelenchus* spp. (nematode), pathogens of pine trees. *Nematologica* **38**, 133-145.

Riga, E., Beckenbach, K. and Webster, J. M. (1992). Taxonomic relationships of *Bursaphelenchus xylophilus* and *B. mucronatus* based on interspecific and intraspecific cross-hybridization and DNA analysis. *Fundamental and Applied Nematology* **15**, 391-395.

Riggs, R. D., Schmitt, D. P. (1988). Complete characterization of the race scheme for *Heterodera glycines*. *Journal of Nematology* **20**, 392-395.

Roberts, P. A. (1993). The future of nematology, Integration of new and improved management strategies. *Journal of Nematology* **25**, 383-394.

Robinson, M. P. (1989). Quantification of soil and plant populations of *Meloidogyne* using immunoassay techniques. *Journal of Nematology* **21**, 583-584. (Abstr.).

Robinson, M. P., Butcher, G., Curtis, R. H., Davies, K. G. and Evans, K. (1993). Characterization of a 34D protein from potato cyst nematodes, using monoclonal antibodies with potential for species diagnosis. *Annals of Applied Biology* **123**, 337-347.

Saiki, R. K., Gelfand, D. H., Stoffel, S., Scharf, S. J., Higuchi, R., Horn, G. T., Mullis, K. B. and Erlich, H. A. (1988). Primer-directed enzymatic amplification of DNA with a thermostable DNA polymerase. *Science* **239**, 487.

Schmitt, D. P. and Shannon, G. S. (1992). Differentiating soybean responses to *Heterodera glycines* races. *Crop Science* **32**, 275-277.

Schmitt, D. P., Barker, K. R., Noe, J. P. and Koenning, S. R. (1990). Repeated sampling to determine the precision of estimating nematode population densities. *Journal of Nematology* **22**, 552-559.

Schomaker, C. H. and T. H. Been. (1992). Sampling strategies for the detection of potato cyst nematodes; developing and evaluating a model. *In* 'Nematology from

Molecule to Ecosystem' (F. Gommers and P.W.Th. Maas, eds.). pp. 182-194. *European Society of Nematologists.* Wageningen, Netherlands.

Schots, A. F., Gommers, F. J. and Egberts, E. (1992). Quantitative ELISA for the detection of potato cyst nematodes in soil samples. *Fundamental and Applied Nematology* **15**, 55-61.

Schots, A. F., Gommers, F. J., Baker, J. and Egberts, E. (1990). Serological differentiation of plant-parasitic nematode species with polyclonal and monoclonal antibodies. *Journal of Nematology* **22**, 16-23.

Seinhorst, J. W. (1965). The relation between nematode density and damage to plants. *Nematologica* **11**, 137-154.

Sidhu, G. S. and Webster, J. M. (1981). The genetics of plant-nematode systems. *Botanical Review* **47**, 387-419.

Sijmons, P. C. (1993). Plant-nematode interactions. *Plant Molecular Biology* **23**, 917-931.

Southey, J. F. (1986). 'Laboratory methods for work with plant and soil nematodes.' Ministry of Agriculture Fisheries and Food. London, HMSO.

Stratford, R., Shields, R., Goldsbrough, A. P. and Fleming, C. (1992). Analysis of repetitive DNA sequence from potato cyst nematodes and their use as diagnostic probes. *Phytopathology* **82**, 881-886.

Tarjan, A. C. and O'Bannon J. H. (1984). Nematode parasites of citrus. *In* 'Plant and Insect Nematodes' (W.R. Nickle, ed.). pp. 395-434. Marcel Dekker, Inc., NY.

Thorne, G. (1961). 'Principles of Nematology'. 553 pp. McGraw-Hill, NY.

Triantaphyllou, A. C. (1985). Cytogenetics, cytotaxonomy and phylogeny of root-knot nematodes. *In* 'An Advanced Treatise on *Meloidogyne*. Vol. I. Biology and Control' (J. N. Sasser and C. C. Carter, eds.). pp. 113-126. North Carolina State University Graphics, Raleigh, NC.

Triantaphyllou, A. C. (1987). Genetics of nematode parasitism of plants. *In* 'Vistas on Nematology' (J.A. Veech and D.W. Dickson, eds.). pp. 354-371. Society of Nematologists. Hyattsville, MD.

Triantaphyllou, A. C. and Esbenshade, P. R. (1990). Demonstration of multiple mating in *Heterodera glycines* with biochemical markers. *Journal of Nematology* **22**, 452-456.

Van Meggelen, J. C., Karssen, G., Janssen, G. J. W., Verkerk-Bakker, B. and Janssen, R. (1994). A new race of *Meloidogyne chitwoodi* Golden, O'Bannon, Santo and Finley, 1980. *Fundamental and Applied Nematology* **17**, 93.

Veech, J. A. (1981). Plant resistance to nematodes. *In* 'Plant-Parasitic Nematodes' (B.M. Zuckerman and R.A. Rohde, eds.). Vol. 3. pp. 377-404. Academic Press, NY.

Viglierchio, D. R. and Schmitt, R. V. (1983). On the methodology of nematode extraction from field samples, Comparison of methods for soil extraction. *Journal of Nematology* **15**, 450-454.

Vrain, T. C. (1993). Restriction fragment length polymorphism separates species of the *Xiphinema americanum* group. *Journal of Nematology* **25**, 361-364.

Vrain, T. C., Wakarchuk, D. A., Levesque, A. C. and Hamilton, R. I. (1992). Intraspecific rDNA restriction fragment length polymorphism in the *Xiphinema americanum* group. *Fundamental and Applied Nematology* **15**, 563-573.

Wallace, H. R. (1963). 'The Biology of Plant-Parasitic Nematodes'. 280 pp. Arnold, London.

Wallace, H. R. (1978). The diagnosis of plant diseases of complex etiology. *Annual Review of Phytopathology* **16**, 379-402.

Wallace, H. R. (1983). Interactions between nematodes and other factors on plants. *Journal of Nematology* **15**, 221-227.

Weingartner, D. P. and Shumaker, J. R. (1990). Effects of soil fumigants and aldicarb on bacterial wilt and root-knot nematodes in potato. *Journal of Nematology Supplement* **22**, 681-688.

Weingartner, D. P., McSorley, R. and Goth, R. W. (1993). Management strategies in potato for nematodes and soil-borne diseases in subtropical Florida. *Nematropica* **23**, 233-245.

Wendt, K., Vrain, T. C. and Webster, J. M. (1993). Separation of three species of *Ditylenchus* and some host races of *D. dipsaci* by restriction fragment length polymorphism. *Journal of Nematology* **25**, 555-563.

Wiggers, R. J., Magill, C. W., Starr, J. L. and Price, H. J. (1991). Evidence against amplification of four genes in giant cells induced by *Meloidogyne incognita*. *Journal of Nematology* **23**, 421-424.

Williamson, V. M. (1991). Methods for collection and preparation of nematodes. Part 4. Molecular techniques for nematode species identification. *In* 'Manual of Agricultural Nematology' (W.R. Nickle, ed.). pp. 107-123. Marcel Dekker, Inc., NY.

Wilson, M. A., Bird, D. Mck. and van der Knapp, E. (1994). A comprehensive subtractive cDNA approach to identify nematode-induced transcripts in tomato. *Phytopathology* **84**, 299-303.

Wood, W. B., ed. (1988). 'The Nematode *Caenorhabditis elegans*'. Cold Spring Harbor Laboratory, NY.

Wyss, U. (1981). Ectoparasitic root nematodes feeding behavior and plant cell responses. *In* 'Plant-parasitic Nematodes, Vol. III' (B.M. Zuckerman and R.A. Rohde, eds.). pp. 325-351. Academic Press, New York.

Xue, B., Baillie, D. L., Beckenbach, K. and Webster, J. M. (1992). DNA hybridization probes for studying the affinities of three *Meloidogyne* populations. *Fundamental and Applied Nematology* **15**, 35-41.

Xue, B., Baillie, D. L. and Webster, J. M. (1993). Amplified fragment length polymorphisms of *Meloidogyne* spp. using oligonucleotide primers. *Fundamental and Applied Nematology* **16**, 481-487.

Yuan, W.-Q., Barnett, O. W., Westcott III, S. W. and Scott, S. W. (1990). Tests for transmission of prunus necrotic ringspot and two nepoviruses by *Criconemella xenoplax*. *Journal of Nematology* **22**, 489-495.

# 6

# POTENTIAL OF PATHOGEN DETECTION TECHNOLOGY FOR MANAGEMENT OF DISEASES IN GLASSHOUSE ORNAMENTAL CROPS

## I.G. Dinesen[1] and A. van Zaayen[2]

[1]*The Danish Plant Directorate, Lyngby, Denmark*
[2]*Inspection Service for Floriculture and Arboriculture (NAKB), Den Haag, The Netherlands*

## I. INTRODUCTION

Growers of ornamental plants in glasshouses are faced with very expensive production facilities which makes the unit cost of producing a potted plant or cut

Advances in Botanical Research Vol. 23
Incorporating Advances in Plant Pathology
ISBN 0-12-005923-1

flower very high. It is essential, therefore, that the plants bring adequate returns by being of high quality, making their cultivation worthwhile. Not only do blemishes in the crop cause reduction in the economic return, but, for instance, lack of uniformity can also be detrimental. Uniformity can be enhanced by growing plants in an optimal environment and under strictly defined and constant growing conditions. Even then, plants may not be entirely uniform due to genetic variation among plants of the same variety.

Crop damage may be caused by one of several different factors. It can be physiological in nature such as wilting caused by water stress, or caused by infectious plant pathogens. This chapter will deal with avoiding damage caused by pathogens including fungi, bacteria, and viruses.


## II. THE NEED FOR HEALTHY PLANT MATERIAL

The interest in using healthy plant material for propagation has increased in most countries because of the requirement for reducing pesticide usage in response to environmental concerns. The use of healthy plant propagative material has also enhanced the health and quality of glasshouse crops.

Several plant viruses that affect ornamental glasshouse crops are highly contagious and their effects on plants are often drastic, seriously reducing crop yield and quality. A great deal of money is spent on preventing virus diseases from becoming disastrous (Bos, 1982). In addition to viruses, fungi and bacteria can cause disease problems in ornamental crops. Under experimental conditions, the losses caused by viruses, in particular, but also by bacteria and fungi are difficult to assess, as it is extremely hard to prevent contamination of the healthy control plants. Besides, artificial infection may not always reflect what happens under natural conditions.

Losses due to plant diseases can be either direct or indirect. Direct losses include yield reduction and decreased quality caused by symptomless infections by plant pathogens. Crop failure also is a typical direct loss. Other direct losses are defects in visual attractiveness such as aberrant size, shape and colour, which reduce quality and market value. The costs of maintaining healthy crops are of the indirect type. Such costs may include indexing of propagation material for freedom of plant pathogens to meet plant passport requirements and inspections to meet import, export and plant quarantine regulations. Breeding for resistance and research are two additional examples of indirect costs associated with plant health.

The damage caused by various viruses, alone or together, can be enormous. Crop loss in the United States due to plant virus infections has been estimated at $ 1.5-2.0 billion annually (Bialy and Klausner, 1986). In 1982 Bos reviewed crop losses due to viruses and discussed the difficulties encountered with damage evaluation. One of the most striking examples is that of 'swollen shoot' in cacao. This virus killed millions of cacao trees in West Africa and caused numerous trees to be destroyed because of a suspected infection. By 1977, 162 million cacao trees had been cut down in Ghana and the rate of destruction had risen to 15 million a year, equivalent to an area of 9400 ha (Thresh, 1980). Similarly, South-America lost 20 million orange trees infected by tristeza virus during a period of ten years (Bos, 1972). Losses from most virus diseases, however, are difficult to measure unless crops are completely destroyed. Because of the difficulties in obtaining accurate information there are few precise estimates on crop losses caused by viruses in monetary terms, and available estimates are confined to losses in the major crops like cereals, rice, sugar beet and potato (Walkey, 1985).

Many ornamental crops are propagated vegetatively. Virus infection of vegetatively propagated plants can be serious and unless special measures are taken all the propagules removed from an infected plant will be infected. Barnett (1988) reviewed virus damage in ornamental crops. There are several examples of losses caused by viruses becoming apparent only after the introduction of virus-free material allowed comparison with the infected clones grown previously (Bos, 1983). Virus infection in the 'Baccara' rose grown in glasshouses is one such example (Pool *et al.*, 1970). The heat-treated and virus-tested clonal line yielded 13.5% more blooms than the commercial stock during a 12-month period. Hakkaart (1964) inoculated plants of a virus-free carnation variety with four carnation viruses. Three of those reduced the number of flowers per plant and decreased the quality of the flowers. A combination of two viruses resulted in more severe symptoms than those produced by each virus alone. He found that even a virus, like carnation mottle virus, which does not cause visible symptoms, may adversely affect yield. Gippert and Schmelzer (1973) eliminated viruses from a number of *Pelargonium zonale* varieties through meristem culture. Although viruses may be symptomless in pelargonium, the plants grown from meristems were more vigorous than control plants and produced 20-30% more cuttings. In addition, rooting of the cuttings was improved, so that total production increased by about 36%. These results have been confirmed in commercial application.

Van Zaayen *et al.* (1992) compared 20 virus-free alstroemeria plants (cv. Snow Queen) from meristems and 20 plants infected with alstroemeria mosaic virus and alstroemeria carla virus in an experiment conducted in a commercial glasshouse. The virus-free plants produced 50% more flower stems than the virus-infected plants and, furthermore, the virus-free plants had more high quality flower stems, healthier leaves, and the flower stems had a longer vase life since the leaves did not yellow as quickly.

In recent years tospoviruses such as the tomato spotted wilt virus and impatiens necrotic spot virus have caused much damage to ornamental glasshouse crops. Cho *et al.* (1987) recorded crop losses of 50-90% in lettuce in Hawaii due to tospoviruses, but figures for crop losses in ornamentals are again difficult to find. As complete crops sometimes have to be rogued to prevent the spread of virus to other crops, the economic loss can be considerable. Tospoviruses are transmitted rapidly over a wide area by some species of thrips.

Ornamental foliage crops can also suffer from virus infections. Ornamental aeroids with severe symptoms due to dasheen mosaic virus are usually unmarketable. In addition, the virus causes yield losses of up to 60% in *Caladium* sp., *Dieffenbachia* sp., *Philodendron* sp. and *Zantedeschia* sp. (Zettler and Hartman, 1987).

Estimates of crop losses due to bacterial diseases are rare. Kennedy and Alcorn (1980) have made some estimates based on questionnaires from different areas within the United States. The losses caused by bacterial diseases were estimated at $206 x $10^6$ in 1976. The pathogen causing the greatest dollar-loss per year ($54.5 x $10^6$) was *Pseudomonas glycinea* in soybean in the state of Iowa. The authors concluded that the annual importance of these pathogens emphasizes the continuous need for research into their control and epidemiology.

When plants are propagated vegetatively, there is a high risk of pathogens being spread along with the new plants, so that in time all plants in a production system become infected. Therefore, regardless of the extent of infection, once it occurs, it is usually necessary to begin production anew with pathogen-free plants in order to maintain a healthy crop with desired characteristics.

In general it is not practical and often impossible to use chemical means to combat the bacteria and fungi which invade the vascular tissue of plants, or the viruses and viroids which move systemically throughout plant tissues. In these cases, the only way to obtain healthy plant material is through selection, heat treatment and/or meristem-tip culture.

## III. PLANT QUALITY

The trade in ornamentals including all kinds of propagating material has increased enormously during the last fifteen years. For example, the turnover of cut flowers and plants at flower auctions in the Netherlands has increased from a value of 1,113 million guilders in 1975 to 5,052 million guilders in 1992 (PVS and PGF, 1993). Ornamentals are now produced in a large number of countries and the number of attractive species with many different varieties continues to grow.

With the increasing interest in production of ornamentals, the competition has also grown and growers attempt to plant the best available varieties. Because breeders and producers of propagating material compete internationally for buyers, it is of vital importance to both buyer and seller that basic planting material meets the highest possible quality standards. High quality plants for propagation provide the best opportunity for producing a healthy crop.

### A. Clonal Selection

Selection of healthy plants is often the first step in obtaining disease-free plant material. The selection process in commercial nurseries should be done from their cleanest and healthiest plants produced the previous season. Early spring is often the optimal time for selection of basic plant material because plants are stressed and disease symptoms are most likely to be apparent under the low light and poor growing conditions that prevail during this time. Moreover, differences in growth rate are most apparent during the early spring which makes it easier to select vigorous plants for propagation.

### B. Quality Control of Propagation Material

To recognize the quality of propagating material, the elements composing the quality of a product must be defined. Quality consists of the following three main elements

(Van Ruiten, 1988):

a. *trueness to type and varietal purity* - the degree to which every cutting or plantlet is of the same identified variety and/or origin (deviations of the type should not occur);

b. *plant health* - freedom of quarantine diseases and freedom of other pests and quality diseases up to reasonable levels or certification standards;

c. *basic quality requirements* - including uniformity, vitality, appropriate root development, and acceptable treatment and packaging.

In addition to plant quality, characteristics such as resistance to certain diseases or tolerance to climatic variation, a certain colour, or abundance of flowers may be important. These properties can be acquired by breeding and selection, which is usually done at research institutes or by private companies. Uniformity, one of the basic quality requirements, can be enhanced by growing plants in an optimal environment and under uniform growing conditions. Even when this is done, however, all plants of one variety may not be entirely uniform, due to genetic heterogeneity. Vegetative propagation through cuttings or by the application of *in vitro* propagation usually leads to a high standard of uniformity. Regardless of the method used, the starting material should, as much as possible, be free from pathogens to prevent their multiplication and impact on uniformity and other quality characteristics.

One of the oldest examples of a virus, eventually leading to decreased quality, is that of the tulip breaking virus (TBV). Tulip flowers with the colour breaking symptoms were once sought because of their beauty (Schenk, 1976). Now it is known that TBV, the cause of this phenomenon, is undesirable as it leads to degeneration of the bulb stocks. Flower colour breaking can be caused by several viruses in a number of ornamental crops, e.g. freesia mosaic virus in purple (Fig. 1), pink and red

*Fig. 1.* Flower colour breaking caused by freesia mosaic virus, an example of a viral infection that at one time was considered desirable.

freesia's (Hakkaart, 1973), alstroemeria streak virus in purple, pink and red alstroe-meria's (Van Zaayen *et al.*, 1994), and turnip mosaic virus or cucumber mosaic virus in numerous crops (Walkey, 1985). Although such symptoms may look very decorative, they are variable and usually are accompanied by several unwanted effects such as crop loss. The yellow vein netting symptoms on leaves of pelargoni-um, once very popular, are also no longer considered desirable (OEPP/EPPO, 1992) because an infectious agent, probably a virus that can be spread through the whole crop, may be the causal agent (Cassells *et al.*, 1982). Virus infection of foliage crops can also cause quality problems. For example, dasheen mosaic virus causes distorted leaves of *Zantedeschia* sp. and *Dieffenbachia* sp., which affect their market quality (Zettler and Hartman, 1987). Nevertheless, in ornamental foliage crops, viruses occasionally are still considered desirable. The *Abutilon* mosaic virus which causes yellow-green mosaic symptoms in *Abutilon* plants, is an acceptable virus infection but is an exception in the ornamental industry.

Bacterial leafspot and stem rot is recognized as the most important disease in pelargonium cultivars. The causal agent is *Xanthomonas campestris* pv. *pelargonii* (Hellmers, 1952). In general, when an ornamental is visually infected with a plant pathogenic bacterium, the quality of the plant is so poor that it has no commercial value.

Another commercial factor connected with quality is the reputation of the producer. Commercial producers of ornamental crops work hard to sell their crops and usually develop a clientele who trust their products. A minor deviation from acceptable quality due to any one of various factors can damage the trust between producer and client.

In the foregoing, factors affecting minimum acceptable quality that a product must fulfil is described. In some European countries such as Denmark and the Nether-lands, independent inspection and certification systems have been developed since the middle 1940's in a collaborative industry effort to recognize and describe the quality of propagation material of ornamental crops (see Section IV). International acceptance and application of such a certification system is one goal of the European and Mediterranean Plant Protection Organization (EPPO) (Lopian, 1993a, 1993b). By introducing inspection schemes on national and international levels, species by species, goals can be realized to improve the quality of planting material and provide a better guarantee for freedom from diseases.

International agreement on required minimum standards, however, will not be sufficient in the future. Agreement also needs to be reached on methods and techniques, applied for regular visual inspection, sampling, testing, etc., in short - quality assurance. Quality assurance in ornamental plants has recently been dealt with by Steed (1993). Quality assurance is defined in terms of meeting customer requirements at the lowest cost. In 1987, the World Standards Organisation (ISO) published its first quality system standard, ISO 9000, which is called BS 5750 in Great Britain. BS 5750 lays down a set of criteria for management within commercial companies through which they can be registered independently by British Standards Institution (BSI) and others, leading to marketing benefits. To meet these criteria, existing sales, purchasing, production, testing, inspection, packing, and despatch procedures must be analyzed and assessed against BS 5750 standards. Written, quantitative quality control procedures must accompany applications to register with

BSI and be available to external assessors. External assessors from BSI or other certification agencies spend a number of days auditing the companies' systems against BS 5750 prior to registration. After the assessments, use of the quality control data should improve quality and progressively reduce costs (Steed, 1993). BS 5750 is applied in practice by a number of horticultural companies including suppliers of ornamental plants (Richardson, 1993). In Denmark, a grower of kalanchoë plants is already working towards a quality certificate under this system (Prins, 1993).

In the near future, quality certificates will also be required by inspection services in Europe, which will have to meet the requirements of EN 45,000. EN 45,000 is a European set of standards meant for plant inspection services and their testing laboratories, but is comparable to ISO 9000.

## IV.  CERTIFICATION SCHEMES FOR ORNAMENTALS

Unfortunately, once pathogens have been eliminated, crops do not automatically remain free from the pathogens. Many measures must be taken to prevent re-infection with viruses and other pathogens. Elimination and control of diseases ought to be an essential element in production schemes of vegetatively propagated ornamentals and the introduction of infected propagating material must be prevented. Goals can be realized to improve the quality of propagating material by the introduction of certification schemes and quality control systems at both the national and international levels. Schemes for the production of certified, pathogen-tested, vegetatively propagated ornamental plants have been and are still being developed. On a national level such certification schemes are operational for some ornamental crops in Denmark, France and the Netherlands (Van Ruiten, 1988, 1993). For the last eight years the possibilities for adopting comparable and compatible certification schemes has been discussed by the EPPO Panel on Certification of Pathogen-tested Ornamentals within the European and Mediterranean Plant Protection Organization (EPPO). Several of such schemes have now been published (e.g. OEPP/EPPO, 1991a, 1992, 1993a) and schemes for several other ornamentals are currently under discussion (Lopian, 1993b).

## A. The OEPP/EPPO General Scheme

The OEPP/EPPO now has a general scheme for the production of certified pathogen-tested vegetatively propagated ornamental plants (OEPP/EPPO, 1991b). The EPPO Council, having examined the proposals of the Working Party on Phytosanitary Regulations, recommends to the member states to take account of the general scheme in their national certification schemes for ornamental plants. The general scheme runs as follows as quoted from the recommendations of the EPPO council (OEPP/EPPO, 1991b):

1. Selection of the best type of *candidate material* (including vigour and agronomic value) to be taken into the scheme (best new cultivars or best clones of existing cultivars).

2. Production of pathogen-tested *nuclear stock* material, either by selective testing

of the candidate material or, if no pathogen-free material of the type required is obtained, by meristem-tip culture followed by testing and reselection of the best types. All candidate material undergoing testing is kept in isolation (quarantine) from the material which has given a negative test result and is maintained as certified nuclear stock.

3. Maintenance of the nuclear stock under conditions ensuring freedom from infection, with regular checks on health status by re-testing. If possible, samples of each selected type may also be maintained by *in vitro* culture.

4. Multiplication of the nuclear stock to produce *propagation stock*, under conditions ensuring freedom from infection or infestation. Propagation stock can be one or a number of generations, to be defined for each crop, and it must be possible to follow the filiation of each plant or group of plants through the generations from nuclear stock. This stage may in some cases be performed *in vitro*. Random checks are made on health status by re-testing.

5. Mass propagation to give material for certification. Regular visual inspection for health status. Random checks on pathogen status by re-testing.

6. Certification of the final product released for sale *(certified stock)*, e.g. cuttings, bulbs, mother plants, *in vitro* plants.

Throughout the whole procedure, care should be taken to check for varietal purity, and to look for possible mutations or back mutations. In general, it is desirable also to check for trueness to type but for many ornamentals this may be difficult in practice.

The scheme should be under the strict control of an official inspection/certification organization, throughout all of its stages. Establishments involved in the production of nuclear, propagation or certified stock should be officially registered. Sampling to determine whether the demands of the certification scheme have been met and thus to issue certificates should be performed by the official organization.

The European and Mediterranean Plant Protection Organization has been given the task by its member states to develop internationally acceptable and harmonized certification schemes for ornamentals (Lopian, 1993b). As was mentioned earlier, certification schemes for several ornamentals, e.g. pelargonium (Fig. 2), chrysanthemum and carnation, have been finalized by the EPPO Panel and other schemes are under way. Tolerance levels for certain pathogens have also been included. Each of the three categories of propagation material must undergo a designated testing schedule to ensure its freedom from a predetermined, crop-specific set of pathogens. Plants which do not fulfil this health standard are rejected. To accompany the crop-specific certification schemes, nursery requirements for establishments participating in the schemes have been devised by EPPO (OEPP/EPPO, 1993b).

The adoption of the EPPO certification schemes, even if not compulsory, should lead to a high standard and thus to a better world-wide acceptance of plant material from nurseries in the EPPO countries (Lopian, 1993b).

## B. National Certification of Ornamentals

For several years a number of ornamental plant species, including the main flori-

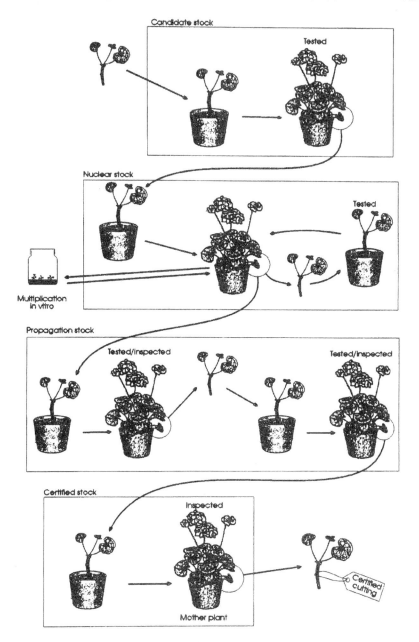

*Fig.* 2. Diagrammatic representation of the OEPP/EPPO certification scheme for pelargonium (Courtesy of OEPP/EPPO, Paris)

cultural crops, have been included in the Dutch inspection system. A general compulsory basic quality control system exists, but companies may also participate voluntarily in a certification system for certain crops. All nurseries producing planting material (meaning: 'propagating material and plants, which are intended to be planted for cultivation purposes'; Van Ruiten, 1993) are registered. The implementing inspection organization is the Inspection Service for Floriculture and Arboriculture (NAKB). The Dutch and Danish systems for certification have served as examples for many of the recent EPPO-recommendations. The 'voluntary' Dutch certification scheme is discussed in this section (Van Ruiten, 1993).

The objective of the Dutch certification schemes for ornamental crops is to ensure that reliable material is produced and marketed. This requires visual inspections for varietal trueness to type, health and quality. Certification, with the ultimate objective to produce and market virus-free or pathogen-free material, is voluntary. The main immediate purpose of certification is the establishment of a propagation system based on individually selected and tested plants, which have been found to be disease free. The first stage of the system is in the hands of the individual companies because of the often varying assortment of varieties at each nursery. The certification system imposes obligations on the companies such as the need for separate glasshouse sections, which meet given isolation (e.g., insect-proof) and hygiene standards. Separate glasshouse sections are required for: entry and pre-testing of candidate material, maintenance of nuclear stock (SEE class-material), multiplication of basic material for propagation stock (EE class), and production of elite certified material (E class). Certification rules require that all plants concerned have been individually tested in the preliminary (candidate) phase for all known virus diseases; that all individual plants in the nuclear stock have been tested at least three times per year for any virus infection that might have been introduced (in this stage, the indicator plant *Saponaria vaccaria* is used in addition to ELISA for carnation, see Section VI); that all nuclear plants have flowered in order to make a varietal check; and that all plants, in every stage, are grown in steamed soil.

The prescribed testing has to be done by the companies, in their own laboratories or in another laboratory under contract. The NAKB inspection service supervises the entire propagation programme. There are regular inspections at the companies involved, and results of self-performed tests must be available. Before material from a company is marketed as certified material, NAKB-inspection officers take samples from all production classes (SEE, EE and E) for virus testing (Table 1). For plants of the SEE and EE classes, every individual plant is sampled, but for testing of E-class material, sampling takes place on the basis of statistically reliable strategies (Van Ruiten, 1988). These check tests are handled at the inspection service's laboratory. Only if the results are satisfactory (i.e., in SEE and EE class plants: 0% virus, and in E class plants: ≤5% virus) and the material appears to satisfy all other inspection criteria, certified planting material can be supplied during the growing season (Van Ruiten 1993). NAKB gives certificates for each batch of cuttings. Further regular testing takes place during the growing season.

The Dutch certification schemes are based on practical experience, i.e. on problems and diseases existing in practice. In the current European Union, however, boundaries are fading away and the possible introduction of 'new' pests and diseases

*Table I.* Tolerance levels set for 1993 for pathogens of some ornamental crop plants by The Netherlands (NAKB).

| Ornamental host plant | Pathogens tested | Tolerance Levels | | |
|---|---|---|---|---|
| | | Nuclear stock (SEE) | Propaga-tion stock (EE) | Elite or certified stock (E) |
| Carnation | viruses | 0% | 0% | 5% |
| Chrysanthemum | viruses and viroids | 0% | 0% | 1% |
| Pelargonium | all pathogens | 0% | 1% | 5%[a] |

[a]An exception is *Xanthomonas campestris* pv. *pelargonii* for which the tolerance is 0% for all classes.

from adjacent countries need to be considered. The OEPP/EPPO-recommendations in general is comprised of tests for all possible diseases of the crop concerned. These recommendations will in principle be incorporated into the Dutch regulations.

In Denmark a board was founded in 1978 to develop a Danish certification scheme. The objectives that were established are still the basis for the Danish program. The scheme involves choosing the best clones by carrying out trials under glasshouse conditions, testing for diseases, and undertaking a clean-up program. Healthy and genetically defined material can be sold following the directive for plants which is controlled by The Danish Plant Directorate. The selection and testing are done by The Danish Institute of Plant and Soil Science, which is an institute under the Ministry of Agriculture. Certified plants can either be grown by a private company or, as many of them are, at The Danish Growers' Elite Plant Station. The Danish program now includes many glasshouse ornamental crops which are tested for a variety of different pathogens (Table 2). Examples of pathogens for which testing must be done and found to be absent before certification is granted is illustrated in a relevant page from the Danish Plant Directive (Fig. 3).

## V. SAMPLING FOR CERTIFICATION

Certification schemes have been proposed or are already in operation for a number of specific ornamental plants (see Section IV). In the early stages of production schemes, where only a limited number of plants are involved, all plants must be tested for the presence of certain pathogens. In later production stages, only random samples are taken for testing for reasons of practicability. Some pathogens can be checked by visual inspection since they normally cause typical symptoms on leaves or other plant parts.

*Table II.* The number of pathogens for which ornamentals must be tested under the Danish certification scheme.

| Ornamental Host | Pathogen type | | |
|---|---|---|---|
| | Fungi | Bacteria | Viruses |
| Aeschynanthus hildebrandii | 0 | 1 | 1 |
| Argyranthemum sp. | 0 | 2 | 2 |
| Campanula isophylla | 1 | 0 | 1 |
| Begonia elatior | 0 | 1 | 1 |
| Dendranthema sp. | 0 | 2 | 4 |
| Dianthus caryophyllus | 1 | 2 | 4 |
| Dieffenbachia maculata | 0 | 1 | 2 |
| Epipremum auremum | 0 | 1 | 2 |
| Euphorbia pulcherima | 0 | 1 | 1 |
| Hibiscus sp. | 1 | 1 | 4 |
| Hoya sp. | 1 | 0 | 1 |
| Hydrangea macrophylla | 0 | 0 | 7 |
| Kalanchoe blossfeldiana | 0 | 1 | 2 |
| Pelargonium sp. | 2 | 1 | 9 |

## A. Sampling for Detection of Bacterial Pathogens

Most bacterial diseases included in certification schemes are vascular diseases so stems can usually be sampled for detecting the pathogens. As the density of pathogenic bacteria is generally higher in older than in younger plant parts, the samples to be tested should be taken from the base of the stem. The optimal time of year for sampling depends on the actual pathogen to be detected. *X. campestris* pv. *begoniae*, for instance, is normally the greatest problem in June in northern Europe. This bacterium may remain latent in *Begonia elatior* for 10-14 days during the spring and summer, and 20-40 days during the winter (Bech and Paludan, 1990).

Pelargonium plants are very often infected by both viruses and bacteria. Symptoms may be either masked or clearly developed depending on the environmental conditions. The bacterium, *X. campestris* pv. *pelargonii*, causes the most damage under conditions of high humidity and temperature. The best time for sampling for this pathogen is on dry, sunny days, when the infected plants loose their turgor pressure and can, therefore, be identified easily. In horticultural practice the cuttings are usually taken from the upper part of the plants to reduce the risk of transmitting pathogens.

A severe disease of *Chrysanthemum frutescens* is crown gall caused by *Agrobacterium tumefaciens*. Spread of *A. tumefaciens* through asymptomatic cuttings may also be influenced by the time of the year at which the cuttings are taken from the mother plant (Bazzi *et al.*, 1989). Experimental trials in Italy clearly indicate that if *A. tumefaciens* is inoculated during November the pathogen can be found 45 cm from the inoculation point after 60 days, whereas the bacterium moved only 5 cm from the inoculation point when inoculated in March. Therefore sampling for healthy plants

| Plant | Pathogen | Years of validity as nuclear stock |
|---|---|---|
| *Narcissus* L. | Arabis mosaic nepovirus<br>Cucumber mosaic cucumovirus<br>Narcissus late season yellows potyvirus<br>Narcissus mosaic potexvirus<br>Narcissus tip necrosis carmovirus<br>Narcissus white streak potyvirus<br>Narcissus yellow stripe potyvirus<br>Raspberry ringspot nepovirus<br>Strawberry latent ringspot nepovirus<br>Tobacco rattle tobravirus<br>*Ditylenchus destructor*<br>*Ditylenchus dipsaci* | 4 |
| *Nephrolepis exaltata* (L.) Schott | Tomato spotted wilt virus<br>*Aphelenchoides* spp. | 1 |
| *Nerium oleander* L. | *Pseudomonas syringae* pv. *savastanoi*<br>Cocumber mosaic cucumovirus<br>Tomato spotted wilt virus | 2 |
| *Pelargonium* L.'Herit. ex Ait. | *Xanthomonas campestris* pv. *pelargonii*<br>*Verticillium albo-atrum*<br>*Verticillium dahliae*<br>Cucumber mosaic cucumovirus<br>Pelargonium flower break virus<br>Pelargonium leaf curl virus<br>Pelargonium line pattern virus<br>(syn. Pelargonium ring pattern virus)<br>Pelargonium ringspot virus<br>Tobacco ringspot nepovirus<br>Tomato black ring nepovirus<br>Tomato ringspot nepovirus<br>Tomato spotted wilt virus | 1 |
| *Philadephus* L. | *Agrobacterium tumefaciens*<br>Raspberry ringspot nepovirus | 10 |

*Fig. 3.* A page from the Danish Plant Directive illustrating the number of pathogens for which ornamental nuclear stock plants must be tested to meet certification standards.

should take place in early spring when the movement of bacteria is very slow and the risk of distributing the bacteria via cuttings is reduced.

## B. Sampling for Detection of Viruses and Viroids

For large-scale routine testing for viruses, viral titers must be high in the plant material at the time of sampling. Factors that influence the success of indexing tests for viruses include the type of plant tissue collected, distribution of virus in the plant, time of year, stage of plant maturity, time and temperature at which plant samples are stored, presence of different virus strains, and the effect of mixed infections (Barnett, 1986). The probability of detecting a single infected plant in composite samples must also be considered (Gibbs and Gower, 1960).

Leaves with obvious symptoms of mosaic, mottle, or chlorosis usually are the best sources of material for virus detection. Where severe necrosis occurs, however, detection of virus may be troublesome or impossible because intact virus particles may be absent from such tissue (Huttinga and Maat, 1987). Probably for the same reason, older leaves usually are not a good virus source and should not be used in virus indexing procedures, although there are exceptions as in the case of pelargonium.

Viruses can be present in such plants as pelargonium without causing visible symptoms, symptoms may be mild, or occur seasonally (Bouwen and Maat, 1992). In carnation, the carnation mottle virus does not lead to visible symptoms at any time during the season (Hakkaart, 1964). Such latent viruses are readily disseminated widely without being noticed, especially in vegetatively propagated crops. Other viruses are present in minute concentrations, too low to yield visible symptoms. Sampling methodologies are most important for these viruses and are often based on experience because experimental results are not always available. The following examples will illustrate the complexity of sampling, for large-scale routine testing.

For detection of alstroemeria mosaic virus, alstroemeria streak virus, alstroemeria carla virus and cucumber mosaic virus (alstroemeria-strain) by ELISA, two to three full-grown leaves per sample must be collected from different shoots of an alstroemeria plant; lower (older) leaves should be avoided (De Blank *et al.*, 1994; Van Zaayen *et al.*, 1994). Full-grown leaves of carnation, just below the top of the plants, but not old leaves, give the best results for detection of carnation mottle virus, carnation etched ring virus and carnation latent virus (Hakkaart and Van Olphen, 1971). Consequently, for certification of carnation nuclear and propagation stock in the Netherlands, one full-grown leaf (not too old and not too young) per carnation plant is picked. In the NAKB diagnostic laboratory, two leaves are combined for testing by ELISA. The same is done for chrysanthemum, one full-grown leaf is picked per plant, and two leaves are tested together with ELISA for chrysanthemum virus B, tomato aspermy virus, and tomato spotted wilt virus. In the EPPO-certification scheme for pathogen-tested material of chrysanthemum, however, freshly collected chrysanthemum roots or leaves (with symptoms, if present) are recommended for use in ELISA (OEPP/EPPO, 1993a).

For detection of chrysanthemum stunt viroid (CSVd) and, occasionally, of chrysanthemum chlorotic mottle viroid (CCMVd) by bi-directional electrophoresis in the NAKB laboratory in the Netherlands, a mixed sample of 15 leaves (three leaves per plant, from top, middle and bottom) is used per test. This is based on the results of Huttinga *et al.* (1987) with potato spindle tuber viroid (PSTVd) and CSVd. These authors suggested that composite samples could be tested successfully based on an experiment with PSTVd in which they were able to detect one diseased potato leaflet among 499 healthy ones. Research in France on the detection of CSVd and CCMVd indicated that, although viroids are detected in any part of infected chrysanthemum plants, their concentration is higher at the top and in the young leaves, which are the best material for indexing (Monsion *et al.*, 1980). The slab gel electrophoresis technique they used was also said to be very suitable for routine mass indexing of plantlets from *in vitro* culture.

For inspection of nuclear and propagation stock of pelargonium in the Netherlands, two full-grown, older leaves with petioles and one mature leaf with petiole, respectively, must be sampled per plant for testing in ELISA for the pelargonium flower-break virus (PFBV) and pelargonium line pattern virus (PLPV). The petioles are tested for PLPV (Bouwen and Maat, 1992) and the leaves for PFBV. For testing of certification stock, only one leaf of a plant is taken at random, which is the common procedure with all certified crops.

Previous season infections with freesia mosaic virus (FreMV) are easy to detect with ELISA in all aboveground plant parts. Testing for the more important primary infections which develop during the current season, however, is much more difficult. Older, symptomatic leaves or flowers showing flower colour breaking are unsatisfactory virus sources. The youngest leaves and the stem give the best result. For optimal detection, leaves should be picked just before plants are lifted. The suitability of freesia corms for testing has not yet been investigated thoroughly. Freesia leaves are also tested with ELISA for bean yellow mosaic virus (BYMV).

Due to the erratic distribution of some plant viruses, leaves are not always the best virus source and other plant parts, if available should be substituted. Flowers of *Bouvardia* sp., for example, are superior to leaves for detection of cucumber mosaic virus by ELISA (J. Schuuring, personal communication), and petioles of full-grown leaves of *Pelargonium peltatum* 'Tavira' for detection of PLPV by ELISA (Bouwen and Maat, 1992). The latter virus could, however, easily be detected in leaves of some *P. zonale* cultivars.

Plantlets derived from meristem culture and other *in vitro* material usually yield adequate material for virus testing. The most efficient way is to collect unused plant parts including the top shoots and leaves which are cut off during the regular transfer of *in vitro* plantlets to fresh nutrient media. Because the virus in question may occur in young plantlets at low concentrations, repeated testing is required.

## C.  Sampling for Detection of Fungal Pathogens

Large-scale routine testing for fungi which cause diseases in ornamentals is not yet a common practice. An exception is the testing of carnation for *Fusarium oxysporum* f.sp. *dianthi* which is occasionally done on about 100 carnation samples at a time in the NAKB laboratory in the Netherlands. Testing for fungi is, however, generally limited to specific instances where symptomatic plant parts (roots, stems, or leaves) are sampled and tested for the presence of fungi, which subsequently need to be identified and, if necessary, tested for pathogenicity.

## D.  Sampling for Detection of MLO's

Mycoplasma-like organisms (MLO's) cause yellowing diseases and other aberrations of various ornamental plants. Because the MLO's are found in the phloem tissue of stems or roots, it is these tissues that are sampled for indexing for MLO's. Occasionally stem sections from chrysanthemums with phyllody are tested for the presence of MLO's in the NAKB laboratory in the Netherlands, where many (ornamental) fruit-tree samples are examined for MLO's causing pear decline.

## VI. TESTING FOR THE PRESENCE OF PATHOGENS

When the selection and sampling procedures have taken place the next step is to test the plant material for various diseases and/or pathogens. The plant part to be tested is that part where the concentration of pathogen is potentially the greatest.

Pathogen-indexing or testing is an important element in certification programmes and for plant quarantine, but is often also required at the request of growers. Methods used in testing laboratories should be rapid, reliable, sensitive, relatively cheap, and applicable to a large number of samples. At the same time, such methods should be simple to undertake and give clear-cut test results. Unfortunately, in many cases a single testing method may not be sufficient to verify the cleanliness of a plant sample. Not only may several pathogens be present, but also different methods sometimes need to be combined to yield optimal pathogen-indexing results. Results can be influenced by plant sampling methods, seasonal variances, and efficiency of testing methods.

During the first stage of plant production under a certification programme, plants are kept separate from commercial production in designated glasshouses. Such plants are called pre-nuclear plants or candidate material. These plants cannot be declared pathogen-free until careful tests have been performed and the results are negative. Because the concentration of pathogens in the plants is often very low, the tests have to be repeated several times. During the period when the plants are kept in isolation, they are grown under conditions which enhance expression of the diseases which are of concern and assessed for symptoms frequently.

## A. Tests for Bacterial Diseases

Pelargonium is chosen for this section as a model plant to illustrate a testing procedure. In pelargonium, pathogenic bacteria are distributed throughout the whole plant via the vascular system. Consequently, as pelargonium cultivars are propagated vegetatively, pathogens are easily transmitted from mother plants to the cuttings (Reuther, 1983).

The conventional bacteriological technique using selective media and physiological tests for the identification of bacteria provides high specificity, but is too laborious for application in practical horticulture (Reuther, 1991). A simple and effective method for testing pelargonium cuttings is the 'broth method'. In this method the lower 2-3 cm of a cutting is taken, surface disinfected, and aseptically transferred to a test tube with nutrient broth. If there is no growth in the broth after 14 days, the original cutting is declared healthy (Hellmers, 1958). Testing of the basal segment of *in vitro*-mother plants provides a high degree of confidence in evaluating the status of contamination with pathogenic bacteria (Reuther, 1991).

The first step, before detection methods such as serology and DNA-based techniques can be applied, is the extraction of bacteria from the plant material to be tested. Many nurseries use an isolation method for extraction of *X. campestris* pv. *pelargonii*, in which basipetal stem portions are ground in buffer and plated on a nutrient medium (Digat, 1987). Bacterial colonies with colour and morphological characteristics typical for the xanthomonads are then tested by a serological method or other procedure. Some laboratories perform a serological test directly on the extract.

Another extraction method is the 'shaking method' (Fig. 4). In this procedure the lower 3 cm of a cutting is cut into small pieces and transferred to a bottle with a small amount of water. The sample is shaken at 130 rpm for 12-20 h and the fluid tested with an appropriate method (Dinesen, 1981). Sensitivity of the 'shaking method' can be increased by concentrating extracted bacteria by centrifugation at 8000 rpm for 20 min. Another advantage of this method is that the bacteria may multiply during incubation on the shaker (Dinesen and DeBoer, 1994). After extraction of the bacteria a reliable and sensitive test method is used. Some of the most advantageous techniques in horticultural practice are discussed below.

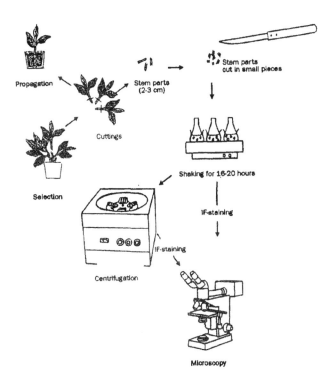

*Fig. 4.* A general extraction procedure for testing ornamental glasshouse plants for bacterial pathogens by immunofluorescence.

*I.G. Dinesen and A. van Zaayen*

*1. Immunofluorescence staining (IF)*
This method has been described by several authors (e.g. De Boer and Hall, 1988; see also Chapter 2). The procedure uses a specific antibody tagged with a fluorescent dye. The size and shape of bacterial cells coated with labelled antibody can be analyzed with a microscope equipped to provide illumination in the near- ultraviolet range and with appropriate filters. The technique can be carried out in the direct mode in which the fluorescent dye is conjugated with the primary antibody, or in the indirect mode in which the fluorescent dye is conjugated with a secondary antibody, which is specific for the primary antibody. The indirect procedure is advantageous in that one commercially available conjugate can be used for any primary antibody preparation (DeBoer, 1990).
  The extracted sample is used undiluted, diluted 1:10, 1:100, and 1:1000 and from each dilution 25 µl is transferred to a microscope slide and fixed. The slide is then treated with the primary antibody, washed and treated with the secondary antibody followed by washing. The preparation is then ready for observation with a microscope. This method is rather sensitive and the detection level is about $10^3$ immunofluorescence forming units (IFU) per ml (DeBoer, 1990). Cross-reaction with saprophytic bacteria, however, may occur and therefore some experience is required to be able to distinguish between cross-reacting bacteria and pathogenic bacteria.
  Cross-reactivity is a problem with all serodiagnostic tests for detection of bacteria, but the specificity of monoclonal antibodies, which react with a single antigenic determinant, is potentially greater in comparison to that of polyclonal antibodies, which react with a large number of different determinants (DeBoer *et al.*, 1988). Immunofluorescence is a very useful method for laboratories, where routine testing is applied to many samples, as part of the process for obtaining healthy plant material.
  In the NAKB laboratory in the Netherlands, where IF is applied for certain bacteria as a pre-screening test of various crops, about 20,000 samples per year are pre-screened for *X. campestris* pv. *pelargonii*. IF-positive samples are subjected to immunofluorescence colony staining (IFC; Van Vuurde, 1990). Isolation of IFC-positive colonies allows confirmation of the target organism or of cross-reacting bacteria. At the Danish Institute for Plant Pathology the following ornamentals are routinely tested for different bacterial diseases by using IF : *Argyranthemum* sp., *Begonia elatior*, *Dendranthema* sp., *Dianthus caryophyllus*, *Dieffenbachia maculata*, *Euphorbia pulcherrima*, *Kalanchoe blossfeldiana* and pelargonium.

*2. Enzyme Linked Immunosorbent Assay (ELISA)*
ELISA, as a serological procedure for detecting pathogens, is discussed in more detail in sub-section B and Chapter 2. Its use for detecting bacterial plant pathogens has been limited. A polyclonal multiwell ELISA has been developed for testing whole cells of *X. campestris* pv. *pelargonii*. This assay made it possible to detect the bacterium in sap from inoculated but asymptomatic pelargoniums in one hour. Cross-reaction with various other non-xanthomonads and fungi did not occur. The antiserum invariably cross-reacted with several other pathovars of *X. campestris*, but this was not a problem for indexing and diagnosis since the host is indicative of the pathovar (Anderson and Nameth, 1990).

*3. Dot immunobinding assay (DIA)*

DIA differs from conventional ELISA primarily in that the solid phase matrix used to adsorb the test material is a plastic or nylon membrane, rather than a well in a polystyrene microtiter plate. Usually DIA involves the application of microliter-quantities of sample to a membrane, which is incubated in a solution containing specific antibodies. The membrane is then incubated with enzyme-conjugated anti-globulin antibodies that react with bound immunoglobulins. The enzymatic reaction is revealed by the addition of an appropriate substrate, which, after conversion, precipitates to form a colored spot (Lazarovits, 1990).

. This system is not very commonly used in the industry, but it's development is in progress. A commercial kit for detection of *Xanthomonas campestris* pv. *pelargonii* by DIA is available.

## B. Testing for Viruses

*1. ELISA*

ELISA is the major method currently in use in the Netherlands for large-scale routine detection of over 40 viruses of ornamental crops mainly grown in glasshouses. The average number of samples tested in the NAKB laboratory in the Netherlands with ELISA is 20,000 per week, but this number increases to 50,000 during busy periods.

ELISA was introduced into plant virology by Voller *et al.* (1976) and modified by Clark and Adams (1977). It has proved to be a reliable, very sensitive and relatively inexpensive detection method, which determines the presence of an antigen such as a plant virus by means of serological and enzymatic reactions. Of the various types of ELISA protocols, the double antibody sandwich DAS-ELISA (Fig. 5) is the one

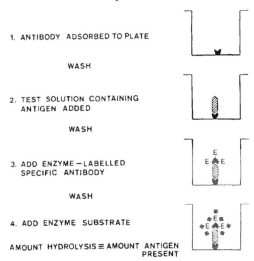

1. ANTIBODY ADSORBED TO PLATE

WASH

2. TEST SOLUTION CONTAINING ANTIGEN ADDED

WASH

3. ADD ENZYME−LABELLED SPECIFIC ANTIBODY

WASH

4. ADD ENZYME SUBSTRATE

AMOUNT HYDROLYSIS ≡ AMOUNT ANTIGEN PRESENT

*Fig. 5.* Diagrammatic representation of the double antibody sandwich ELISA procedure for determining the presence of viral antigens in ornamental plants.

most often used for large-scale routine detection of viruses in ornamental plants. The test is performed in 96-well, flat-bottom polystyrene microtiter plates. Filling of ELISA plates for coating with antibody, adding plant samples, conjugate and substrate has been fully automatized in many diagnostic laboratories.

To conduct the procedure, antibodies specific for a particular virus, are first added to wells in the plate to coat the well surfaces. After incubation, unabsorbed antibodies are washed from the wells and sap samples are added, one sample per well. If the virus is present in the sap, it will react with the adsorbed antibodies and thus be bound to the plate. Next, after washing, a solution containing antibody-enzyme conjugate is added to the wells. Wherever the virus is bound in a well, the enzyme will also be bound by means of a second serological reaction. Nonspecific binding of enzyme is prevented by special buffer components. After another washing to remove the unbound conjugate, a suitable enzyme substrate is put into each well. The enzyme activity is measured after incubation on the basis that the amount of hydrolyses is directly proportional to the amount of virus present. In plant virology, alkaline phosphatase is generally used as the enzyme of choice because of its colourless substrate p-nitrophenyl phosphate. Conversion of the substrate to a yellow coloured product indicates a positive test result. A sample is considered positive if the absorbance value is higher than the average absorbance value of the non-infected control samples plus two or three times the standard deviation of the controls (Clark and Bar-Joseph, 1984).

Possibilities, advantages, and modifications of ELISA have been reviewed elsewhere (e.g. Clark and Bar-Joseph, 1984), whereas Barnett (1986) and Lange (1986) dealt with advantages and disadvantages in practical application. The quality of the antiserum is of utmost importance. Most antisera are commercially available. In the NAKB laboratory that tests ornamental plants in the Netherlands, mainly antisera prepared by D.Z. Maat of the DLO-Research Institute for Plant Protection in Wageningen are used. The preparation of antibodies and enzyme conjugates has been described by Tóbiás et al. (1982). For routine detection polyclonal antisera are preferred, in most cases, over monoclonal antibodies. The former generally are cheaper, have a broader detection range, and are less vulnerable to changes in test conditions.

Negative and positive control samples must be incorporated into each test to enable distinction between healthy and diseased plants and to serve as a control for reagent and test conditions. Positive controls may consist of fresh sap from plants infected with the virus(es) in question and maintained in the glasshouse for that purpose, frozen sap from such plants, infected plant sap diluted with glycerol and stored in the freezer, or a purified virus solution. For viruses that are difficult to maintain in plants in the glasshouse and for which other storage methods are also not feasible, *in vitro* cultures of infected plants can be maintained.

Samples to be tested usually are prepared by squeezing leaves in extraction buffer, with a press or a power-driven crusher (e.g. Pollähne roller press, Fig. 6). The optimum ratio of plant material to buffer should be determined experimentally. Appropriate suspensions usually contain 5-10% (w/v) plant material. For routine testing of viruses of ornamentals in the Netherlands, 2-3 leaves are used and the amount of extraction buffer is adjusted in accordance with the weight of plant tissue. If the background (i.e., non-specific adsorption of enzyme) needs to be reduced, substances like Tween 20, polyvinyl pyrrolidone, or egg-albumin may be added to

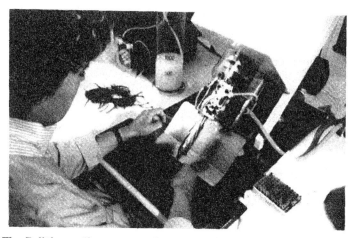

*Fig. 6.* The Pollähne roller press used to crush plant tissue to obtain plant sap for testing by ELISA for viral infections.

extraction and conjugate buffers (D.Z. Maat, personal communication). If proper cleaning of the roller press is difficult between samples, samples are homogenized in extraction buffer in separate plastic bags with a hand press or in a homogenizer.

Buffer composition, incubation periods and temperatures of the ELISA procedure all may need to be modified to give better results in certain cases. The best extraction buffer for pelargonium, for instance, contains five times as much Tween 20 as the standard extraction buffer, no sodium phosphate, but much more potassium phosphate than is normally used. The pH of plant extracts is very important and should be at least 6, but preferably around 7 to obtain adequate results.

The slowest, but in some cases the optimum DAS-ELISA procedure, involves overnight incubations for coating, plant sample, and enzyme conjugate. At the NAKB laboratory in the Netherlands, this is only done for viruses that are difficult to detect. This could be due to low virus concentrations in the plant, erratic virus distribution, or weak antigenic properties of a virus, e.g. alstroemeria mosaic and streak viruses, carnation etched ring virus and cucumber mosaic virus. In most other cases, the faster method of about 3 hrs incubation of the conjugate at 30 °C is applied. A simplified modification of DAS-ELISA for viruses that occur in very low concentrations, in which plant sample and enzyme conjugate are incubated simultaneously (Flegg and Clark, 1979), has also been tried in the Dutch laboratory for a strain of cucumber mosaic virus and the carnation etched ring virus. The results were unreliable, however, and the method was abandoned and replaced by the procedure in which the conjugate was incubated overnight at 4 °C.

Commercially available kits for detecting groups of viruses, such as the potyvirus group, can be a great help in the testing of plant samples with unknown virus infections. Here again, polyclonal group-specific antibodies are preferred over monoclonal antibodies, as the former can detect a greater number of different viruses (Wong *et al.*, 1994; Van Zaayen *et al.*, 1994).

## 2. *Bioassay*

Inoculation of indicator plants with sap extracted in buffer from test plants is a very effective and sensitive detection method. Local lesions may develop on inoculated leaves of indicator hosts beginning at three days after inoculation, possibly followed by systemic symptoms such as mosaic or mottle on newly emerged leaves. Bioassays are generally not suited for large-scale routine testing as they are time consuming, require a lot of glasshouse space, and are not very fast. Sometimes symptoms only appear several weeks after inoculation. Bioassays, however, can be combined with ELISA tests to check for the presence of viruses in 'new' crops, e.g. crops that are going to be certified; or to check for the presence of viruses in 'unknown' crops, e.g. symptomless *Petunia* sp. in which potato virus Y sometimes occurs. Bioassays are also useful for confirming equivocal positive readings in ELISA as sometimes occurs when testing cyclamen for tomato spotted wilt virus. *Kalanchoe* sp. also often express virus symptoms poorly and tend to give erratic results in ELISA and electron microscopy (I. Bouwen, personal communication). A potyvirus in *Kalanchoe* sp. was detected with monoclonal group-specific antibodies in leaves of *Chenopodium quinoa*, inoculated with sap from *Kalanchoe* plants with suspected viral infections, but not directly in *Kalanchoe* leaves themselves (I. Bouwen and J. Schuuring, personal communication).

Negative ELISA results with symptomless plants also sometimes need to be confirmed with a bioassay test because virus concentrations may be too low to detect with serological techniques. This may be true for the dasheen mosaic virus in *Dieffenbachia maculata*. According to Paludan and Begtrup (1982), a screening programme for *Dieffenbachia* should be based on sap inoculation to young plants of *Philodendron* varieties, as this test was the most effective one.

Bioassays are also used in the Netherlands to improve viral detection in general for certification procedures for certain crops such as carnations. All carnation nuclear stock plants are not only tested by ELISA for carnation mottle virus (CarMV) and carnation etched ring virus (and, in the near future, also for carnation necrotic fleck virus (CNFV) and three other carnation viruses), but a bioassay using *Saponaria vaccaria* 'Pink Beauty' as the indicator plant is also conducted (Hakkaart and Van Olphen, 1971). This indicator host was more sensitive for detection of CarMV than *Chenopodium amaranticolor* or serological diagnosis. Apart from being able to detect very low levels of CarMV, *S. vaccaria* also reacts to several other viruses (i.e., carnation etched ring virus, carnation latent virus, carnation ringspot virus and carnation vein mottle virus), for which *C. amaranticolor* is less susceptible.

In the OEPP/EPPO certification schemes for pathogen-tested material of pelargonium and chrysanthemum (OEPP/EPPO, 1992, 1993a) the use of either indicator plants or ELISA is dictated for testing of nuclear stock. Nuclear stock and candidate nuclear stock of carnation, however, should be checked by inoculation on test plants, but ELISA may be used in parallel with biological testing. This applies to all carnation viruses except for CNFV, which should be tested by both ELISA and visual inspection (OEPP/EPPO, 1991a). CNFV is an exception because this virus is not mechanically transmissible.

## 3. *Electron microscopy*

Electron microscopes (EM) are invaluable for detection of viral infections in plants.

When symptomatic leaves of a crop have not reacted positively in ELISA with antisera to known viruses, examination of crude sap, stained with phosphotungstic acid or uranyl acetate, may indicate whether virus particles are present and their possible taxonomic status. A negative result, however, does not exclude a virus infection since labile viruses may not be visualized in standard electron microscopic procedures. Preparation and viewing of samples is rather time-consuming, which renders this method unfit for large-scale routine application.

Immunosorbent electron microscopy (ISEM) combines direct visualisation of virus particles with the specificity of serological reactions (Derrick, 1973; Milne and Luisoni, 1977). For this procedure plant sap is placed on an electron microscope grid and treated with the appropriate virus antiserum. If the plant virus is present, antibodies decorate it, making the virus particles easier to see. Variations and applications of the procedure are reviewed by Milne (1986). Although ISEM is not suited for large-scale screening programmes as each grid must be handled separately, it has definite advantages. It is very sensitive and crude antisera, even with low titres, can be used successfully. Moreover, the specificity of ISEM is broad and this lack of strain specificity is usually preferred in routine testing as mentioned above. ISEM can be used to confirm ELISA results and to develop positive/negative threshold levels for ELISA. This was done by De Blank *et al.* (1994) to determine whether the detection of alstroemeria mosaic potyvirus (AlMV) needed improvement. The results of ISEM indicated the presence of a second potyvirus, which later proved to be alstroemeria streak virus (Wong *et al.*, 1994; Van Zaayen *et al.*, 1994). The results also explained the problems previously encountered with the detection of AlMV.

## C. Testing for Viroids

Bi-directional electrophoresis (Huttinga *et al.*, 1987) is used at the NAKB laboratory in the Netherlands for testing of chrysanthemums for chrysanthemum stunt viroid (CSVd) and, occasionally, for chrysanthemum chlorotic mottle viroid (CCMVd). The method was developed after Schumacher *et al.* (1983) found the means to lower the background staining of gels when sensitive reagents (e.g. $AgNO_3$) were used. Prior to its adoption by the diagnostic laboratories, the bi-directional electrophoresis was compared with molecular hybridization methods, which are very sensitive, but the former method proved to be more practical and it does not need the hazardous chemicals of the other methods (Huttinga *et al.*, 1987). The presence of viroid RNA can readily be detected in symptomless infected plants long before the appearance of symptoms even in susceptible varieties. The technique is, however, rather laborious, only 32 samples can be handled at a time. As one sample may represent 15 leaves from 5 plants, in principle 160 plants can be indexed at a time. Annually, about 1500 samples are being tested for CSVd in the Netherlands.

## D. Testing for Fungi

Currently detection of pathogenic fungi is based on visual inspection and identification of pure cultures obtained by transfer of diseased tissue to sterile nutrient agar media in.Petri dishes. Sometimes a bioassay is required to test pathogenicity of isolates particularly when closely related saprophytic species are common. *Fusarium oxysporum*, for example, can best be distinguished from saprophytic *Fusarium* spp. by inoculation onto host plants. For bioassays, young, potted test plants are inoculated by pouring a spore suspension of the isolated fungus on the soil. Symptoms may appear after two weeks in the case of *F. oxysporum* f.sp. *cyclaminis*, but only after 6-12 weeks when *F. oxysporum* f.sp. *dianthi* is involved.

## E. Testing for MLO's

MLO's in the phloem of chrysanthemums (Bertaccini *et al.*, 1990) can be detected by fluorescence microscopy (Davies *et al.*, 1986). Stem sections are stained with 4'-6, diamidino-2-phenylindole (DAPI), as described by Seemüller (1976). Prior to sectioning, the stems may be fixed in 5% (v/v) glutaraldehyde, pH 7.0, and stored until use. A freeze microtome is preferred for sectioning. After staining for 20-30 min in a 0.1% DAPI-solution, the sections can be viewed with a UV-light microscope. About eighty samples can be handled per day.

## F. Future Developments

### 1. Viruses and Bacteria

It was predicted many years ago that modern methods based on nucleic acid hybridisation techniques with cDNA probes and dot-blot immunoassays will take over the well-known assays for routine detection of plant viruses. Recently, the polymerase chain reaction (PCR) was invented and proves to be a promising technique. Although all these methods are being applied in research on viral and bacterial pathogens, the link to large-scale routine screening has scarcely been made. A dot-blot immunoassay or dot-ELISA (on nitrocellulose paper) is in use at the laboratory of the Seed Pathology Institute in Denmark, where a modified simple procedure suitable for routine use was developed (Lange, 1986).

The purpose of the application, however, must always be kept in mind when choosing which diagnostic technique to use or when interpreting results (Barnett, 1986). As long as excellent antisera are available, 'old shoes should not be thrown away before adequate new ones are available'. This obviously does not hold for those viruses which can not be detected or detected only with great difficulty using the common procedures. Moreover, detection of group-specific nucleic acids would be a great help for large-scale screening programmes.

## 2. *Viroids and MLO's*

As serological methods have no prospect for the detection of viroids and MLO's, and the techniques available so far are rather laborious, a great deal may be expected from the application of techniques like PCR (e.g., Namba *et al.*, 1993). It may take some time before these techniques are ready for routine testing, but their availability is eagerly awaited.

## 3. *Pathogenic fungi*

New techniques, which could be used for rapid large-scale routine detection, are in great need for fungi such as *Fusarium oxysporum* f.sp. *dianthi* and *cyclaminis*. Both fungi may occur in plants long before the appearance of symptoms and early detection could prevent spread of the disease. Detection of *F. oxysporum* f.sp. *cyclaminis*, based on isozyme patterns (Kerssies *et al.*, 1994), is ready for testing in practice. Another exciting development is the detection of fungal propagules in soil, with the aim of preventing future infections with destructive pathogens such as *Verticillium dahliae*. This pathogen can now be detected in rose stems by means of monoclonal antibodies (Van de Koppel and Schots, 1994), but here again testing on a large scale in practice has not yet been done. Fungal detection, however, certainly is progressing.

## VII. ELIMINATION OF PLANT PATHOGENS

### A. Meristem Culture

Over 40 years ago, meristem-tip culture was applied for the first time to free virus-infected dahlia plants from viruses (Morel and Martin, 1952). The healthy shoots grown from the meristems had to be grafted on young, virus-free seedlings since techniques for *in vitro* rooting had not yet been developed. Tissue culture techniques have since been improved and dahlia meristems can now be readily rooted (Mullin and Schlegel, 1978). Another ten years were required, however, for the development of a more refined meristem propagation procedure to eradicate dahlia mosaic virus, which was combined with specific methods to detect the virus in very low quantities (Wang *et al.*, 1988).

Meristem culture has now been applied to many crops with the intent to eliminate viral infections. This subject has been reviewed by Hollings (1965), Mori *et al.* (1969), Quak (1977, 1982), Wang and Hu (1980), and Kartha (1986). These review articles deal with various topics including location and dimensions of meristems, degree of virus elimination (some viruses are more readily eliminated than others), speculations about possible causes of virus elimination by meristem culture, the dependence on reliable and sensitive detection methods, and the danger of re-infection. The ornamental crops that have been dealt with in these reviews are carnation, *Dianthus barbatus*, chrysanthemum, *Cymbidium* sp., and narcissus (Hollings, 1965); iris, lily, and petunia (Mori *et al.*, 1969; in Japanese); amaryllus, *Caladium hortulanum*, freesia, gladiolus, hyacinth, iris, lily, *Narcissus tazetta*, *Nerine* sp., pelargonium, petunia and *Ranunculus asiaticus* (Quak, 1977); and *Budleia davidii*, *Daphne* sp. and *Hydrangea macrophylla* (Wang and Hu, 1980).

During the last 10-15 years only a few new cases were reported of successful trials to free ornamental crops from specific viruses. Much more information has been gathered, however, on viruses and the detection methods havé been improved and simplified. Large-scale testing for commercial purposes is more feasible now than before. To the list of ornamentals that were made virus-free previously have been added: *Aeschynanthus hildebrandii*, of the difficult to eliminate tobacco mosaic virus (Paludan, 1985b); *Alstroemeria* spp., of alstroemeria mosaic virus and alstroemeria carla virus (Hakkaart and Versluijs, 1985, 1988); various members of the *Araceae* (ornamental foliage crops) from dasheen mosaic virus (Zettler and Hartman, 1987); *Cymbidium* sp., not only from Cymbidium mosaic virus, which had been reported earlier by others, but also from Odontoglossum ringspot virus, a strain of tobacco mosaic virus, by immersion of the freshly excised meristems in antiserum to this virus (Inouye, 1984); *Euphorbia pulcherrima*, of poinsettia mosaic virus and poinsettia cryptic virus (Paludan and Begtrup, 1986); freesia, 25 varieties, from freesia mosaic virus, bean yellow mosaic virus and cucumber mosaic virus (Bertaccini *et al.*, 1989); *Kalanchoe* sp., from Kalanchoe latent virus (Paludan, 1985c); *Laeliocattleya areca*, from Cymbidium mosaic virus but not from Odontoglossum ringspot virus (Ishii, 1974); and *Delphinium* sp., from cucumber mosaic virus (Van Zaayen *et al.*, 1992).

Only a few publications deal with meristem culture to free plants from non-viral pathogens. It is generally accepted, however, that the application of meristem culture not only yields virus-free plants, but also plants free from bacteria and fungi (Pierik, 1987), viroids (Lizarraga *et al.*, 1980; Paludan, 1985a) and mycoplasmas (Green *et al.*, 1989). Bech and Paludan (1990) did not detect the bacterium X. *campestris* pv. *begoniae* in plantlets grown from ca. 0.25 mm long meristems, excised from severely infected mother plants, over a period of 12 months (Fig. 7). Chrysanthemum stunt viroid (CSVd) could be eliminated by a cold treatment of plants at 5°C for six months, followed by meristem culture. Paduch-Cichal and Kryczynski (1987) applied this combined treatment to chrysanthemums and potatoes and obtained 18-80% viroid-free plants, depending on the type of viroid and on the host plant species.

Nematodes, particularly those that live inside the roots like *Pratylenchus* species, most probably can also be eliminated through meristem culture, but there are no references on the subject.

It should be noted that the term 'pathogen-free' plants is often not used correctly. Plants can be said to be free from only those pathogens for which testing has been done. Normally the term 'pathogen-tested' would be a more accurate designation.


## B. Other Treatments


The effectiveness of meristem culture for eliminating viruses may be improved by prior heat treatment of the mother plants (Nyland and Goheen, 1969), although not all plant species and cultivars can withstand such treatment. The combination of heat treatment and meristem culture was initially used to free carnations from viruses (Quak, 1957), and is still used occasionally for carnations at the NAKB laboratory in the Netherlands. Heat treatment for six weeks at 30-36 °C is used for carnations.

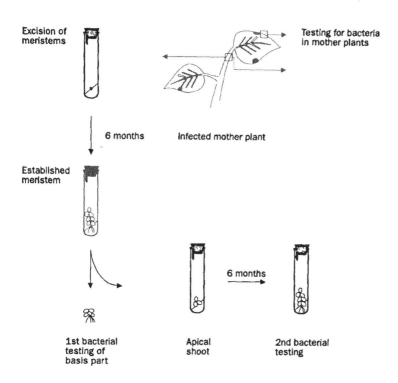

*Fig. 7.* Testing strategy for bacterial pathogens in a meristem culture scheme for ornamental glasshouse crop plants. (Adapted from Bech and Paludan, 1990)

Meristem culture has also been combined with chemotherapy which involved addition of therapeutic agents such as ribavirin to the nutrient medium on which meristems are grown (Lerch, 1987; Long and Cassells, 1986; Witkowski *et al.*, 1972). Recently the use of ribavirin was reported for eradication of Odontoglossum ringspot virus from *in vitro* cultures from *Cymbidium* sp. meristems (Toussaint *et al.*, 1993).

## C. Final Remarks

Meristem culture is an effective means for elimination of viruses and certain other pathogens, which cannot be avoided or controlled in other ways. Availability of reliable tests for those pathogens to be eliminated in this way is, however, of utmost importance. Accurate tests are required to index the host plants from which meristems are taken as well as plants grown from the meristematic tissue to ensure that the plants are indeed free from those pathogens. Testing is preferably done on waste plant tissue cut from *in vitro* plantlets during regular transfer to fresh nutrient media, but sometimes the pathogen concentration in such tissue is too low and mature plants need to be tested as a final check (Van Zaayen *et al.*, 1992). The resulting pathogen-free crops are usually of high quality, are no longer a source of contamination and may be stored *in vitro* to avoid re-infection. Today, many ornamental crops are propagated vegetatively by tissue culture and meristem culture is an essential part of this large-scale, rapid multiplication process. When plant stocks are multiplied by stem cuttings as is done with carnation, chrysanthemum, and pelargonium, for example, pathogen-free mother stock must first be established from meristem cultures.

A remaining problem is that of re-infection of pathogen-tested plants once they are grown outside of the laboratory. Certain measures, if well implemented, can prevent re-infection. These measures differ for various pathogens and depend largely on how they are spread. Resistance is perhaps the ultimate solution to plant diseases, but it will be a long time before crops resistant to the most harmful organisms and viruses are developed.

## REFERENCES

Anderson, M.J. and Nameth, S.T. (1990). Development of a polyclonal antibody-based serodiagnostic assay for the detection of *Xanthomonas campestris* pv. *pelargonii* in geranium plants. *Phytopathology* **80**, 357-60.

Barnett, O.W. (1986). Application of new test procedures to surveys: merging the new with the old. *In*: 'Developments and Applications in Virus testing' (R.A.C. Jones and L. Torrance, eds.), pp. 247-267. The Lavenham Press Ltd., Sudbury, United Kingdom.

Barnett, O.W. (1988). Virus damage evaluation. *Acta Horticulturae* **234**, 489-496.

Bazzi, C., Gozzi, R. and Mazzucchi, U. (1989). Translocation of *Agrobacterium tumefaciens* biovar 1 in stems of *Chrysanthemum frutescens* L. *Phytopathologia Mediterranea* **28**, 28-32.

Bech, K. and Paludan, N. (1990). Is *Xanthomonas campestris* pv. *begoniae* transmitted in meristem culture? Inoculation trials with *Begonia elatior* and testing for bacteria in meristem plants. *Tidsskrift for Planteavl* **94**, 337-344.

Bertaccini, A., Bellardi, M.G. and Rustignoli, E. (1989). Virus-free *Freesia* corms produced by meristem-tip culture. *Advances in Horticultural Science* **3**, 133-137.

Bertaccini, A., Davis, R.E., Lee, I.M., Conti, M., Dally, E.L. and Douglas, S.M. (1990). Detection of chrysanthemum yellows mycoplasmalike organism by dot hybridization and southern blot analysis. *Plant Disease* **74**, 40-43.

Bialy, H. and Klausner, A. (1986). A new route to virus resistance in plants. *Biotechnology* **4**, 96.

Bos, L. (1972). Hoe en waarom vormen plantevirussen een probleem. *Bedrijfsontwikkeling* **3**, 611-616.

Bos, L. (1982). Crop losses caused by viruses. *Crop Protection* **1**, 263-282.

Bos, L. (1983). Plant virus ecology; the role of Man, and the involvement of governments and international organizations. *In* 'Plant Virus Epidemiology' (R.T. Plumb and J.M. Thresh, eds.), pp. 7-23. Blackwell Scientific Publications, Oxford, United Kingdom.

Bouwen, I. and Maat, D.Z. (1992). Pelargonium flower-break and pelargonium line pattern viruses in the Netherlands; purification, antiserum preparation, serological identification, and detection in pelargonium by ELISA. *Netherlands Journal of Plant Pathology* 98, 141-156.

Cassells, A.C., Minas, G. and Bailiss, K.W. (1982). Pelargonium net vein agent and pelargonium petal streak as beneficial infections of commercial pelargoniums. *Scientia Horticulturae* 17, 89-96.

Cho, J.J., Mitchell, W.C., Mau, R.F.L. and Sakimura, K. (1987). Epidemiology of tomato spotted wilt virus disease on crisphead lettuce in Hawaii. *Plant Disease* 71, 505-508.

Clark, M.F. and Adams, A.N. (1977). Characteristics of the microplate method of enzyme-linked immunosorbent assay for the detection of plant viruses. *Journal of General Virology* 34, 475-483.

Clark, M.F. and Bar-Joseph, M. (1984). Enzyme immunosorbent assays in plant virology. *In:* 'Methods in Virology', Volume 7, (K. Maramorosch and H. Koprowski, eds.), pp. 51-85. Academic Press, New York.

Davies, D.L., Clark, M.F. and Adams, A.N. (1986). A mycoplasma-like organism associated with a decline-like disease in English pears. *Acta Horticulturae* 193, 329-332.

De Blank, C.M., Van Zaayen, A. and Bouwen, I. (1994). Towards a reliable detection of alstroemeria mosaic virus. *Acta Horticulturae* 377, 199-208.

De Boer, S.H. (1990). Immunofluorescence for bacteria. *In:* 'Serological Methods for Detection and Identification of Viral and Bacterial Plant Pathogens' (R. Hampton, E. ball and S. De Boer, eds.), pp. 295-298. APS Press, St. Paul, Minnesota.

De Boer, S.H. and Hall, J.W. (1988). An automated microscope system for estimating the population of *Clavibacter michiganensis* subsp. *sepedonicus* cells labelled with monoclonal antibodies in immunofluorescence. *Canadian Journal of Plant Pathology* 10, 215-220.

De Boer, S.H., Wieczorek, A. and Kummer, A. (1988). An ELISA test for bacterial ring rot of potato with a new monoclonal antibody. *Plant Disease* 72, 874-878.

Derrick, K.S. (1973). Quantitative assay for plant viruses using serologically specific electron microscopy. *Virology* 56, 652-653.

Digat, B. (1987). Methodologie de la detection des bacteries pathogenes du pelargonium et organisation generale de la certification sanitaire de cette culture en France. *Bulletin OEPP/EPPO Bulletin* 17, 281-286.

Dinesen, I.G. (1981). Production of plants free of pathogenic bacteria. *Proceedings of the Fifth International Conference on Plant Pathogenic Bacteria 1981*, 518-522. Centro Internacional de Agricultura Tropical, Cali, Columbia.

Dinesen, I.G. and De Boer, S.H. (1994). Extraction of *Clavibacter michiganensis* subsp. *sepedonicus* from composite samples of potato tubers. *American Potato Journal* 72, 133-142.

Flegg, C.L. and Clark, M.F. (1979). The detection of apple chlorotic leafspot virus by a modified procedure of enzyme-linked immunosorbent assay (ELISA). *Annals of*

*Applied Biology* **91**, 61-65.

Gibbs, A.J. and Gower, J.C. (1960). The use of a multiple-transfer method in plant virus transmission studies - some statistical points arising in the analysis of results. *Annals of Applied Biology* **48**: 75-83.

Gippert, R. and Schmelzer, K. (1973). Erfahrungen mit Spitzenmeristemkulturen von Pelargonien (*Pelargonium zonale*-Hybriden). *Archiv fur Phytopathologie und Pflanzen schutz* **9**, 353-362.

Green, S.K., Luo, C.Y. and Lee, D.R. (1989). Elimination of mycoplasma-like organisms from witches' broom infected sweet potato. *Journal of Phytopathology* **126**, 204-212.

Hakkaart, F.A. (1964). Description of symptoms and assessment of loss caused by some viruses in the carnation cultivar 'William Sim'. *Netherlands Journal of Plant Pathology* **70**, 53-60.

Hakkaart, F.A. (1973). Some results of experiments with leaf necrosis of freesia. *Plant Virology - Proceedings of the Seventh Conference of the Czechoslovak Plant Virologists,* High Tatras, 1971. Publishing House of the Slovak Academy of Sciences, Bratislava, 231-234.

Hakkaart, F.A. and Van Olphen, M. (1971). *Saponaria vaccaria* 'Pink Beauty', a new test plant for carnation mottle virus. *Netherlands Journal of Plant Pathology* **77**, 127-133.

Hakkaart, F.A. and Versluijs, J.M.A. (1985). Viruses of *Alstroemeria* and preliminary results of meristem culture. *Acta Horticulturae* **164**, 71-75.

Hakkaart, F.A. and Versluijs, J.M.A. (1988). Virus elimination by meristem-tip culture from a range of *Alstroemeria* cultivars. *Netherlands Journal of Plant Pathology* **94**, 49-56.

Hellmers, E. (1952). Bacterial leafspot of *Pelargonium* (*Xanthomonas pelargonii*) in Denmark. *Transaction of the Danish Academy of Technical Sciences* **4**, 1-35.

Hellmers, E. (1958). Four wilt diseases of perpetual-flowering carnations in Denmark. *Dansk botanisk Arkiv* **18**, 1-200.

Hollings, M. (1965). Disease control through virus-free stock. *Annual Review of Phytopathology* **3**, 367-396.

Huttinga, H. and Maat, D.Z. (1987). Purification. *In*: 'Viruses of Potatoes and Seed-Potato Production' (J.A. de Bokx and J.P.H. van der Want, eds.), pp. 25-32. Pudoc, Wageningen, the Netherlands.

Huttinga, H., Mosch, W.H.M. and Treur, A. (1987). Comparison of bi-directional electrophoresis and molecular hybridization methods to detect potato spindle tuber viroid and chrysanthemum stunt viroid. *Bulletin OEPP/EPPO Bulletin* **17**, 37-43.

Inouye, N. (1984). Effect of antiserum treatment on the production of virus-free *Cymbidium* by means of meristem culture. *Nogaku Kenkyu* **60**, 123-133.

Ishii, M. (1974). Partial elimination of virus from doubly infected orchids by meristem explant culture. *Acta Horticulturae* **36**, 229-233.

Kartha, K.K. (1986). Production and indexing of disease-free plants. *In*: 'Plant Tissue Culture and its Agricultural Applications' (L.A. Withers and P.G. Alderson, eds.), pp. 219-238. Butterworths, London, United Kingdom.

Kennedy, B.W. and Alcorn, S.M. (1980). Estimates of U.S. crop losses to procaryote plant pathogens. *Plant Disease* **64**, 674-676.

Kerssies, A., Everink, A., Hornstra, L. and Van Telgen, H.J. (1994). Electrophoretic detection method for *Fusarium oxysporum* species in cyclamen and carnation by using isoenzymes. *In*: 'Modern Assays for Plant Pathogenic Fungi: Identification, Detection and Quantification' (A. Schots, F.M. Dewey and R. Oliver, eds.), pp. 57-62. CAB International, Oxford, United Kingdom.

Lange, L. (1986). The practical application of new developments in test procedures for the detection of viruses in seed. *In*: 'Developments and Applications in Virus Testing' (R.A.C. Jones and L. Torrance, eds.), pp. 269-281. The Lavenham Press Ltd., Sudbury, United Kingdom.

Lazarovits, G. (1990). The dot immunobinding assay (DIA) - bacteria. *In*: 'Serological Methods for Detection and Identification of Viral and Bacterial Plant Pathogens'. (R. Hampton, E. ball and S. De Boer, eds.) pp. 249-261. APS Press, St. Paul, Minnesota.

Lelliot, R.A. (1988). Bacteria I: Pseudomonas and Xanthomonas. *In*: 'European Handbook of Plant Diseases' (I.M. Smith, J. Dunez, R.A. Lelliot, D.H. Phillips and S.A. Archer, eds.), pp. 136-75. Blackwell, Oxford, United Kingdom.

Lerch, B. (1987). On the inhibition of plant virus multiplication by ribavirin. *Antiviral Research* **7**, 257-270.

Lizarraga, R.E., Salazar, L.F., Roca, W.M. and Schilde-Rentschler, L. (1980). Elimination of potato spindle tuber viroid by low temperature and meristem culture. *Phytopathology* **70**, 754-755.

Long, R.D. and Cassells, A.C. (1986). Elimination of viruses from tissue cultures in the presence of antiviral chemicals. *In*: 'Plant Tissue Culture and its Agricultural Applications' (L.A. Withers and P.G. Alderson, eds.), pp. 239-248. Butterworths, London, United Kingdom.

Lopian, R. (1993a). European and Mediterranean Plant Protection Organization - profile of a regional plant protection organization. *In*: BCPC Monograph No 54: 'Plant Health and the European Single Market' (D. Ebbels, ed.), pp. 267-272. BCPC/AAB/BSPP, United Kingdom.

Lopian, R. (1993b). EPPO's contribution to a harmonized Europe: certification schemes to improve the health status of ornamental plants. *In*: BCPC Monograph No 54: 'Plant Health and the European Single Market' (D. Ebbels, ed.), pp. 273-278. BCPC/AAB/BSPP, United Kingdom.

Milne, R.G. (1986). New developments in electron microscope serology and their possible applications. *In*: 'Developments and Applications in Virus Testing' (R.A.C. Jones and L. Torrance, eds.), pp. 179-191. The Lavenham Press Ltd., Sudbury, United Kingdom.

Milne, R.G. and Luisoni, E. (1977). Rapid immune electron microscopy of virus preparations. *Methods in Virology* **6**, 265-281.

Monsion, M., Macquaire, G., Bachelier, J.C., Faydi, C. and Dunez, J. (1980). Detection of chrysanthemum stunt and chlorotic mottle viroids by slab gel electrophoresis. *Acta Horticulturae* **110**, 321-325.

Morel, G.J. and Martin, C. (1952). Guérison de dahlias atteints d'une maladie à virus. *Comptes Rendus* **235**, 1324-1325.

Mori, K., Hamaya, E., Shimomura, T. and Ikegami, Y. (1969). Production of virus-free plants by means of meristem culture. *Journal of the Central Agricultural Experiment Station* No. 13, Japan.

Mullin, R.H. and Schlegel, D.E. (1978). Meristem-tip culture of dahlia infected with dahlia mosaic virus. *Plant Disease Reporter* **62**, 565-567.

Namba, S., Kato, S., Iwanami, S., Oyaizu, H., Shiozawa, H. and Tsuchizaki, T. (1993).

Detection and differentiation of plant-pathogenic mycoplasmalike organisms using polymerase chain reaction. *Phytopathology* **83**, 786-791.

Nyland, G. and Goheen, A.C. (1969). Heat therapy of virus diseases of perennial plants. *Annual Review of Phytopathology* **7**, 331-354.

OEPP/EPPO. (1991a). Certification scheme. Pathogen-tested material of carnation. *Bulletin OEPP/EPPO Bulletin* **21**, 279-290.

OEPP/EPPO. (1991b). General scheme for the production of certified pathogen-tested vegetatively propagated ornamental plants. *Bulletin OEPP/EPPO Bulletin* **21**, 757.

OEPP/EPPO. (1992). Certification scheme. Pathogen-tested material of pelargonium. *Bulletin OEPP/EPPO Bulletin* **22**, 285-296.

OEPP/EPPO. (1993a). Certification scheme. Pathogen-tested material of chrysanthemum. *Bulletin OEPP/EPPO Bulletin* **23**, 239-247.

OEPP/EPPO. (1993b). Certification scheme. Nursery requirements - recommended requirements for establishments participating in certification of fruit or ornamental crops. *Bulletin OEPP/EPPO Bulletin* **23**, 249-252.

Paduch-Cichal, E. and Kryczynski, S. (1987). A low temperature therapy and meristem-tip culture for eliminating four viroids from infected plants. *Journal of Phytopathology* **118**, 341-346.

Paludan, N. (1985a). Inactivation of viroids in chrysanthemum by low-temperature treatment and meristem-tip culture. *Acta Horticulturae* **164**, 181-186.

Paludan, N. (1985b). Inactivation of tobacco mosaic virus in *Aeschynanthus hildebrandii* by means of heat treatment, chemotherapy and meristem-tip culture. *Tidsskrift for Planteavl* **89**, 273-278.

Paludan, N. (1985c). Inactivation of *Kalanchoë* latent virus strain 1 in *Kalanchoë* using heat treatment and meristem-tip culture. *Tidsskrift for Planteavl* **89**, 191-195.

Paludan, N. and Begtrup, J. (1982). *Dieffenbachia maculata* (Lodd.) G. Don. Virus attack in Danish cultures, survey and diagnosis. *Tidsskrift for Planteavl* **86**, 399-404.

Paludan, N. and Begtrup, J. (1986). Inactivation of poinsettia mosaic virus and poinsettia cryptic virus in *Euphorbia pulcherrima* using heat-treated mini-cuttings and meristem-tip culture. *Tidsskrift for Planteavl* **90**, 283-290.

Pierik, R.L.M. (1987). 'In vitro Culture of Higher Plants'. Martinus-Nijhoff Publishers, Dordrecht, Boston, Lancaster.

Pool, R.A.F., Wagnon, H.K. and Williams, H.E. (1970). Yield increase of heat-treated 'Baccara' roses in a commercial greenhouse. *Plant Disease Reporter* **54**, 825-827.

Prins, M. (1993). ISO 9000-serie: kwaliteit stapsgewijs opbouwen. *Vakblad voor de Bloemisterij* **47**, 26-31.

PVS and PGF (1993). Tuinbouwstatistiek 1992. The Hague, the Netherlands.

Quak, F. (1957). Meristeemcultuur, gecombineerd met warmtebehandeling, voor het verkrijgen van virusvrije anjerplanten. *Tijdschrift over Plantenziekten* **63**, 13-14.

Quak, F. (1977). Meristem culture and virus-free plants. *In*: 'Applied and Fundamental Aspects of Plant Cell, Tissue and Organ Culture' (J. Reinert and Y.P.S. Bajaj, eds.), pp. 598-615. Springer-Verlag, Berlin, Germany.

Quak, F. (1982). Weefselkweek en virussen. *Gewasbescherming* **13**, 77-88.

Reuther, G. (1983). Propagation of disease-free *Pelargonium* cultivars by tissue culture. *Acta Horticulturae* **131**, 311-319.

Reuther, G. (1991). Elimination of *Xanthomonas campestris* pv. *pelargonii* and endophytic bacteria from *Pelargonium*. *In*: 'Pelargonium Micropropagation and

Pathogen Elimination'. Cost 87, 17-30. A printed report from the Commission of the European Communities, Brussels, Belgium.

Richardson, J.M. (1993). Quality assurance in ornamental plants, BS5750 in practice. *In*: BCPC Monograph No 54: 'Plant Health and the European Single Market' (D. Ebbels, ed.), pp. 95-100. BCPC/AAB/BSPP, United Kingdom.

Schenk, P.K. (1976). Virologists and ornamentalists. *Acta Horticulturae* 59, 17-23.

Schumacher,J., Randles, J.W. and Riesner, D. (1983). A two-dimensional technique for the detection of circular viroids and viroids. *Analytical Biochemistry* 135, 288-295.

Seemüller, E. (1976). Investigations to demonstrate mycoplasma-like organisms in diseased plants by fluorescence microscopy. *Acta Horticulturae* 67, 109-111.

Steed, P. (1993). Quality assurance in ornamental plants. *In*: BCPC Monograph No 54: 'Plant Health and the European Single Market' (D. Ebbels, ed.), pp. 89-94. BCPC/AAB/BSPP, United Kingdom.

Thresh, J.M. (1980). The origin and epidemiology of some important plant virus diseases. *In* 'Applied Biology', Volume 5 (T.H. Coaker, ed.), pp. 1-65. Academic Press, London, New York, San Francisco.

Tóbiás, I., Maat, D.Z. and Huttinga, H. (1982). Two Hungarian isolates of cucumber mosaic virus from sweet pepper (*Capsicum annuum*) and melon (*Cucumis melo*): identification and antiserum preparation. *Netherlands Journal of Plant Pathology* 88, 171-183.

Toussaint, A., Kummert, J., Maroquin, C., Lebrun, A. and Roggemans, J. (1993). Use of VIRAZOLE[R] to eradicate odontoglossum ringspot virus from *in vitro* cultures of *Cymbidium* Sw. *Plant Cell, Tissue and Organ Culture* 32, 303-309.

Van de Koppel, M.M. and Schots, A. (1994). A double (monoclonal) antibody sandwich ELISA for the detection of *Verticillium* species in roses. *In*: 'Modern Assays for Plant Pathogenic Fungi: Identification, Detection and Quantification' (A. Schots, F.M. Dewey and R. Oliver, eds.), pp. 57-62. CAB International, Oxford, United Kingdom.

Van Ruiten, J.E.M. (1988). Production and trade of healthy propagating material: a need for a quality control system. *Acta Horticulturae* 234, 523-528.

Van Ruiten, J.E.M. (1993). Quality inspection and certification schemes for ornamental plants and fruit crops in the Netherlands. *In*: BCPC Monograph No 54: 'Plant Health and the European Single Market' (D. Ebbels, ed.), pp. 83-88. BCPC/A-AB/BSPP, United Kingdom.

Van Vuurde, J.W.L. (1990). Immunofluorescence colony staining. *In*: 'Serological Methods for Detection and Identification of Viral and Bacterial Plant Pathogens' (R. Hampton, E. Ball and S. De Boer, eds.), pp. 299-305. The American Phytopathological Society, St. Paul.

Van Zaayen, A., De Blank, C.M. and Bouwen, I. (1994). Differentiation between two potyviruses in *Alstroemeria*. *European Journal of Plant Pathology* 100, 85-90.

Van Zaayen, A., Van Eijk, C. and Versluijs, J.M.A. (1992). Production of high quality, healthy ornamental crops through meristem culture. *Acta Botanica Neerlandica* 41, 425-433.

Voller, A., Bartlett, A., Bidwell, D.E., Clark, M.F. and Adams, A.N. (1976). The detection of viruses by enzyme-linked immunosorbent assay (ELISA). *Journal of General Virology* 33, 165-167.

Walkey, D.G.A. (1985). 'Applied Plant Virology'. Heinemann, London, United

Kingdom.

Wang, P.J. and Hu, C.Y. (1980). Regeneration of virus-free plants through *in vitro* culture. *Advances in Biochemical Engineering* **18**, 61-99.

Wang, W.C., Tronchet, M., Larroque, N., Dorion, N. and ʼAlbouy, J. (1988). Production of virus-free dahlia by: meristem-culture and virus detection through cDNA probes and ELISA. *Acta Horticulturae* **234**, 421-428.

Witkowski, J.T., Robins, R.K., Sidwell, R.W. and Simon, L.M. (1972). The design, synthesis and broad spectrum antiviral activity of 1-b-D-ribafuranosyl-1,2,4,-triazole-3-carboxamide and related nucleosides. *Journal of Medicinal Chemistry* **15**, 1150-1154.

Wong, S.M., Chng, C.G., Reiser, R.A. and Horst, R.K. (1994). Further characterization of a potyvirus in Alstroemeria Endowment Series. *Acta Horticulturae* **377**, 63-71.

Zettler, F.W. and Hartman, R.D. (1987). Dasheen mosaic virus as a pathogen of cultivated aroids and control of the virus by tissue culture. *Plant Disease* **71**, 958-963.

# 7

# INDEXING SEEDS FOR PATHOGENS

C.J. Langerak, R.W. van den Bulk and A.A.J.M. Franken

*DLO Centre for Plant Breeding and Reproduction Research,
Wageningen, The Netherlands*

## I. INTRODUCTION

A worldwide increase of agricultural land destined for production of seed together with an expansion of international trade and tighter restrictions for movement of seed and germplasm have increased the interest in seed health testing. Increased knowledge about the effect of seed-borne pathogens on seed quality has contributed to this interest. Moreover, recent developments in seed technology have increased the requirement for information about the health status of seed lots. Such information is of direct relevance to seed treatments directed to the improvement of the uniformity in sowing, germination, emergence and protection against seed-borne and soil-borne pathogens.

The first records on the occurrence of pathogenic microorganisms on seeds can be dated back to the beginning of this century. In 1918, the first attempts to isolate fungi from seeds were made by Doyer (Agarwal and Sinclair, 1987). The initial objective was to explain poor germination results in the official seed testing

Advances in Botanical Research Vol. 23
Incorporating Advances in Plant Pathology
ISBN 0-12-005923-1

programmes. The first seed health testing method involved stimulation of specific disease symptoms in a germination test by adapting the incubation conditions to enhance development of seed quality-affecting fungi. Studies on the existence of such host-parasite relationships made clear that seeds often carry pathogens. These pathogens may not affect the plant in the seedling stage, but cause disease much later, sometimes even at the maturation stage of the plant. This phenomenon required another approach than the seedling test for detection of the target pathogen. Numerous plant pathologists have since worked on the development and standardization of seed health testing methods, mostly in concerted actions within the framework of the technical Plant Disease Committee (PDC) of the International Seed Testing Association (ISTA). Until the 1970s, most activities of this group of seed pathologists were focussed on the development of methods for detection of seed-borne fungi and nematodes. Later on, attention was also paid to the detection of seed-borne viruses and bacteria, which can usually cause serious disease problems at much lower infection levels. Detection of these microorganisms became possible after the development of selective isolation methods and immunoassays, the latter already in use at that time in the clinical and veterinary sectors (Van Vuurde et al., 1988).

The occurrence of seed-borne pathogens has been described by many authors. The most recent updated compilation of this information was presented by Richardson (1990) as section 1.1 of the Handbook on Seed Health Testing of ISTA. Detection methods have been described in the literature for a wide range of seed-borne pathogens. Detailed descriptions for the detection of several seed-borne fungi and some bacteria, viruses and nematodes can be found in the Working Sheets of the ISTA Handbook on Seed Health Testing (Anonymous, 1981-1994). A number of methods for the detection of fungi in economically important crops were intensively evaluated by members and co-workers of the PDC and have been incorporated in the International Rules of ISTA (Anonymous, 1993). Excellent technical information for detection of bacteria in seeds is given in Saettler et al. (1989), and for their identification by Schaad (1988a). Serological tests which can be used for detection and identification of viral and bacterial plant pathogens were described by Hampton et al. (1990). Handbooks, reviewing various aspects of seed pathology including the detection of seed-borne pathogens, have been compiled by Neergaard (1977) and Agarwal and Sinclair (1987). Compendia dealing with descriptions of seed-borne diseases in seeds of rice (Agarwal et al., 1989), chickpea (Haware et al., 1986) and wheat (Mathur and Cunfer, 1993) also provide useful information about the available seed health testing methods.

There can be various reasons to analyse seeds for the presence or absence of plant pathogens. In most cases these reasons are related to a particular stage of the seed production process, which starts the moment seeds are formed and ends when delivered to the farmer. The purpose of testing, the nature of the pathogen, and the level of contamination that can be tolerated should determine the methodology to be followed (Langerak and Franken, 1994).

The users of seed health testing methods are often not aware of the fact that analytical data do not always provide the information which actually is needed in a particular situation. Although many methods have been described in the literature and in handbooks, hardly any information is available regarding the correct way test results can or should be interpreted. The aim of this chapter is to indicate the important features of seed health testing, the various purposes of testing, which testing methods have been described in the literature, and the implications of a seed health testing method. Finally, some examples will be given regarding interpretation, reporting and evaluation of test results.

## II. IMPORTANT FEATURES OF SEED HEALTH TESTING METHODS

A seed health testing method chosen for a particular purpose (Langerak and Franken, 1994) consists of a clear protocol which describes the way concrete information can be obtained about the health or disease status of a seed lot. Such a protocol may consist of one or more individual tests which may give either a positive result indicating the presence of a pathogen in the test sample or a negative one, indicating that it is absent.

It is obvious that one test may give strictly qualitative information, whereas another one may provide quantitative or semi-quantitative data (Sheppard, 1993). The steps required in various methods usually include sampling, incubation, separation and identification. Thereafter verification may be needed to confirm the identity. Verification of the identity of a bacterium isolated from an extract by dilution plating on an agar medium, for example, may involve the application of a particular serological technique, performing one or several biochemical reactions, or testing the pathogenicity by inoculation of plant parts or whole plants grown under standardized conditions.

A seed health testing method in practice may involve the use of one single test which provides a quantitative measure of the target pathogen for which no further confirmation is needed. Generally, tests do not meet all required characteristics (Table I). It is, therefore, often more efficient and reliable to combine two or three tests, one having the best characteristics for sensitivity and suitability for mass application (screening), and the other(s) for its specificity, precision and accuracy (verification).

Minor differences in test protocol, use of equipment, and source or quality of chemicals can strongly influence the accuracy and precision of a test. It should be noted that the term 'analytical specificity' is used, for example, to define the degree of immunological cross-reactivity between similar antigens (Trigalet et al., 1978). This definition does not relate to the ability of a test to predict the level of disease in a seed lot nor to the performance of a seed lot in the field. Neither does the 'analytical sensitivity', which describes the capability of the method to detect small numbers of target cells or virus particles (Sheppard *et al*, 1986; Sheppard, 1993). Sheppard *et al.* (1986) also discussed the terms 'diagnostic specificity' and 'diagnostic sensitivity', with reference to the utility of seed health testing to measure the actual health or disease status of the seed lot or crop. Diagnostic specificity is a measure of the ability of a negative test to correctly identify healthy seed lots. Diagnostic sensitivity refers to the ability of a positive test to correctly identify a seed lot with a specified disease. These characteristics must be determined by testing seed lots with known disease status. Values for 'true posi-

*Table I.* Relative importance of laboratory test factors.

| Factor | Screening test used in combination with a verification test | Verification test used in combination with a screening test |
|---|---|---|
| Costs | +++ | + |
| Simplicity | +++ | + |
| Mass application | +++ | ++ |
| Precision | ++ | ++++ |
| Accuracy | ++ | +++ |
| Analytical specificity | ++ | ++++ |
| Analytical sensitivity | ++++ | ++ |
| Diagnostic sensitivity | ++++ | +++ |
| Diagnostic specificity | ++ | ++++ |

+ = minor, ++ = moderate, +++ = major, ++++ = crucial
After Sheppard *et al.* (1986) and Franken and Van Vuurde (1990).

tive' (TP) and 'true negative' (TN) as well as 'false negative' (FN) and 'false positive' (FP) test results must be obtained and acceptable levels must be agreed upon. Detailed descriptions of the calculation and the use of these characteristics in test method evaluations have been given by Sheppard *et al.* (1986) and Sheppard (1993) and were used by Franken (1993).

The relative importance of test factors described in Table I can be used to determine the possible combination of tests for detection of seed-borne bacteria. Examples of such combinations may include: a (semi-) selective enrichment method combined with a serological method (Taylor *et al.*, 1979), a serological method followed by dilution plating (Van Vuurde *et al.*, 1983), and application of dilution-plating in combination with DNA hybridization (Schaad *et al.*, 1989). The approach of combining two or more tests for a seed health assay is not common practice, because the crucial characteristics of the available methods such as the diagnostic specificity and sensitivity are often not known. Moreover, good screening or confirmation tests are not always available, are too expensive, or are not applicable for technical reasons. An example of such a confirmatory test is immunosorbent-electron microscopy (ISEM) as described by Van Vuurde *et al.* (1988) and Franken *et al.* (1990) for confirmation of doubtful results obtained with the enzyme-linked immunosorbent assay (ELISA) applied for the detection of seed-borne viruses. Although efforts are undertaken to develop alternative methods for detection of seed-borne fungi (Blakemore *et al.*, 1994; Husted, 1994), not many are available at this time for application on a routine basis.

## III. PURPOSE OF TESTING

Seed health testing can take place in several stages of the seed production chain as illustrated in Figure 1. The reason for testing will differ at each stage as has been described by Langerak and Franken (1994). In the production phase, information about the health or disease status of maturing seeds might be useful to support or to justify curative or preventative control measures to be undertaken prior to, or following harvest. Moreover, knowledge about the health status obtained before harvest might be used to adapt the post-harvest storage conditions and to give advice regarding processing of the seed lot. For example, one might prevent unwanted cross contamination between healthy and diseased seed lots, or one might avoid contamination of cleaning equipment.

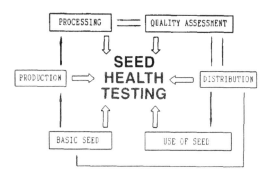

*Fig. 1.* Scheme representing the various stages of the seed production chain at which seed health testing can be useful. Each stage determines which seed health test is the most appropriate. (After Langerak and Franken, 1994).

Checks on the health status of a seed lot prior to or during the various substages of processing such as cleaning, grading, storage and seed treatment are not always made. However, such information could obviously contribute to an indirect or even direct improvement of the seed quality and is therefore important from the economic point of view as well. Having information about the level of external contamination, one could exclude certain lots from a liquid cleaning processing-step to avoid smearing of pathogens from seed to seed or from seed lot to seed lot. For some pathogens, e.g. squash mosaic virus (Middleton, 1944), it is known that their presence may be linked to the size or weight of the seed. On the basis of such knowledge one might adapt grading to sort out lighter seeds that may have a higher chance of being infected (Ryder and Johnson, 1974).

The most obvious stage for seed health testing is when the seed lot is ready for delivery and normally quality assessment takes place. The way the seed is distributed to the user and the final destination affect the choice of the seed health testing method. It is possible that based on health tests an additional chemical, physical or biological treatment is required to meet national quality standards or those of the potential user. In such cases, a subsequent test is sometimes needed to verify or to measure the efficacy of the treatment. When biologically active compounds, e.g. fungicides or bactericides, have been applied to the seed, the traditional microbiological techniques will generally not be suitable. In these cases serological or other modern molecular techniques based on DNA or RNA analysis are not suitable either, as they generally do not distinguish between dead and live cells or active and inactivated virus. A bioassay or culture method may give the best information about the viability of the seed-borne inoculum after being treated.

Often different requirements for specific pathogens are needed for seed lots which will be exported. The method of processing and the treatment that was applied will then determine whether testing for the presence of pathogens is still possible or reliable. Data from tests carried out at an early stage during seed production and processing may be helpful to convince the client that the seed lot meets all the official requirements included in phytosanitary or quarantine regulations of the importing country.

Health tests are normally not necessary once the seed has reached its final destination. Nevertheless, difficulties may arise when the seed is considered to be the source of the problem in a disease outbreak or epidemic, and it cannot be proven that the seed has been treated in such a way that the risk of disease transmission is negligible. In this respect it is often not realized that the methods based on serology (e.g. ELISA) and molecular techniques available are not suitable for application in all situations, as they do not indicate the viability of the target pathogen. A reliable method could then be based on a culture test for the pathogen, a growing-on test to examine seedlings and older plants for symptoms, or a host inoculation test showing specific symptoms of the disease (Krinkels, 1992).

In case seed is intended to be used for further multiplicaton or for long term storage in gene banks, requirements set for disease 'freedom' should be as stringent as possible. Very sensitive and, if possible, non-destructive methods have to be carried out to confirm the absence of the target pathogens.

## IV.  CRUCIAL FACTORS IN SEED HEALTH TESTING METHODS

The choice of a suitable seed health testing method to analyse a seed lot or sample depends primarily on the type of target pathogen(s), the selected optimal sampling procedure for detection of the pathogen(s), the defined tolerance levels, and the analytical and diagnostic features of a test. It is also relevant to know in advance how data obtained can be evaluated and interpreted, and what actions have to be taken on the basis of test results. Several of these factors will be discussed.

### A.  Sample Size

Samples drawn from a seed lot must be representative. Directions for correct sampling procedures are given in the ISTA International Rules for Seed Testing (Anonymous, 1993). Each seed health testing method requires decisions in advance on either the number of individual seeds to be tested or number and size of subsamples, taking into account the established tolerance level and confidence intervals. It is important in this respect to establish the probability of accepting a seed lot with disease incidence higher than the tolerable disease level, and the probability of rejecting a seed lot with disease incidence less then the tolerable disease level.

The distribution of the pathogen in a sample, the diagnostic sensitivity and specificity, and the tolerable disease levels, including 'zero tolerance', are factors which determine the correct sample size. The mathematical implications have been discussed by Geng *et al.* (1983) and Masmoudi *et al.* (1994), who based their calculations on the Poisson Distribution for a test in which the number of positive samples indicate the presence of the pathogen. Recently, Roberts *et al.* (1993) and Ridout (1994) combined both the mathematical studies based on binomial distributions and practical considerations in the development of a rational design and interpretation of seed tests for bacterial diseases. This approach might equally be applicable to fungal and viral diseases at low levels of seed infestation. It is emphasized that this approach also needs a thorough understanding of the relationship between inoculum, disease transmission, and detection thresholds.

## B. Technical and Biological Factors

Logistics and technical facilities to carry out the planned tests in a seed health testing laboratory have been described by Langerak (1988). Several factors may cause variation in test results. For instance, it is very important to dry samples with a high moisture content when the period between sampling and testing exceeds several days. Otherwise, moulds associated with such seeds may become active and interfere with the detection of the target pathogen. Seeds and associated microorganisms may respond differently under variable incubation conditions.

Instrument settings for temperature, moisture, light, and pressure must be supervised according to well defined schedules. General guidelines for materials and instruments used in microbiological studies are given in ISO 7218 of the International Organization of Standardization (Anonymous, 1985). Materials such as water, filter paper, sand, soil, nutrient media, chemicals, plates, glassware etc. must be of a standard quality and need regular checks for purity, cleanliness, and toxicity for both seeds and seed-borne organisms. Equipment needed for seed health testing must be clean and sometimes even sterile to avoid cross-contamination with the material to be incubated. Because the quality of tap water may vary over the year and will differ among various locations, it is recommended that deionized or distilled water be used. Test material of a biological nature such as antisera, test plants, reference seed samples, and pathogen isolates need special attention, because there is considerable risk of quality loss over time. Again regular checks have to be carried out according to a fixed scheme.

One of the most important factors in performing seed health tests is the human one. The choice of a test may depend on the skill of the technicians performing the tests. The skills of a technician determine to a large extent the success of test standardization. This especially holds for methods in which subjective assessment is required for characteristics of a biological, and thus variable, nature, such as symptoms, colonies of bacteria and fungi, and fructifications of fungi. It is recommended that such tests are carried out at least in duplicate and are assessed by two technicians. Criteria should be established with respect to accuracy, defined as the degree of similarity between the measurement and the true value of the measured quality, and the performance of the test according to ISO/DIS 6107-8 (Anonymous, 1991a).

## C.  Standardization

The specificity and sensitivity of a seed health assay are of major importance. It is necessary that methods be performed with the same accuracy and precision within and between laboratories to obtain comparable results in repeated testing. However, it is known that application of standardized methods in various laboratories tends to generate different results. This was for instance shown in comparative tests for detection of *Xanthomonas campestris* pv. *campestris* (Schaad, 1988b) and *Pseudomonas syringae* pv. *phaseolicola* (Van Vuurde and Van den Bovenkamp, 1984). The differences found between, but also within laboratories may be due to several factors such as sample handling, heterogeneity of the sample, human factors, technical factors, and the analysis of data (Franken and Sheppard, 1993).

The precision of a method is determined by its reproducibility (between different laboratories) and repeatability (within a laboratory), expressed as the closeness of agreement between mutually independent test results obtained under stipulated conditions, as described in ISO 5725 (Anonymous, 1986). To study the repeatability, the operator and the environment are considered constant factors which do not contribute to the variability. When studying the reproducibility, the operator and environment contribute to the variability  according to ISO 5725. The reproducibility of test methods may vary and some methods may become more or less sensitive to test conditions. It is essential therefore to repeat testing on well-stored reference samples in order to evaluate the precision of the test results. The cause of variation should be found and eliminated if reproducibility is lower than originally described and established for the method. This will lead to improvement  and further standardization of the method. Disagreements on whether a method is optimal for use in seed health testing are generally not due to lack of either analytical or diagnostic sensitivity and specificity, but due to a lack of precision.

## D.  Relevance of Disease

The status of a seed-borne disease, viz. quarantine or quality disease, also determines the technical and biological characteristics needed for a seed health assay.  Acceptance of a pre-defined level of a pathogen in a sample of a certain size is a typical feature of a quality disease. Rejection on the basis of a 'zero-tolerance' generally deals with tests applied for quarantine seed-borne diseases. In the latter case, the presence of dead or viable propagules of the pathogen is not tolerated in the sample submitted for testing. A high diagnostic specificity of the seed health assay is required in the situation of quality diseases, because false positive results cannot be tolerated. When testing for quality diseases, it is important to be informed about the relation between the results of the laboratory method and those obtained in glasshouse or field tests. On the basis of such information one should be able to control the disease by adequate measures. For a quarantine disease, there is no need for high diagnostic specificity, because false positive results are acceptable. The seed lot will be rejected regardless of whether positive results are false (not giving disease in the field or glasshouse) or true (giving disease in the field or glasshouse), and regardless of the percentage of samples giving disease. However, a high diagnostic sensitivity is required for a quarantine disease, because false negative results are not acceptable. Because the risk attributed to quarantine diseases is generally larger, the necessity of looking for small numbers of virus particles, bacterial cells or fungal spores is also higher. This means that the analytical sensitivity often needs to be higher.

## E. Interpretation of Test Results

Test results have to be interpreted and reported, depending on the purpose of the seed health assay (Langerak and Franken, 1994). Most seed health testing methods for fungi, such as plating methods, allow individual inspection of seeds in a representative sample drawn from the seed lot and provide the user directly with qualitative information about the identity of the target pathogen and quantitative information of the percentage of seed infection. Washing tests for fungi, involving the examination of soaking fluids from seed samples for the presence of fungal propagules, and almost all methods described for detection of bacteria and viruses do not allow an assessment of individual seeds, but require an analysis of a certain number of sub-samples, each consisting of an equal number of seeds. Most Probable Number (MPN) or preferably 'maximum likelihood' methods are then needed to estimate percentage of infection as has been described by Taylor and Phelps (1984) and Taylor *et al.* (1993). Ridout (1994) and Roberts *et al.* (1993) improved this method by using the concept of a binomial distribution for occurrence of the pathogen in the sub-samples. The advantage of the latter approach is that higher infection levels can be more accurately estimated with narrower confidence intervals.

Test results can be expressed in several ways, such as the percentage infected or contaminated seeds, the number of fungal spores or colony-forming units per gram or number of seeds tested, the presence or absence of the pathogen in a working sample or number of subsamples, etc. Data reported as: "no .... has been found in the sample tested", "the pathogen has been found in the sample tested" or "the estimated percentage of seed infection is ....%" mostly provide information about whether the pathogen is present as seed-borne inoculum and not about the percentage of seed-transmission. It is difficult to give an absolute indication of the predictive value of a seed health method as such values depend on the prevalence of a disease, given a specific diagnostic sensitivity and specificity of the method applied. The prevalence of disease has been defined as: 'the amount of disease in a specific population at a given point of time' and has a direct relation to the field incidence of a disease according to Sheppard *et al.* (1986) and Sheppard (1993).

## F. Comparison of Diagnostic Values

An example to illustrate diagnostic values of seed tests was worked out by Franken (1993) on the basis of data obtained by van Vuurde *et al.* (1991) for an immunofluorescence microscopy (IF) test and a dilution plating (DP) test applied for detection of *P. syringae* pv. *phaseolicola* in 22 bean samples, each divided in five sub-samples of 1000 seeds (Table II). The disease incidence in the field was determined for the 22 seed lots, assuming that the appearance of symptoms in the field was indicative for true positive seed lots. For the IF-test it was found that 7 samples gave a true negative result (TN=7); ten samples gave a true positive result (TP=10) and five samples gave a false positive result (FP=5), implying that no false negatives were obtained (FN=0). For the DP-test the results were: TN=12, TP=7, FP=0 and FN=3.

The values for the diagnostic sensitivity and diagnostic specificity were calculated using the formulae given by Sheppard (1993; see Appendix). It is clear that Van Vuurde *et al.* calculated higher values for the diagnostic specificity of IF and the diagnostic specificity of DP. Their calculations were based on an expected 10% disease transmission for the disease according to Taylor *et al.* (1979), and the number of infected seeds present in the seed samples estimated with MPN tables (Taylor and Phelps, 1984). In such situations only the heavily infected seed lots tested would give halo blight in 10 000 plants.

Based on the field incidence of halo blight, a prevalence of 45.5% was calculated by using Sheppard's (1993) formula. Predictive values of the IF-test and the DP-test for detecting *P. syringae* pv. *phaseolicola* (Table III) were calculated for several levels of prevalence on the basis of the diagnostic values given in Table II. Higher predictive values of positive IF tests and of negative DP tests were obtained using the data presented by Van Vuurde *et al.* (1991) as a result of their higher diagnostic values. Nevertheless, the conclusion remains the same, i.e. at

*Table II.* The diagnostic sensitivity and specificity of an immunofluorescence test and a dilution plating test, calculated from data of van Vuurde *et al.* (1991).

|                        | Immunofluorescence | Dilution plating |
| ---------------------- | ------------------ | ---------------- |
| Diagnostic specificity | 58.3 (89) %[a]     | 100 (100) %      |
| Diagnostic sensitivity | 100 (100) %        | 70 (78) %        |

[a]Between brackets: values given by Van Vuurde *et al.* (1991) on basis of an expected 10% disease transmission in the field.

*Table III.* The predictive values of an immunofluorescence (IF) test and a dilution plating (DP) test for detecting *P. syringae* pv. *phaseolicola* in bean seeds, in relation to disease development in the field for 10,000 plants per sample (calculated from Van Vuurde *et al.*, 1991).

| Prevalence (%) | Predictive value (%) of a positive test result (halo blight) | | Predictive value (%) of a negative test result (no symptoms) | |
|---|---|---|---|---|
| | IF | DP | IF | DP |
| 1.0 | 2.4  (8.4)[b] | 100  (100) | 100  (100) | 99.7  (99.8) |
| 20.0 | 37.5  (69.4) | 100  (100) | 100  (100) | 92.3  (94.8) |
| 45.5[a] | 66.7  (88.4) | 100  (100) | 100  (100) | 78.3  (84.5) |
| 50.0 | 70.6  (90.0) | 100  (100) | 100  (100) | 75.0  (82.0) |

[a]Prevalence in paper by Van Vuurde *et al.* (1991) based on symptoms in the field
[b]Between brackets: the values based on the values for diagnostic sensitivity and specificity, given by Van Vuurde *et al.*(1991). See also Table II.

low prevalence of disease (1%), the predictive value of a positive IF test is relatively low, whereas at a high prevalence the predictive value of a negative DP test is relatively low. A negative IF result has a high predictive value, whereas a positive DP result has a high predictive value. In practice this means that IF is a good method to identify 'healthy' samples, and DP is a good method to predict disease in the field.

Sheppard *et al.* (1986) demonstrated that the predictive value of a positive result of a seed health method in the development stage may be relatively good, because experiments conducted in this stage mostly deal with artificially created situations of high disease prevalence. In practice, however, the method will often have lower predictive values of a positive result, because of a much lower natural prevalence of the disease. Therefore, the principles of diagnostic sensitivity, diagnostic specificity, and prevalence of a disease should be applied to any test method under development in order to measure the predictive value of such a test.

As mentioned before, the interpretation of test results may be more difficult when a combination of two tests is used. The problem of dealing with a negative result obtained by one test and a positive result obtained by another test was also studied for the detection of *P. syringae* pv. *phaseolicola* using IF and DP. The probability of isolating the target bacterium from bean seeds after dilution plating increases with increasing numbers of fluorescent cells observed in IF (Table IV). This probability will increase further when two or more plating media are used. A similar correlation was found for detecting *Xanthomonas campestris* pv. *campestris* in crucifer seeds (Franken, 1992a).

In the Netherlands, risk classes were established for IF-positive and DP-negative bean seed lots tested for the presence of *P. syringae* pv. *phaseolicola* (Van Vuurde *et al*, 1991). This classification per seed lot was based on the number of IF-positive subsamples per lot (giving an estimation of the percentage of infested

*Table IV.* The correlation between immunofluorescence (IF) and dilution plating (DP) methods for detecting *P. syringae* pv. *phaseolicola* in bean seed (adapted from Van Vuurde *et al.*, 1991).

| | Number of IF-positive cells in 100 fields[a] | | | | | |
|---|---|---|---|---|---|---|
| | 1-4 | 5-20 | 21-100 | 102-500 | 501-2500 | >2500 |
| Percent positive in IF and DP | 8.8 | 16.7 | 30.0 | 35.4 | 65.2 | 76.2 |

[a] For detecting *P. syringae* pv. *phaseolicola* a specificity-tested polyclonal antiserum was used.

seeds), the number of fluorescent cells found in IF for each subsample, and the number of saprophytes found on King's medium B for those positive-IF subsamples. At high numbers of fluorescent cells, the probability of isolating the pathogen is known (Table IV). The number of saprophytes found on King's medium B gives an indication of the probability of failure of the dilution-plating method caused by overgrowth of the target bacterium by saprophytes. The probability that viable cells of the target pathogen are present in a seed lot is low, when low numbers of fluorescent cells in IF, e.g. less than 4 cells per 100 microscope fields, and no colonies on King's medium B are found. The probability that viable cells of the target bacterium are present in a seed lot is much higher, when both high numbers of fluorescent cells in IF (e.g. more than 500 cells/100 microscope fields) and many saprophytes on King's medium B (e.g. more than 150 colonies/plate in the undiluted extract) are found. Similar approaches can be used for other verification tests. It has to be emphasized that only a relatively high or low risk can be indicated. Whether the risk is acceptable is up to the user of the test result. When the user prefers a high level of confidence for detecting the pathogen, the diagnostic sensitivity of the test should be increased, for example by testing two or more media in DP or increasing the number of subsamples to be tested (Geng *et al.*, 1983).

## V. SEED HEALTH TESTING METHODS

Seed health testing methods have been developed for different purposes, which consequently influence their applicability for routine testing. For example, a detection method to study the mechanism of disease transmission via seed for a particular pathogen may be useful for that research, but will not automatically be suitable for large scale application in a seed health testing programme. Attention has been paid to this aspect in reviewing and selecting the literature for methods which might be useful for routine indexing of seeds for seed-borne pathogens.

Guidelines for international standardization of seed health testing methods were drafted at the first Workshop on Seed Health Testing of the technical Plant Disease Committee (PDC) of ISTA, held in Cambridge in 1958. The methods in use at that time dealt only with the detection of seed-borne fungi. Since this meeting, various seed health testing methods have been evaluated on the basis of data obtained in comparative testing programmes of the PDC. The first compilation of evaluated methods was presented by Noble (1965). Descriptions of these methods have been revised since then, and are, together with a number of newly evaluated methods, published by ISTA in so-called Working Sheets (Anonymous, 1981 - 1994). These Working Sheets, describing seed health testing methods for individual pathogens separately for each host, constitute Section 2 of the ISTA Handbook on Seed Health Testing. The seed health testing methods dealing with detection of fungi, bacteria, viruses and nematodes have been summarized in Table V. Prerequisite for inclusion of a test is a thorough evaluaton of the test with regard to its reproducibility within and between laboratories, and publication in either a scientific journal or in technical reports of PDC Workshops or Seminars.

The Working Sheets provide the user with detailed descriptions of one or more tests for each host-pathogen combination and references to the literature. The descriptions generally include an indication of the number of seeds, size or weight of working samples which are normally analyzed, and recommendations regarding chemical or physical treatment of the seed prior to incubation. Detailed instructions are given with regard to the most important incubation factors, such as the substrate, chemicals, incubation conditions, incubation time, and the need for exposure to light. On the basis of experience obtained in comparative testing programmes, special attention is paid to those factors which are important in reducing development of non-target organisms, thus providing the best conditions for examination of the incubated seed. Photographs and drawings are added on a number of sheets in order to facilitate the examination as much as possible. In the near future, attention will be paid in the Working Sheets to the diagnostic features and precision of the tests, especially in those cases where more than one test is described for a host-pathogen combination. In addition to the seed health methods mentioned in the Working Sheets, many publications describe tests for detection of fungi (Table VI), bacteria (Table VII) and viruses (Table VIII). Only the most useful publications have been included.

## A. Detection of Seed-Borne Fungi

It is evident from Tables V and VI that several approaches are available for detection of seed-borne fungi in seeds. The most simple procedure is inspection of the seed for characteristic visible signs of the presence of a target pathogen. Such signs can be a characteristic discoloration, malformation, or presence of typical fruiting bodies. Because there is a risk that symptomless seeds also may be infected, an additional test may be required. When direct inspection results in the observation of typical structures of the pathogen, such as pycnidia of *Septoria apiicola* on the seed coat of celery, it may also be necessary to verify the identity of the fungus. Furthermore, it might be necessary to assess the viability of conidia or other fungal propagules present on the seed. For example, an additional test for *Septoria* on celery to determine the viability of conidia released during seed

Table V. List of available Working Sheets from the ISTA Handbook on Seed Health Testing (Anonymous, 1981-1994), describing detection methods for seed-borne fungi, bacteria and viruses.

| Host | Pathogen | Method | Author(s), year | ISTA Number |
|---|---|---|---|---|
| Avena sativa | Pyrenophora avenae | agar and blotter test | Rennie and Tomlin, 1984 | 3 |
| Beta vulgaris | Pleospora betae | agar test | Mangan, 1982 | 49 |
| Cicer arietinum | Ascochyta rabiei | blotter test | Mathur, 1981 | 38 |
| Coriandrum sativum | Protomyces macrosporus | direct inspection, washing test | Mathur, 1981 | 41 |
| Cruciferae | Xanthomonas campestris | dilution plating of seed washings | Schaad, 1982 | 50 |
| Daucus carota | Alternaria dauci | agar and blotter test | Gambogi, 1987 | 4 |
| | Alternaria radicina | agar and blotter test | Gambogi, 1987 | 5 |
| Eleusine coracana | Cochliobolus nodulosa | blotter test | Mathur, 1981 | 36 |
| | Pyricularia grisea | blotter test | Mathur, 1981 | 37 |
| Fabaceae | Pseudomonas phaseolicola | IF combined with dilution plating | van Vuurde and van den Boyenkamp, 1987 | 65 |
| Glycine max | Diaporthe phaseolorum | agar test | Kulik, 1985 | 27 |
| | Cercospora kikuchi | agar test | Kulik, 1985 | 34 |
| | Peronospora manshurica | direct inspection, washing test | Hansen and Mathur, 1987 | 64 |
| Gossypium spp. | Colletotrichum gossypii | growing-on test | Meiri, 1981 | 42 |
| | Xanthomonas malvacearum | growing-on test | Meiri and Volcani, 1981 | 43 |
| Gramineae | Acremonium coenophialum | NaOH-soak test | Welty and Rennie, 1985 | 55 |
| Helianthus annuus | Botrytis cinerea | blotter test | Anselme and Champion, 1981 | 44 |
| Hordeum vulgare | Pyrenophora graminea | agar, blotter, growing-on test | Rennie and Tomlin, 1984 | 6 |
| | Ustilago nuda | embryo test | Rennie, 1982 | 48 |
| Lactuca sativa | Lettuce mosaic virus | growing-on test, indicator plant test | Rohloff and Marrou, 1981 | 9 |
| Linum usitatissimum | Botrytis cinerea | agar test | Anselme and Champion, 1981 | 10 |
| | Alternaria linicola | agar and blotter test | Malone, 1982 | 46 |
| | Phoma exigua | agar and blotter test | Malone, 1982 | 47 |
| Lolium spp. | Gloeotinia temulenta | soaking, droplet test | Matthews, 1981 | 35 |
| Lycopersicon esculentum | Fusarium oxysporum | agar test | Gambogi, 1987 | 58 |

| Host | Pathogen | Test | Reference | No. |
|------|----------|------|-----------|-----|
| Oryza sativa | Pyricularia oryzae | blotter test | Mathur, 1981 | 12 |
| | Monographella albescens | blotter test | Mia et al., 1985 | 62 |
| Pennisetum thyphoïdes | Sclerospora graminicola | washing and NaOH-soak test | Shetty and Mathur, 1985 | 40 |
| Phaseolus vulgaris | Colletotrichum lindemuthianum | blotter test | Anselme and Champion, 1981 | 45 |
| Picea spp. | Caloscypha fulgens | agar test | Sutherland, 1987 | 63 |
| Pinus spp. | Fusarium moniliforme | agar and blotter test | Anderson, 1987 | 56 |
| Pisum sativum | Ascochyta pisi | agar test | Hewett, 1987 | 16 |
| Sesamum indicum | Alternaria spp. | blotter test | Yu et al., 1987 | 59-61 |
| | Macrophomina phaseolina | blotter test | Meiri, 1981 | 26 |
| Triticum aestivum | Leptosphaeria nodorum | agar and blotter test | Kietreiber, 1984 | 19 |
| | Tilletia spp. | washing test | Kietreiber, 1984 | 53 |
| | Ustilago nuda (tritici) | embryo (staining) test | Rennie, 1982 | 48 |
| Valerianella locusta | Perenospora valerianellae | washing test | Champion, 1984 | 51 |
| Zea mais | Cochliobolus heterosthrophus | blotter test | Mathur, 1981 | 39 |
| Zinnia spp. | Alternaria zinniae | blotter test | Gambogi, 1982 | 32 |

Table VI. Detection methods used in research or routine testing of seeds for seedborne fungi on major crops.

| Pathogen | Crop | Method[a] | Reference(s)[b] |
|---|---|---|---|
| Acremonium coenophialum | tall fescue | NaOH-soak/staining test<br>ELISA<br>immunoblot test | Welty and Rennie, 1985 (WS-55); Welty et al., 1986<br>Johnson et al., 1983<br>Gwinn et al., 1991 |
| Alternaria brassicicola | crucifers | agar test<br>blotter test | Wu and Lu, 1984<br>Bassey and Gabrielson, 1983 |
| A. dauci<br>A. linicola<br>A. padwickii<br>A. radicina | carrot<br>flax, linseed<br>rice<br>carrot | agar and blotter test<br>agar and blotter test<br>agar and blotter test<br>agar and blotter test | De Tempe, 1968; Gambogi, 1987 (WS-4)<br>De Tempe, 1963; Malone, 1982 (WS-46)<br>Shetty and Shetty, 1988; Agarwal et al., 1989<br>De Tempe, 1968; Gambogi, 1987 (WS-5);<br>Pryor et al., 1994 |
| A. longissima, sesami, sesamicola<br>A. zinniae | sesame<br>Zinnia spp. | blotter test<br>blotter test | Yu et al., 1982, 1987 (WS 59-61)<br>Gambogi et al., 1976; Gambogi, 1982 (WS-32) |
| Ascochyta fabae<br>A. pisi<br>A. rabiei | lentil<br>pea<br>chickpea | agar test<br>agar test<br>blotter test | Morrall and Beauchamp, 1988<br>Hewett, 1987; Hewett, 1987 (WS-16)<br>Maden et al., 1975; Mathur, 1981 (WS-38) |
| Botrytis allii<br>B. cinerea | onion<br>flax, linseed | agar test<br>agar and blotter test | Maude and Presly, 1977; Kritzman and Netzer, 1978<br>De Tempe, 1963; Anselme and Champion, 1981<br>(WS-10) |
|  | sunflower | blotter test | Anselme and Champion, 1975, 1981 (WS-44) |
| Cercospora kikuchi | soybean | agar test, blotter test | Kulik, 1985 (WS-34); McGee et al., 1980 |
| Cochliobolus sativus<br>C. nodulosus (Bipolaris nodulosa)<br>C. miyabeanus (Bipolaris oryzae)<br>C. heterostrophus | barley<br>finger millet<br>rice<br>maize | blotter test<br>blotter test<br>blotter test<br>blotter test | De Tempe, 1970a<br>Mathur, 1981 (WS-36)<br>Agarwal et al., 1989<br>Singh et al, 1974; Mathur, 1981 (WS-39) |
| Colletotrichum gossypii | cotton | agar and growing-on test | Veliri, 1981 (WS-42); Da Cruz Machado<br>and Langerak, 1993 |
| C. lindemuthianum | bean | blotter test | Anselme and Champion, 1981 (WS-45) |
| Diaporthe phaseolorum | soybean | agar test | Kulik, 1985 (WS-27) |

| | | |
|---|---|---|
| *Didymella bryoniae* | cucurbits | agar test, blotter test | Lee *et al.*, 1984 |
| *Fusarium* spp. | barley, chickpea, lentil | agar and blotter test | Jorgensen, 1983; Diekmann and Assad, 1989 |
| | cereals | agar and blotter test | De Tempe, 1970b |
| *F. moniliforme* | wheat | agar and blotter test | Shaarawy *et al.*, 1991 |
| | sorghum | agar and blotter test | Wu and Mathur, 1987 |
| *F. moniliforme* var. *subglutinans* | onion | agar and blotter test | Vannacci, 1981 |
| *F. oxysporum* | tomato | agar test | Gambogi, 1987 (WS-58) |
| *Gloeotinia temulenta* | grasses | soaking test, agar test | Matthews, 1981 (WS-35); Alderman, 1991 |
| *Leptosphaeria maculans* | crucifers | blotter test | Maguire *et al.*, 1978; Maguire and Gabrielson, 1983 |
| *L. nodorum* | wheat | agar, blotter, growing-on and fluorescence test | Kietreiber, 1984 (WS-19); Rennie and Tomlin, 1984; Jorgensen, 1989 |
| | barley, wheat | agar-fluorescence test | Manandhar and Cunfer, 1991; Cunfer and Manandhar, 1992 |
| *Macrophomina phaseolina* | sesame | blotter test | Meiri, 1981 (WS-26) |
| *Monographella albescens* | rice | blotter test | Mia *et al.*, 1985, 1987 (WS-62); Agarwal *et al.*, 1989 |
| *M. nivalis* | wheat | agar and blotter test | De Tempe, 1970b; Cristani, 1992 |
| *Peronospora manshurica* | soybean | washing test | Hansen and Mathur, 1987 (WS-64) |
| *P. valerianellae* | corn salad | washing test | Champion and Mecheneau, 1979; Champion, 1984 (WS-51) |
| *P. farinosa* | spinach | washing test | Inaba *et al.*, 1983 |
| *Perenosclerospora sorghi* | maize, sorghum | NaOH-maceration | Prabhu *et al.*, 1984; Rao *et al.*, 1985 |
| *Phoma exigua* var. *linicola* | flax, linseed | agar and blotter test | De Tempe, 1963; Malone, 1982 (WS-47) |
| *Phomopsis* spp. | soybean | blotter test ELISA, immunoblot test | McGee *et al.*, 1980 Gleason *et al.*, 1987 |
| *Pleospora betae* | sugar beet | agar and blotter test | De Tempe, 1978; Mangan, 1982 (WS-49); Mangan, 1983 |

| | | | |
|---|---|---|---|
| *Pyrenophora avenae*, *P. graminea*, *P. teres* | oats, barley | agar and blotter test, agar, blotter and growing-on test | Rennie and Tomlin, 1984 (WS-3); De Tempe, 1964; Jorgensen, 1980; Jorgensen 1982a,b; Rennie and Tomlin, 1984 (WS-6) |
| *Pyricularia grisea* | finger millet | blotter test | Rarganathaiah and Mathur, 1978; Mathur, 1981 (WS-37); |
| *P. oryzae* | rice | blotter test | Mathur, 1981 (WS-12); Agarwal *et al.*, 1989 |
| *Sclerospora graminicola* | pearl millet | NaOH-soak test | Shetty *et al.*, 1978; Shetty and Mathur, 1985 (WS-40) |
| *Septoria apiicola* | celery | seedling inoculation test | Hewett, 1968 |
| *Tilletia Indica* | rice, wheat | NaOH soak test, NaOH soak test and washing test | Agarwal *et al.*, 1989; Agarwal *et al.*, 1993 |
| *T. caries*, *T. controversa* | wheat | washing test | Kietreiber, 1984 (WS-53); Castro *et al.*, 1994 |
| *Ustilago nuda*, *U. nuda* (*U. tritici*) | barley, wheat | embryo test, embryo (staining) test | Hewett and Damgaci, 1986; Rennie, 1988 (WS-25) Rennie, 1982 (WS-48) |

[a]ELISA, Enzyme-linked immunosorbent assay

[b]References followed by a WS (Working Sheet) number between parentheses refer to the ISTA Handbook on Seed health testing, section 2: Working sheets, each dealing with one pathogen on one host (Anonymous, 1981-1994).

washing is recommended by Hewett (1968).

Seed washing may be included in a direct test and involves placing individual seeds or portions of seeds in water or water plus detergent to promote release of spores or conidia. Staining techniques may be needed to aid distinction between closely related species such as *Tilletia spp.* in wheat (Trione and Krygier, 1977; Stockwell and Trione, 1986). Another direct principle used for detection of seed-borne fungi that invade the seed, such as *Ustilago nuda* and *Acremomium coenophialum*, deals with visual inspection of internal parts of the seed after separation, clarification and, if necessary, staining of mycelium fragments in the seed tissue. These tests are in general time consuming, but give rather good information about the presence of the pathogen. However, no information is obtained about the viability of the target pathogen. Again, additional tests, such as a growing-on test, will offer such information. For infections of wheat and barley with *Ustilago nuda*, such a growing-on test takes several months in order to get the first loose smut symptom indicating disease transmission. For *Acremonium coenophialum*, viability may also be estimated by the use of a tissue culture assay (Conger and McDaniel, 1983).

Indirect detection methods are predominantly used for detection of seed-borne fungi. They involve incubation of individual seeds on a substrate such as filter paper (blotter test) or agar media (agar test), or, more rarely, in a substrate such as sand or soil (growing-on test or soil/sand test).

In one variant of the blotter test, seeds are placed on top of or between several layers of moist paper and incubated for some time either in darkness or exposed to a 12 hour photoperiod. After incubation, either specific symptoms develop on the seedlings, provided the seed is able to germinate, or presence of the target pathogen is assessed on the basis of characteristic fructifications developing on infected tissue or on the seed coat. In the latter case, microscopic inspection will aid the final examination and identification.

Alternatively, germination of the seed may be prevented in the blotter test, in order to create conditions for development of mycelium or spores of the pathogen on the seed itself. Germination may be prevented by addition of a herbicide such as 2,4-D to the substrate, or by freezing the moist blotter with seeds at - 20°C after pre-imbibition at 20°C. In both cases development of characteristic mycelium and/or fructifications will occur directly on the seed surface, especially when during the final incubation period the seeds are exposed to alternating light (fluorescent white or near ultra violet) and dark periods. Microscopic inspection at magnifications of 12-50 x will aid the final examination and identification. In both blotter tests, the regime of temperature and light may be varied, the degree of saturation of the substrate with water may be optimized, and selective agents can be added.

Various agar tests also offer an indirect way of testing. As described for the blotter tests, seeds are plated directly on solid agar media containing nutrients and additives to promote outgrowth of the target pathogen from the seed. Incubation conditions, such as temperature and exposure to light, generally determine the extent of characteristic colonization and sporulation of the pathogenic fungi, allowing either direct recognition by visual inspection or microscopic examination. Agar tests, in contrast to blotter tests, often require pretreatment of the seeds with a hypochlorite solution or heat treatment to reduce fast growing saprophytes from colonizing the substrate and overgrowing the pathogens. Such pretreatments may enhance the specificity of the test, but may also reduce sensitivity as superficial inoculum of the pathogen may be killed.

## B. Detection of Seed-Borne Bacteria

Direct inspection of seeds for the presence of seed-borne bacteria is possible in some cases, as has been described for beans infected with *Pseudomonas syringae* pv. *phaseolicola* showing fluorescing spots on the seed coat under ultraviolet light (Wharton, 1967). However, the reliability is generally low and only applicable for bean cultivars with white seed coats. Other direct and indirect tests as described for fungi, such as microscopic examination of seed washings and blotter or agar tests are not suitable for detection of bacteria. One exception described by Agarwal *et al.* (1989) for *Pseudomonas fuscovaginae* on rice seed involves the plating of washed seeds on King's medium B and the detection of fluorescent colonies under UV light after 1-2 days.

The detection of pathogenic bacteria first involves separation of the bacteria from seed, which can be achieved by soaking, washing, or extraction after crushing or maceration of the seed. Once the bacteria are released from the seed, the seed extract can be analyzed for the presence of pathogenic bacteria. To increase the sensitivity of a method, pre-enrichment can be used, for example in combination with IF or DP (Bashan and Assouline, 1983; Naumann *et al.*, 1986; Wong, 1991).

Several approaches have been described (Table VII) for routine seed health testing. IF or ELISA tests on seed extracts have the advantage of being quick and rather cost effective, however, a fluorescence microscope is required for IF, and antisera with high specificity and sensitivity are essential. A disadvantage of IF and ELISA is that these tests do not differentiate between dead and living bacterial cells. An alternative test, which is often applied, is dilution plating (DP), in which case an aliquot of undiluted or diluted seed extract is plated onto a (semi)-selective agar medium. Such DP tests have the advantage that living bacterial cells are detected as colony-forming units. Disadvantages of plating tests are the duration of the test, and possible growth inhibition of the target bacterium by other bacteria present in the seed extract or by toxicity of the extract itself. In addition, verification of suspect colonies is almost always necessary. Verification may include serological, biochemical or pathogenicity tests. For pathogenicity tests, bacteria from suspect colonies are used to inoculate seedlings or plant parts, which show specific symptoms after incubation. Pathogenicity tests may also be used directly as a detection method (seedling inoculation test). The seed extract is then used, either directly or after enrichment, for inoculation of the host plant or plant parts (Fett, 1979; Webster *et al.*, 1983; Mehta, 1990). Growing-on tests involve the incubation of seeds in sand or soil and examination of the resulting seedlings and plants for specific symptoms. This method does give information about the transmission of the pathogen from seed to seedling/plant, but is rather time-consuming and may be replaced in the future by faster methods as described above.

Table VII. Detection methods used in research or routine testing of seeds for seedborne bacteria on major crops.

| Pathogen | Crop | Method[a] | Reference(s)[b] |
|---|---|---|---|
| Clavibacter michiganensis subsp. insidiosus | lucerne | IF combined with dilution plating | Nemeth et al., 1991 |
| Clavibacter michiganensis subsp. michiganensis | tomato | IF combined with seedling inoculation test<br>dilution plating | van Vaerenbergh and Chauveau, 1987<br>Fatmi and Schaad, 1989; Kritzman, 1991;<br>Gitaitis, 1990 |
| | | IF combined with dilution plating | Franken et al., 1993; Anonymous, 1992a |
| Curtobacterium flaccumfaciens | bean | IF, seedling inoculation test<br>growing-on test combined with IF | Calzolari et al., 1987<br>Diatloff et al., 1993 |
| Erwinia stewartii | maize | ELISA + IF + dilution plating (combination)<br>ELISA | EPPO Quarantine Procedure 91/3066<br>Lamka et al., 1991 |
| Pseudomonas avenae | rice | growing-on (= seedling symptom) test | Agarwal et al., 1989 |
| Pseudomonas fuscovaginae | rice | plating whole seeds on selective medium | Agarwal et al., 1989 |
| Pseudomonas glumae | rice | dilution plating | Agarwal et al., 1989 |
| Pseudomonas syringae pv. glycinea | soybean | growing-on test<br>enrichment followed by plant inoculation | Parashar and Leben, 1972<br>Fett, 1979 |
| Pseudomonas syringae pv. phaseolicola | bean | IF combined with dilution plating<br>dilution plating | van Vuurde and van den Bovenkamp, 1987<br>Mohan and Schaad, 1987 (WS-65), 1989;<br>van Vuurde et al., 1991 |
| | | dilution plating, confirmation with<br>toxin bioassay | Jansing and Rudolph, 1990 |
| Pseudomonas syringae pv. pisi | pea | IF combined with dilution plating<br>dilution plating | Franken and van den Bovenkamp, 1990<br>Taylor, 1984; Ball and Reeves, 1991 |
| Pseudomonas syringae pv. tomato | tomato | growing-on test<br>plating enriched seed extract | Jones et al., 1989<br>Bashan and Assouline, 1983; Naumann et<br>al., 1986 |

| Organism | Host | Method | Reference[b] |
|---|---|---|---|
| Xanthomonas campestris pv. campestris | cabbage | dilution plating | Kritzman, 1991 |
| | | dilution plating | Schaad, 1982 (WS-50), 1989; Chang et al., 1991; Franken et al., 1991 Franken, 1992a,b |
| | | dilution plating combined with IF | |
| Xanthomonas campestris pv. carotae | carrot | dilution plating | Kuan, 1989 |
| Xanthomonas campestris pv. malvacearum | cotton | growing-on test | Meiri and Volcani, 1981 (WS-43); Halfon-Meiri and Volcani, 1977 |
| Xanthomonas campestris pv. oryzae, pv. oryzicola | rice | growing-on test<br>direct IF<br>dilution plating | Agarwal et al., 1989<br>Unnamalai et al., 1988<br>Ming et al., 1991 |
| Xanthomonas campestris pv. phaseoli (including var. fuscans) | bean | dilution plating<br><br>IF<br>IF combined with dilution plating<br>IF with enriched seed extract | Sheppard et al., 1989; Dhanvantari and Brown, 1993<br>Malin et al., 1983<br>van Vuurde et al., 1983<br>Wong, 1991 |
| Xanthomonas campestris pv. translucens | wheat, barley | dilution plating<br>dilution plating, confirmation by DIBA | Schaad and Forster, 1989; Duveiller, 1989<br>Claflin and Ramundo, 1987 |
| Xanthomonas campestris pv. undulosa | wheat, barley, rye | seedling inoculation test<br>dilution plating<br>IF<br>ELISA with enriched seed extract | Mehta, 1990<br>Duveiller, 1990<br>Duveiller and Bragard, 1992<br>Frommel and Pazos, 1994 |
| Xanthomonas campestris pv. vesicatoria | tomato, pepper | dilution plating<br><br>IF combined with dilution plating<br>plating enriched seed extract | McGuire and Jones, 1989; Kritzman, 1991; Sijam et al., 1991, 1992; Gitaitis et al., 1991<br>Anonymous, 1992b<br>Bashan and Assouline, 1983 |

[a] DIBA, Dot-immunobinding assay; ELISA, Enzyme-linked immunosorbent assay; IF, Immunofluorescence
[b] References followed by a WS (Working Sheet) number between parentheses refer to the ISTA Handbook on Seed health testing, section 2: Working sheets, each dealing with one pathogen on one host (Anonymous, 1981-1994).

## C. Detection of Seed-Borne Viruses

Routine indexing of seed lots for virus infections has a rather short history when compared with testing for fungi. Detection of viruses in seeds generally differs from detection of fungi and bacteria, because viruses need living plant cells to multiply. This implies that isolation media, as described for fungi and bacteria, cannot be used. It is logical in this respect that the first routine methods developed were growing-on tests and indicator plant tests. For example, for lettuce mosaic virus such a test has been described (Rohloff and Marrou, 1981), and has been standardized internationally via the PDC of ISTA.

A replacement of these rather laborious methods took place in the seventies when the first serological tests were introduced. The applicability of the latex agglutination test for detection of barley stripe mosaic virus (Lundsgaard, 1976) was one of the first agglutination tests which was evaluated within the PDC of ISTA, and appeared to be a useful routine method. Development of other serological methods took place in the last two decades and resulted in the introduction of the Ouchterlony double diffusion test (ODD), the enzyme-linked immunosorbent assay (ELISA) in various modifications, and the dot immunobinding test (DIB), which are all suitable for routine testing (Table VIII). Tests described in this table such as immunosorbent electron microscopy (ISEM) or the similar serological specific electron microscopy (SSEM), and radio immunosorbent assay (RISA) are methods which are not very suitable for routine application, because of the facilities required. However, they are very useful for research purposes. For instance, ISEM or SSEM are good tests for confirmation of positive test results obtained in ELISA because they visualize the morphology of the virus particles.

A variant of ELISA, the enzyme-linked fluorescent assay (ELFA), can be a useful routine test as well. For the detection of lettuce mosaic virus and soybean mosaic virus in seeds, it was shown that the application of ELFA enhanced detection when compared with the standard DAS-ELISA and may be equivalent to biotin-avidin ELISA tests (Dolores-Talens *et al.*, 1989; Hill and Durand, 1986). For all immunochemical techniques, however, it is important to consider that they do not distinguish infective from non-infective virus particles. As a consequence, the correlation between percentage of infection and disease incidence in field or greenhouse should be established. This is especially of importance for quality diseases, for which a certain level of infected seeds can be tolerated. For quarantine diseases which have a zero tolerance this is of less importance, because the presence of infective or non-infective particles results in rejection of the seed lot.

For all these methods, the general principle is that the virus has to be released from the seed before it can be detected. Two ways can be followed to extract the virus. Seeds are ground and suspended in a buffer, or seeds are germinated first, after which either the seedling alone or both the seedling plus remaining seed remnants are crushed and then extracted with a buffer. In the latter case, where only the seedling is extracted, one is in fact proving disease transmission via the seed when a positive reaction is obtained. Use of seedlings for extraction of the virus has the advantage that the virus may multiply during germination, which might increase the sensitivity of the method. Clarification of extracts may be nec-

Table VIII  Detection methods used for research or routine testing of seeds for seedborne viruses on major crops.

| Pathogen | Crop | Method[a] | Reference(s) |
|---|---|---|---|
| Alfalfa mosaic virus | alfalfa | ELISA<br>ELISA, ISEM | Pesic and Hiruki, 1986; Bailiss and Offei, 1990<br>Lange et al., 1983 |
| Barley stripe mosaic virus | barley | latex agglutination test<br>immuno-diffusion test<br>ISEM<br>ELISA | Lundsgaard, 1976<br>Slack and Shepherd, 1975; Carroll et al., 1979<br>Bransky and Derrick, 1979; Lange et al., 1983<br>Lister et al., 1981; Huth, 1988; Anon. 1991b, |
| | | DIBA | Lange and Heide, 1986 |
| Bean common mosaic virus | bean | ELISA<br>Microprecipitin test<br>DIBA<br>ISEM | afarpour et al., 1979; Klein et al., 1992<br>Agarwal et al., 1977<br>Lange and Heide, 1986<br>Lundsgaard, 1983; Raizada et al., 1990 |
| Broad bean stain virus | faba bean, pea, lentil | ELISA | Makkouk and Azzam, 1986; Makkouk et al., 1987; Haack, 1990 |
| Blackeye cowpea mosaic virus | cowpea | immuno-diffusion test | Lima and Purcifull, 1980 |
| Cucumber mosaic virus | lupin | ELISA | Jones and Proudlove, 1991 |
| Cucumber green mottle virus | cucumber | ELISA | Faris-Mukhayyish and Makkouk, 1983. |
| Lettuce mosaic virus | lettuce | indicator plants, growing-on test; ELISA | Rohloff and Marrou, 1981<br>van Vuurde and Maat, 1983, 1985;<br>Falk and Purcifull, 1983 |
| | | ELFA<br>ISEM<br>RISA | Dolores-Talens et al., 1989<br>Bransky and Derrick, 1979<br>Ghabrial et al., 1982 |
| Melon necrotic spot virus | melon | ELISA | Avgelis and Barba, 1986 |

| Virus | Method | Reference |
|---|---|---|
| Pea seedborne mosaic virus | ODD<br>ELISA, ISEM<br>ELISA<br>DIBA | Zimmer, 1979<br>Hamilton and Nichols, 1978<br>Maury et al., 1987; Ding et al., 1992; Wang et al., 1993; Masmoudi et al., 1994<br>Ligat and Randles, 1993 |
| Pea early browning virus | ELISA | van Vuurde and Maat, 1985 |
| Peanut stripe virus | ELISA | Demski and Warwick, 1986; Culver and Sherwood, 1988 |
| Peanut mottle virus | ELISA | Bharathan et al., 1984; Hobbs et al., 1987 |
| Prune dwarf virus,<br>Prunus necrotic ringspot virus | ELISA | Mink and Aichele, 1984 |
| Soybean mosaic virus | ISEM<br>immunodiffusion test<br>Solid-phase RISA<br>ELISA<br>ELFA | Briansky and Derrick, 1979<br>Lima and Purcifull, 1980<br>Bryant et al., 1983<br>Maury et al., 1985; Hill and Durand, 1986<br>Benner et al., 1990 |
| Squash mosaic virus | ELISA | Nolan and Campbell, 1984; Franken et al., 1990 |
| Tobacco mosaic virus | indicator plants | Broadbent, 1965; Paludan, 1982; Green et al., 1987 |
| Tobacco ringspot virus | ELISA<br>ISEM | Lister, 1978<br>Briansky and Derrick, 1979 |

aDIBA, Dot immunobinding assay; ELISA, Enzyme-linked immunosorbent assay; ELFA, Enzyme-linked fluorescent assay; ISEM, Immunosorbent electron microscopy; RISA, Radio immunosorbent assay.

essary in order to avoid unwanted cross reactions between antiserum and certain host proteins in some serological tests. This is not necessary in the ODD test, in which contaminating proteins have a different diffusion rate than the virus resulting in different precipitation bands in the agar. Cross adsorption of antiserum with extracts of healthy seeds may also reduce this problem (Lange and Heide, 1986).

For extensive descriptions of the tests for virus indexing, including materials needed, procedures, additional comments and applications, we refer to the excellent laboratory manual of Hampton *et al.* (1990).

## D. Development of New Detection Methods

In addition to the general methods discussed, new methods have been or are being developed during the last decade (Table IX). Most of these methods are based on DNA techniques, such as hybridization of DNA extracted from seeds with pathogen-specific DNA probes or detection of a specific DNA fragment after application of the polymerase chain reaction (PCR). Extensive information about the use of DNA probes for the detection of plant pathogenic bacteria, including seedborne ones, was given by Rasmussen and Reeves (1992). The DNA techniques can be used either for detection or for confirmatory testing. The advantage of these techniques lies in their sensitivity and specificity. Particularly the application of PCR may result in a high sensitivity, because of the amplification of target DNA. For the detection of *P. syringae* pv. *phaseolicola* in bean seeds, the application of DNA hybridization alone lacked the sensitivity needed for the direct detection of low numbers of bacterial cells (Schaad *et al.*, 1989), but with an amplification step via PCR a higher sensitivity could be obtained (Prosen *et al.*, 1993).

The use of higly specific DNA probes will enable the differentiation at subspecies and, possibly, even at the pathotype level. Such probes are being developed, for example, to enable discrimination between *Pyrenophora graminea* and *P. teres* on wheat seed (Husted, 1994). Also, DNA techniques are relatively fast, compared to other detection techniques.

A disadvantage of DNA techniques is the detection of both live and dead cells. If information about the viability of the pathogen is needed, these techniques could be used as an initial test. A second test, e.g. dilution plating, could then be performed on positive samples to obtain this information. The relevance of DNA techniques for routine indexing of seed lots should become clear in the coming years. The availability of non-radioactive probes certainly increases their value for routine testing.

Detection techniques based on quantification of ergosterol, for fungal infections, or secondary metabolites produced by the pathogen, are not yet used for routine testing. Ergosterol is a common metabolite in a number of fungi, and therefore its detection will not be very specific. However, if produced by the pathogen, the detection of specific metabolites may be incorporated in a seed health testing method. For example, the production of phaseolotoxin by *Pseudomonas syringae* pv. *phaseolicola* can be used to confirm the identity of the bacterium in a toxin bio-

Table IX. New methods, in use or in development, for the detection of seedborne pathogens.

| Pathogen | Crop | Method | Reference(s) |
|---|---|---|---|
| *Erwinia stewartii* | maize | DNA hybridization, development of DNA probe via PCR | Blakemore *et al.*, 1992 |
| *Pseudomonas syringae* pv. *phaseolicola* | bean | plating and DNA hybridization PCR with seed extract | Schaad *et al.*, 1989; Tourte & Manceau, 1994 Tourte and Manceau, 1991; Prosen *et al.*, 1993 |
| *Pseudomonas syringae* pv. *pisi* | pea | DNA hybridization with seed extract | Rasmussen and Wulff, 1990; Rasmussen and Reeves, 1992 |
| | | PCR with seed extract conductimetric assay | Rasmussen and Wulff, 1991 Fraaije *et al.*, 1993 |
| *Cercospora kikuchii* | soybean | direct inspection using image analysis | Paulsen, 1990; Cassady *et al.*, 1992 |
| *Fusarium moniliforme, Stenocarpella maydis* | maize | PCR, development of primers | Blakemore *et al.*, 1994 |
| *Leptosphaeria maculans* | brassicas | PCR with DNA extract from seeds | Taylor, 1993 |
| *Peronosclerospora sacchari* and *sorghi* | maize, sorghum | DNA hybridization with DNA from seeds | Yao *et al.*, 1990, 1991 |
| *Phomopsis/Diaporthe* | soybean | PCR, development of primers | Blakemore *et al.*, 1994 |
| *Pyrenophora graminea* P. *graminea*, P. *teres* | barley | ergosterol quantification DNA hybridization and probe development PCR, development of primers | Gordon and Webster, 1984 Husted, 1994 Blakemore *et al.*, 1994 |
| *Ustilago nuda* | wheat | ergosterol quantification | Shrivastava and Yadav, 1990 |
| Cucumber mosaic virus | lupin | PCR on DNA extracts of ground seeds | Wylie *et al.*, 1993 |
| Pea seedborne mosaic virus | pea | PCR on viral RNA extracted from seed parts | Kohnen *et al.*, 1992 |
| Peanut mottle and peanut stripe virus | peanut | DNA hybridization on DNA extracted from seeds | Bijaisoradat and Kuhn, 1988 |

assay (Jansing and Rudolph, 1990).

Recently, the potential of conductimetric assays, based on the measurement of conductance changes of a semi-selective medium due to bacterial metabolism, was tested for the detection of *Pseudomonas syringae* pv. *pisi* in pea seed lots (Fraaije *et al.*,1993). Selectivity of the medium needs to be improved, however, because saprophytic microorganisms interfered with the conductimetric assay. This assay is relatively fast and almost fully automated and, therefore, is highly suited for large-scale routine indexing.

Another recent technique, image analysis, may be used for objective inspections of seeds, as was shown for soybean (Paulsen, 1990; Casady *et al.*, 1992). However, this technique only identifies seeds that are different in size, form or colour. This limits its application for routine seed health testing because infected seeds that do not show visible symptoms will not be recognized. Still, investigation of new techniques, such as image analysis or conductimetric assays, is worthwhile, because they give new insights and may result in improved seed health testing methods.

## E. Seed Health Testing in Practice

An example to what extent and for which reasons the available seed health testing methods are applied in practice is given in Table X. Data are given for the most important pathogens for which testing was performed in recent years in The Netherlands by the former Government Seed Testing Station, the General Netherlands Inspection Services and private seed companies. Due to changes in the status of a pathogen with regard to importance, regulations, and an increasing awareness of quality control, the numbers of samples tested will vary annually. However, the data presented give an impression of how much seed testing is being done and which methods are actually being used.

## VI. CONCLUDING REMARKS

Major drawbacks in the application of seed health testing methods are the lack of knowledge of the principles and technical features of the methods, lack of standardization, and lack of training. Insufficient insight in the purpose for which a method was developed may lead to its incorrect use and often to misinterpretation of test results. Moreover, the application of some seed health testing methods is sometimes limited by the facilities and investments needed. This especially holds for developing countries, for which seed health testing could be very beneficial for the reliable production of food and fodder. The establishment of well-equipped laboratories and training of personnel is a matter of concern to the Plant Disease Committee (PDC) of ISTA and the Danish International Institute of Seed Pathology for Developing Countries in Copenhagen, Denmark. Both the PDC and the Danish institute regularly organize training courses to improve the skills of seed pathologists all over the world.

The enormously increasing seed trade calls for good seed health tests. On the

Table X. Seed health testing methods for a number of host-pathogen combinations applied for different purposes in the Netherlands.

| Crop | Pathogen | Method | Test centre[a] | Purpose[b] | Number of samples | Number of replicates x seeds | |
|---|---|---|---|---|---|---|---|
| Cereals | Fusarium spp. | seedling symptom test | NAK | ncs | 2600 | 4 x | 50 |
|  | Pyrenophora spp. | deepfreezing blotter test | RPvZ | survey | 160 | 4 x | 50 |
| Beta vulgaris | Phoma betae | agar test | RPvZ | nhse | 25 | 40 x | 5 |
| Brassica spp. | Phoma lingam | deepfreezing blotter test | companies, NAK-G | qasc, hrfc | >500 | 2 x | 1,000 |
|  | Alternaria brassicicola | seedling symptom test | RPvZ | nhse | 50 | 2 x | 100 |
|  | Xanthomonas campestris pv. campestris | IF + dilution plating + pathogenicity test | companies, NAK-G | qasc, hrfc | >500 | 3 x | 10,000 |
| Daucus carota | Alternaria dauci, A. radicina | deepfreezing blotter test | companies, NAK-G | qasc | >500 | 2 x | 100 |
| Lactuca sativa | Lettuce mosaic virus | ELISA | RPvZ | qasc | 160 | 20 x | 100 |
| Linum usitatissimum | Botrytis cinerea | seedling symptom test | NAK | ncs | 500 | 8 x | 50 |
| Lolium spp. | Gloeotinia temulenta | soaking test | RPvZ | nhse, hrfc | 545 | 3 x | 100 |
| Lycopersicon esculentum | Clavibacter michiganensis subsp. michiganensis | IF + dilution plating + pathogenicity test | companies, NAK-G | qasc, hrfc | 200-400 | 2 x | 6,250 |
| Phaseolus vulgaris | Pseudomonas syringae pv. phaseolicola | IF + dilution plating + pathogenicity test | RPvZ | qasc, hrfc | 250 | 5 x | 1,000 |
| Pisum sativum | Pseudomonas syringae pv. pisi | IF + dilution plating + pathogenicity test | RPvZ | qasc, hrfc | 235 | 5 x | 1,000 |
|  | Pea seedborne mosaic virus | ELISA | RPvZ | qasc, hrfc | 230 | 16 x | 100 |
|  | Ascochyta spp. | seedling symptom test | NAK | nsc | 450 | 4 x | 75 |

[a] NAK, General Netherlands Inspection Service for Agricultural Seed and Seed Potatoes; RPvZ, former Government Seed Testing Station; NAK-G, general Netherlands Inspection Service for Vegetable and Flower Seeds.

[b] ncs, national certification scheme; nhse, national health standard for export of seed lots; qasc, quality assesment for private seed companies; hrfc, health requirement for issuance of the International Phytosanitary Certificate.

other hand, the fast developments in molecular biology are providing quite a number of new and alternative detection techniques and methods. Nevertheless, the problems are at the moment bigger than ever before, since no generally accepted agreement exists on which method is the most optimal for use in practice. The main reason being that important information about the value of seed health testing methods is mostly lacking, i.e. data on precision and accuracy, diagnostic sensitivity and specificity. More information is needed on the prevalence of the various diseases, and rates of disease transmission under different conditions in practice. As has been demonstrated, calculation of predictive values for methods on the basis of such data would support the choice of one or two methods to be applied in combination. Such an approach might optimize disease risk analysis for seed lots. Moreover, it will lead to a better understanding of the economical importance of a seed-borne disease, and by that to more realistic phytosanitary regulations.

Considering the fact that worldwide standardization and recommendations for seed health methods are needed, data which will help to indicate the value of a method are also essential for clear analysis of discrepancies between and within laboratories. To achieve a high degree of standardization and quality assurance in seed health testing, it must be possible to check the performance of seed health testing laboratories. For this purpose reference materials are needed, e.g. to check antisera, media, buffers for isolation, molecular probes, etc. Such reference materials are not available, yet. In the long term, the development of suitable reference materials will help to solve problems with regard to standardization and improvement of seed health testing methods.

When new methods for routine indexing of seed samples for pathogens are published, data on precision and diagnostic sensitivity and specificity should preferably be included. The PDC of ISTA has already taken initiatives in this direction by evaluating the Working Sheets of the ISTA Handbook on Seed Health Testing. Working groups of this PDC, which develop and standardize seed health testing methods, will follow protocols as described by Sheppard in "Guidelines for collaborative study procedures to validate characteristics of a method of analysis" (unpublished), which was derived from the "Guidelines for collaborative study procedures in chemical analysis" (Anonymous, 1989). This should lead to reviewed Working Sheets that will give all essential information for worldwide application of routine testing of seeds.

## APPENDIX

For the calculation of values as discussed by Sheppard (1993), the following assumptions were made:

* The sum of true positive (TP) and false negative (FN) test results, (TP+FN), refers to the real number of diseased samples.
* The sum of true negative (TN) and false positive (FP) test results, (TN+FP) refers to the real number of healthy samples.
* The ratio between the number of true positive test results (TP) and the real

number of diseased samples (TP+FN), TP/(TP+FN) x 100, refers to the diagnostic sensitivity.

*   The ratio between the number of true negative test results (TN) and the real number of healthy samples (TN+FP), TN/(TN+FP) x 100, refers to the diagnostic specificity.

Sheppard (1993) also gave calculations for the predictive values of test results. The 'predictive value' in this context means a measure of the probability that, given a negative result, the sample is truly healthy or conversely that, given a positive result, the sample is truly diseased. These values can be formulated as follows:

*   the ratio between the number of true positive test results (TP) and the actual number of samples found positive (TP+FP), TP/(TP+FP) x 100, is designated the predictive value of a positive result, which refers to the probability that given a positive result the sample is truly positive.
*   the ratio between the number of true negative test results (TN) and the actual number of samples found negative (TN+FN), TN/(TN+FN) x 100, is designated the predictive value of a negative result, which refers to the probability that given a negative result the sample is truly negative.

Unlike the diagnostic sensitivity and specificity, the true disease status of the sample is not known beforehand. The predictive value of a positive test can vary strongly with the prevalence of the disease.

'Prevalence' is defined as: the amount of disease in a specified population (eg. one seed in 10 000) at a given point of time and differs from incidence, which is the rate of occurrence of a disease over a stated period of time (Galen and Gambino, 1975). Sheppard (1993) stated that 'prevalence' is equal to the percentage of samples yielding disease in the field or:

*   the ratio between the real number of diseased samples (TP+FN), and the total number of samples tested (TP+FN+TN+FP), (TP+FN)/(TP+FN+TN+FP) x 100, is designated prevalence.

Bayes' Formula or Bayes' Theorem was used by Sheppard (1993), in order to transfer this concept of predictive values from clinical pathology to seed pathology. On basis of the following formulae the prevalence can be taken into account when calculating the predictive values:

*   The predictive value of a positive result, (p.a)/((p.a) + (1-p)(1-b)), which gives a prediction of disease.
*   The predictive value of a negative result, (1-p)b/((1-p)b + p(1-a)), which gives a prediction of health.

Where:  p = prevalence of the disease
        a = diagnostic sensitivity of the test
        b = diagnostic specificity of the test
        p, a and b are expressed as proportions in decimal fractions

# REFERENCES

Agarwal, V.K. and Sinclair, J.B. (1987). 'Principles of seed pathology, Vol. I and II'. CRC Press, Boca Raton, Florida, USA.

Agarwal, .V.K., Nene, Y.L and Beniwal, S.P.S. (1977). Detection of bean common mosaic virus in urdbean (*Phaseolus mungo*) seeds. *Seed Science and Technology* 5, 619-625.

Agarwal, P.C., Mortensen, C.N. and Mathur, S.B. (1989). 'Seed-borne diseases and seed health testing of rice'. Phytopathological Papers No. 30, Danish Government Institute of Seed Pathology for Developing Countries, Copenhagen and CAB International Mycological Institute, Kew.

Agarwal, V.K., Singh, D.V. and Mathur, S.B. (1993). Karnal bunt (*Tilletia indica*). In 'Seedborne Diseases and Seed Health Testing of Wheat' (S.B. Mathur and B.M. Cunfer, eds), pp. 31-43, Danish Government Institute of Seed Pathology for Developing Countries, Copenhagen, Denmark.

Alderman, S.C. (1991). Assessment of ergot and blind seed diseases of grasses in the Willamette Valley of Oregon. *Plant Disease* 75, 1038-1041.

Anonymous (1981, 1994). 'ISTA Handbook on seed health testing, section 2, Working sheets, each dealing with one pathogen on one host'. International Seed Testing Association, Zürich, Switzerland.

Anonymous (1985). 'ISO 7218, Microbiology - General guidance for microbiological examinations'. International Organization for Standardization, Geneva, Switzerland.

Anonymous (1986). 'ISO 5725, Precision of test methods - Determination of repeatability and reproducibility for a standard test method by inter-laboratory tests'. International Organization for Standardization, Geneva, Switzerland.

Anonymous (1989). Guidelines for collaborative study procedures in chemical analysis. Journal of the Association of Official Analytical Chemists 72, 694-704.

Anonymous (1991a). Water quality - vocabulary - part 8, ISO/DIS 6107-8. International Organization for Standardization, Geneva, Switzerland.

Anonymous (1991b). Quarantine procedure no. 34. Barley stripe mosaic hordei virus - inspection and test methods for barley seeds. *EPPO Bulletin* 21, 257-259.

Anonymous (1992a). Quarantine procedure no. 39. *Clavibacter michiganensis* subsp. *michiganensis* - test methods for tomato seeds. *EPPO Bulletin* 22, 219-224.

Anonymous (1992b). Quarantine procedure no. 45. *Xanthomonas campestris* pv. *vesicatoria* - test methods for tomato seeds. *EPPO Bulletin* 22, 247-252.

Anonymous (1993). International rules of Seed Testing. Seed Science and Technology 21, supplement.

Anselme, C. and Champion, R. (1975). Etude de la transmission du *Botrytis cinerea* par les semences de tournesol (*Helianthus annuus*). *Seed Science and Technology* 3, 711-717.

Avgelis, A. and Barba, M. (1986). Application of ELISA test for indexing melon necrotic spot virus in melon seeds. *Annali dell' Istituto Sperimentale per la Patologia Vegetale* 11, 107-111.

Bailiss, K.W. and Offei, S.K. (1990). Alfalfa mosaic virus in lucerne seed during seed maturation and storage and in seedlings. *Plant Pathology* 39, 539-547.

Ball, S.F.L. and Reeves, J.C. (1991). The application of new techniques in the rapid testing for seed-borne pathogens. *Plant Varieties and Seeds* 4, 169-176.

Bashan, Y. and Assouline, I. (1983). Complementary bacterial enrichment techniques for the detection of *Pseudomonas syringae* pv. *tomato* and *Xanthomonas campestris* pv. *vesicatoria* in infected tomato and pepper seeds. *Phytoparasitica* 11, 187-193.

Bassey, E.O. and Gabrielson, R.L. (1983). Factors affecting accuracy of 2,4-D assays of crucifer seed for *Alternaria brassicicola* and relation of assays to seedling disease potential. *Seed Science and Technology* 11, 411-420.

Benner, H.I., Hill, J.H. and Durand, D.P. (1990). Detection of soybean mosaic virus by enzyme-linked fluorescent assay (ELFA). *Seed Science and Technology*, 18, 23-31.

Bharathan, N., Reddy, D.V.R., Rajeshwari, R., Murthy, V.K., Rao, V.R. and Lister, R.M. (1984). Screening peanut germ plasm lines by enzyme-linked immunosorbent assay for seed transmission of peanut mottle virus. *Plant Disease* 68, 757-758.

Bijaisoradat, M. and Kuhn, C.W. (1988). Detection of two viruses in peanut seeds by complementary DNA hybridization tests. *Plant Disease* 72, 956-959.

Blakemore, E.J.A., Reeves, J.C. and Ball S.F.L. (1992). Polymerase chain reaction used in the development of a DNA probe to identify *Erwinia stewartii*, a bacterial pathogen of maize. *Seed Science and Technology* 20, 331-335.

Blakemore, E.J.A., Jaccoud Filho, D.S. and Reeves, J.C. (1994). PCR for the detection of *Pyrenophora* species, *Fusarium moniliforme*, *Stenocarpella maydis* and the *Phomopsis/Diaporthe* complex. In 'Modern Assays for Plant Pathogenic Fungi: Identification, Detection and Quantification' (A. Schots, F.M. Dewey and R. Oliver, eds), pp. 205-213. CAB International, Wallingford, UK.

Brlansky, R.H. and Derrick, K.S. (1979). Detection of seed-borne plant viruses using serologically specific electron microscopy. *Phytopathology* 69, 96-100.

Broadbent, L. (1965). The epidemiology of tomato mosaic. XI. Seed transmission of TMV. *Annals of Applied Biology* 56, 177-205.

Bryant, G.R., Durand, D.P. and Hill, J.H. (1983). Development of a solid-phase radioimmunoassay for detection of soybean mosaic virus. *Phytopathology* 73, 623-629.

Calzolari, A., Tomesani, M. and Mazzucchi, U. (1987). Comparison of immunofluorescence staining and indirect isolation for the detection of *Corynebacterium flaccumfaciens* in bean seeds. *EPPO Bulletin* 17, 157-163.

Carroll, T.W., Gossel, P.L. and Batchelor, D.L. (1979). Use of sodium dodecylsulfate in serodiagnosis of barley stripe mosaic virus in embryos and leaves. *Phytopathology* 69, 12-14.

Casady, W.W., Paulsen, M.R., Reid, J.F. and Sinclair, J.B. (1992). A trainable algorithm for inspection of soybean seed quality. *Transactions of the ASAE* 35, 2027-2034.

Castro, C., Schaad, N.W. and Bonde, M.R. (1994). A technique for extracting *Tilletia indica* teliospores from contaminated wheat seeds. *Seed Science and Technology* 22, 91-98.

Chang, C.J., Donaldson, R., Crowley, M. and Pinnow, D. (1991). A new semiselective medium for the isolation of *Xanthomonas campestris* pv. *campestris* from crucifer seeds. *Phytopathology* 81, 449-453.

Champion, R. and Mecheneau, H. (1979). Méthode de détection de *Peronospora valerianellae*, agent du Mildiou, sur les semences de Mâch (Valerianellae locusta). Comparaison en culture des résultats obtenus. *Seed Science and Technology* 7, 259-263.

Claflin, L.E. and Ramundo, B.A. (1987). Evaluation of the dot immunobinding assay for detecting phytopathogenic bacteria in wheat seeds. *Journal of Seed Technology* 11, 52-61.

Conger, B.V. and McDaniel, J.K. (1983). Use of callus cultures to screen tall fescue samples for *Acremonium coenophialum*. *Crop Science* 23, 172-174.

Cristani, C. (1992). Seed-borne *Microdochium nivale* (Ces. ex Sacc.) Samuels (= *Fusarium nivale* (Fr.) Ces.) in naturally infected seeds of wheat and triticale in Italy.

Culver, J.N. and Sherwood, J.L. (1988). Detection of peanut stripe virus in peanut seed by an indirect enzyme-linked immunosorbent assay using a monoclonal antibody. *Plant Disease* 72, 676-679.

Cunfer, B.M. and Manandhar, J.B. (1992). Use of a selective medium for isolation of *Stagonospora nodorum* from barley seed. *Phytopathology* 82, 788-791.

Da Cruz Machado, J. and Langerak, C.J. (1993). Improvement of a blotter method to detect economically important fungi associated with seeds of cotton. *In* 'Proceedings of the 1st ISTA *Plant Disease* Committee Symposium on Seed health testing, Ottawa, Canada' (J.W. Sheppard, ed.), pp. 48-58. International Seed Testing Association, Zürich, Switzerland.

de Tempe, J. (1963). Health testing of flax seed. *Proceedings of the International Seed Testing Association* 28, 107-131.

de Tempe, J. (1964). *Helminthosporium* spp. in seeds of wheat, barley, oats and rye. *Proceedings of the International Seed Testing Association* 29, 117-140.

de Tempe, J. (1968). The quantitative effect of light and moisture on carrot seed infections in blotter medium. *Proceedings of the International Seed Testing Association* 33, 547-553.

de Tempe, J. (1970a). The determination of *Drechslera sativa* in barley seed in The Netherlands. *Proceedings of the International Seed Testing Association* 35, 185-191.

de Tempe, J. (1970b). Testing cereal seeds for *Fusarium* infection in the Netherlands. *Proceedings of the International Seed Testing Association* 35, 193-206.

de Tempe, J. (1978). Testing sugar beet seed for *Phoma betae*. *Seed Science and Technology* 6, 927-933.

Demski, J.W. and Warwick, D. (1986). Testing peanut seeds for peanut stripe virus. *Peanut Science* 13, 38-40.

Dhanvantari, B.N. and Brown, R.J. (1993). YSSM-XP medium for *Xanthomonas campestris* pv. *phaseoli*. *Canadian Journal of Plant Pathology* 15, 168-174.

Diatloff, A., Wong, W.C. and Wood, B.A. (1993). Non-destructive methods of detecting *Curtobacterium flaccumfaciens* pv. *flaccumfaciens* in mungbean seeds. *Letters in Applied Microbiology* 16, 269-273.

Diekmann, M. and Assad, S. (1989). Comparison of agar and freezing blotter test for detection of *Fusarium* spp. in seeds of lentil, chickpea and barley. *Zeitschrift für Pflanzenkrankheiten und Pflanzenschutz* 96, 134-139.

Ding, X.S., Cockbain, A.J. and Govier, D.A. (1992). Improvements in the detection of pea seed-borne mosaic virus by ELISA. *Annals of Applied Biology* 121, 75-83.

Dolores-Talens, A.C., Hill, J.H. and Durand, D.P. (1989). Application of enzyme-linked fluorescent assay (ELFA) to detection of lettuce mosaic virus in lettuce seeds. *Journal of Phytopathology* **124**, 149-154.

Duveiller, E. (1989). Research on *'Xanthomonas translucens'* of wheat and triticale at CIMMYT. *EPPO Bulletin* **19**, 97-103.

Duveiller, E. (1990). Seed detection of *Xanthomonas campestris* pv. *undulosa* using a modification of Wilbrink's agar medium. *Parasitica* **46**, 3-17.

Duveiller, E. and Bragard, C. (1992). Comparison of immunofluorescence and two assays for detection of *Xanthomonas campestris* pv. *undulosa* in seeds of small grains. *Plant Disease* **76**, 999-1003.

Falk, B.W. and Purcifull, D.E. (1983). Development and application of an enzyme-linked immunosorbent assay (ELISA) test to index lettuce seeds for lettuce mosaic virus in Florida. *Plant Disease* **67**, 413-416.

Faris-Mukhayyish, S. and Makkouk, K.M. (1983). Detection of four seed-borne plant viruses by the enzyme-linked immunosorbent assay (ELISA). *Phytopathologische Zeitschrift* **106**, 108-114.

Fatmi, M. and Schaad, N.W. (1989). Detection of *Clavibacter michiganense* ssp. *michiganense* in tomato seed. In 'Detection of Bacteria in Seed and Other Planting Material' (A.W. Saettler, N.W. Schaad and D.A. Roth, eds), pp. 45-49. The American Phytopathological Society, St. Paul, Minnesota.

Fett, W.F. (1979). Survival of *Pseudomonas glycinea* and *Xanthomonas phaseoli* var. *sojensis* in leaf debris and soybean seed in Brazil. *Plant Disease Reporter* **63**, 79-83.

Fraaije, B.A., Franken, A.A.J.M., van der Zouwen, P.S., Bino, R.J. and Langerak, C.J. (1993). Serological and conductimetric assays for the detection of *Pseudomonas syringae* pathovar *pisi* in pea seeds. *Journal of Applied Bacteriology* **75**, 409-415.

Franken, A.A.J.M. (1992a). Comparison of immunofluorescence microscopy and dilution plating for the detection of *Xanthomonas campestris* pv. *campestris* in crucifer seeds. *Netherlands Journal of Plant Pathology* **98**, 169-178.

Franken, A.A.J.M. (1992b). Application of polyclonal and monoclonal antibodies for the detection of *Xanthomonas campestris* pv. *campestris* in crucifer seeds using immunofluorescence microscopy. *Netherlands Journal of Plant Pathology* **98**, 95-106.

Franken, A.A.J.M. (1993). Characteristics of immunofluorescence microscopy and dilution plating for predicting halo blight of beans and black rot of crucifers. In 'Proceedings 1st ISTA *Plant Disease* Committee Symposium on Seed Health Testing, Ottawa, Canada' (J.W. Sheppard, ed.), 150-155. International Seed Testing Association, Zürich, Switzerland.

Franken, A.A.J.M. and Sheppard, J.W. (1993). The need for accurate guidelines in comparative testing between seed health laboratories. In 'Proceedings 1st ISTA *Plant Disease* Committee Symposium on Seed Health Testing, Ottawa, Canada' (J.W. Sheppard, ed.), 143-148. International Seed Testing Association, Zürich, Switzerland.

Franken, A.A.J.M. and van den Bovenkamp, G.W. (1990). The application of the combined use of immunofluorescence microscopy and dilution plating to detect *Pseudomonas syringae* pv. *pisi* in pea seeds. In 'Plant Pathogenic Bacteria - Proceedings 7th International Conference on Plant Pathogenic Bacteria' (Z. Klement, ed.), pp. 871-875. Akadémiai Kiadó, Budapest.

Franken, A.A.J.M. and van Vuurde, J.W.L. (1990). Problems and approaches in the use of serology for seed-borne bacteria. *Seed Science and Technology* **18**, 415-426.

Franken, A.A.J.M., Maat, D.Z. and Kamminga, G.C. (1990). Detection of squash mosaic virus in seeds of melon (*Cucumis melo*) by enzyme-linked immunosorbent assay (ELISA). *Netherlands Journal of Plant Pathology* **96**, 91-102.

Franken, A.A.J.M., van Zeijl, C., van Bilsen, J.G.P.M., Neuvel, A., de Vogel, R., van Wingerden, Y., Birnbaum, Y.E., van Hateren, J. and van der Zouwen, P.S. (1991). Evaluation of a plating assay for *Xanthomonas campestris* pv. *campestris*. *Seed Science and Technology* **19**, 215-226.

Franken, A.A.J.M., Kamminga, G.C., Snijders, W., van der Zouwen, P.S. and Birnbaum, Y.E. (1993). Detection of *Clavibacter michiganensis* ssp *michiganensis* in tomato seeds by immunofluorescence microscopy and dilution plating. *Netherlands Journal of Plant Pathology* **99**, 125-137.

Frommel, M.I. and Pazos, G. (1994). Detection of *Xanthomonas campestris* pv. *undulosa* infested wheat seeds by combined liquid medium enrichment and ELISA. *Plant Pathology* **43**, 589-596.

Gambogi, P., Triolo, E. and Vannacci, G. (1976). Experiments on the behaviour of the seed-borne fungus *Alternaria zinniae*. *Seed Science and Technology* **4**, 333-340.

Geng, S., Campbell, R.N., Carter, M. and Hills, F.J. (1983). Quality control programs for seed-borne pathogens. *Plant Disease* **67**, 236-242.

Ghabrial, S.A., Li, D. and Shepherd, R.J. (1982). Radio-immunosorbent assay for detection of lettuce mosaic virus in lettuce seed. *Plant Disease* **66**, 1037-1040.

Gitaitis, R.D. (1990). Survey of phytopathogenic bacteria recovered from commercial tomato seed and their secondary spread within tomato transplant fields. *In* 'Plant Pathogenic Bacteria - Proceedings 7th International Conference on Plant Pathogenic Bacteria' (Z. Klement, ed.), pp. 293-298. Akadémiai Kiadó, Budapest.

Gitaitis, R.D., Chang, C.J., Sijam, K. and Dowler, C.C. (1991). A differential medium for semiselective isolation of *Xanthomonas campestris* pv. *vesicatoria* and other cellulolytic xanthomonads from various natural sources. *Plant Disease* **75**, 1274-1278.

Gleason, M.L., Ghabrial, S.A. and Ferriss, R.S. (1987). Serological detection of *Phomopsis longicolla* in soybean seeds. *Phytopathology* **77**, 371-375.

Gordon, T.R. and Webster, R.K. (1984). Evaluation of ergosterol as an indicator of infestation of barley seed by *Drechslera graminea*. *Phytopathology* **74**, 1125-1127.

Green, S.K., Hwang, L.L. and Kuo, Y.J. (1987). Epidemiology of tomato mosaic virus in Taiwan and identification of strains. *Zeitschrift für Pflanzenkrankheiten und Pflanzenschutz* **94**, 386-397.

Gwinn, K.D., Collins-Shepard, M.H. and Reddick, B.B. (1991). Tissue print-immunoblot, an accurate method for the detection of Acremonium coenophialum in tall fescue. *Phytopathology* **81**, 747-748.

Haack, I. (1990). Nachweis von samenubertragbaren Viren in Saatgut von Acker-bohnen und Erbsen mit Hilfe des ELISA. Detection of seed transmissible viruses in seeds of faba beans and peas by means of ELISA. *Archiv für Phytopathologie und Pflanzenschutz* **26**, 337-342.

Halfon-Meiri, A. and Volcani, Z. (1977). A combined method for detecting *Colletotrichum gossypii* and *Xanthomonas malvacearum* in cotton seed. *Seed Science Technology* **5**, 129-139.

Hamilton, R.I. and Nichols, C. (1978). Serological methods for detection of pea seed-borne mosaic virus in leaves and seeds of *Pisum sativum*. *Phytopathology* 68, 539-543.

Hampton, R., Ball, E. and De Boer, S. (1990). 'Serological methods for detection and identification of viral and bacterial plant pathogens - a laboratory manual'. The American Phytopathological Society, St. Paul, Minnesota.

Haware, M.P., Nene, Y.L. and Mathur, S.B. (1986). 'Seedborne diseases of chickpea'. Technical Bulletin no. 1, Danish Government Institute of Seed Pathology for Developing Countries, Copenhagen, Denmark.

Hewett, P.D. (1968). Viable *Septoria* spp. in celery seed samples. *Annals of Applied Biology* 61, 89-98.

Hewett, P.D. (1987). Detection of seed-borne *Ascochyta pisi* Lib. and test agreement within and between laboratories. *Seed Science and Technology* 15, 271-283.

Hewett, P.D. and Damgaci, E. (1986). A new procedure to detect a low incidence of *Ustilago nuda* in seed barley. *Plant Pathology* 35, 377-379.

Hill, J.H. and Durand, D.P. (1986). Soybean mosaic virus. Methods of enzymatic analysis, antigens and antibodies 2, vol. 11, 455-474.

Hobbs, H.A., Reddy, D.V.R., Rajeshwari, R. and Reddy, A.S. (1987). Use of direct antigen coating and protein A coating ELISA procedures for detection of three peanut viruses. *Plant Disease* 71, 747-749.

Husted, K. (1994). Development of species-specific probes for identification of *Pyrenophora graminea* and *P. teres* by dot-blot or RFLP. *In* 'Modern Assays for Plant Pathogenic Fungi: Identification, Detection and Quantification' (A. Schots, F.M. Dewey and R. Oliver, eds), pp. 191-197. CAB International, Wallingford, UK.

Huth, W. (1988). Use of ELISA for detection of barley stripe mosaic virus in barley seeds. Einsatz von ELISA zum Nachweis von Barley Stripe Mosaic Virus in Gerstensamen. *Nachrichtenblatt des Deutschen Pflanzenschutzdienstes* 40, 128-132.

Inaba, T., Takahashi, K. and Morinaka, T. (1983). Seed transmission of spinach downy mildew. *Plant Disease* 67, 1139-1141.

Jafarpour, B., Shepherd, R.J. and Grogan, R.G. (1979). Serologic detection of bean common mosaic and lettuce mosaic viruses in seed. *Phytopathology* 69, 1125-1129.

Jansing, H. and Rudolph, K. (1990). A sensitive and quick test for determination of bean seed infestation by *Pseudomonas syringae* pv. *phaseolicola*. *Zeitschrift für Pflanzenkrankheiten und Pflanzenschutz* 97, 42-55.

Johnson, M.C. anderson, R.L, Kryscio, R.J. and Siegel, M.R. (1983). Sampling procedures for determining endophyte content in tall fescue seed lots by ELISA. *Phytopathology* 73, 1406-1409.

Jones, R.A.C. and Proudlove, W. (1991). Further studies on cucumber mosaic virus infection of narrow leafed lupin (*Lupinus angustifolius*): seed-borne infection, aphid transmission, spread and effects on grain yield. *Annals of Applied Biology* 118, 319-329.

Jones, J.B., McCarter, S.M. and Gitaitis, R.D. (1989). Detection of *Pseudomonas syringae* pv. tomato in tomato. *In* 'Detection of Bacteria in Seed and Other Planting Material (A.W. Saettler, N.W. Schaad and D.A. Roth, eds), pp. 50-58. American Phytopathological Society, St. Paul, Minnesota.

Jorgensen, J. (1980). Comparative testing of barley seed for inoculum of *Pyreno-*

phora graminea and P. teres in greenhouse and field. Seed Science and Technology 8, 377-381.

Jorgensen, J. (1982a). The freezing blotter method in testing barley seeds for inoculum of Pyrenophora graminea and Pyrenophora teres. Repeatability of test results. Seed Science and Technology 10, 639-646.

Jorgensen, J. (1982b). The freezing blotter method in testing barley seed for inoculum of Pyrenophora graminea. Varietal resistance and predictive value of test results. Seed Science and Technology 10, 647-650.

Jorgensen, J. (1983). Disease testing of barley seed and application of test results in Denmark. Seed Science and Technology 11, 615-624.

Jorgensen, J. (1989). Comparison of the fluorescence method with other methods in testing wheat seed for infection with Septoria nodorum. Seed Science and Technology 17, 249-253.

Klein, R.E., Wyatt, S.D., Kaiser, W.J. and Mink, G.I. (1992). Comparative immunoassays of bean common mosaic virus in individual bean (Phaseolus vulgaris) seed and bulked bean seed samples. Plant Disease 76, 57-59.

Kohnen, P.D., Dougherty, W.G. and Hampton, R.O. (1992). Detection of pea seed-borne mosaic potyvirus by sequence specific enzymatic amplification. Journal of Virological Methods 37, 253-258.

Krinkels, M. (1992). De zaak Almeria - toekomstgerichte aanpak bezweert conflict (in Dutch). Prophyta 5, 10-13.

Kritzman, G. (1991). A method for detection of seed-borne bacterial diseases in tomato seeds. Phytoparasitica 19, 133-141.

Kritzman, G. and Netzer, D. (1978). A selective medium for isolation and identification of Botrytis spp. from soil and onion seed. Phytoparasitica 6, 3-7.

Kuan, T.L. (1989). Detection of Xanthomonas campestris pv. carotae in carrot. In 'Detection of Bacteria in Seed and Other Planting Material (A.W. Saettler, N.W. Schaad and D.A. Roth, eds), pp. 63-67. American Phytopathological Society, St. Paul, Minnesota.

Lamka, G.L., Hill, J.H., McGee, D.C. and Braun, E.J. (1991). Development of an immunosorbent assay for seed-borne Erwinia stewartii· in corn seeds. Phytopathology 81, 839-846.

Lange, L. and Heide, M. (1986). Dot immuno binding (DIB) for detection of virus in seed. Canadian Journal of Plant Pathology 8, 373-379.

Lange, L., Tien, P. and Begtrup, J. (1983). The potential of ELISA and ISEM in seed health testing. Seed Science and Technology 11, 477-490.

Langerak, C.J. and Franken, A.A.J.M. (1994). Diagnostic methods for the detection of plant pathogens in vegetable seeds. British Crop Protection Council Monograph 57, 169-178.

Langerak, C.J., Merca S.D. and Mew, T.W. (1988). Facilities for seed health testing and research. Proceedings of the International Workshop on Rice Seed Health, IRRI, Manila, Philippines, 1987, pp. 235-246.

Lee, D.H., Mathur, S.B. and Neergaard, P. (1984). Detection and location of seed-borne inoculum of Didymella bryoniae and its transmission in seedlings of cucumber and pumpkin. Phytopathologische Zeitschrift 109, 301-308.

Ligat, J.S. and Randles, J.W. (1993). An eclipse of pea seed-borne mosaic virus in vegetative tissue of pea following repeated transmission through the seed.

*Annals of Applied Biology* **122**, 39-47.

Lima, J.A.A. and Purcifull, D.E. (1980). Immunochemical and microscopical techniques for detecting blackeye cowpea mosaic and soybean mosaic viruses in hypocotyls of germinated seeds. *Phytopathology* **70**, 142-147.

Lister, R.M. (1978). Application of the enzyme-linked immunosorbent assay for detecting viruses in soybean seed and plants. *Phytopathology* **68**, 1393-1400.

Lister, R.M., Carroll, T.W. and Zaske, S.K. (1981). Sensitive serologic detection of barley stripe mosaic virus in barley seed. *Plant Disease* **65**, 809-814.

Lundsgaard, T. (1976). Routine seed health testing for barley stripe mosaic virus in barley seeds using the latex test. *Zeitschrift für Pflanzenkrankheiten und Pflanzenschutz* **83**, 278-283.

Lundsgaard, T. (1983). Immunosorbent electron microscopy in testing bean seed for bean common mosaic virus (BCMV). *Seed Science and Technology* **11**, 515-521.

Maden, S., Singh, D., Mathur, S.B. and Neergaard, P. (1975). Detection and location of seed-borne inoculum of *Ascochyta rabiei* and its transmission in chickpea (*Cicer arietinum*) *Seed Science and Technology* **3**, 667-681.

Maguire, J.D. and Gabrielson, R.L. (1983). Testing techniques for *Phoma lingam*. *Seed Science and Technology* **11**, 599-605.

Maguire, J.D., Gabrielson, R.L., Mulanax, M.W. and Russell, T.S. (1978). Factors affecting the sensitivity of 2,4 D assays of crucifer seed for *Phoma lingam*. *Seed Science and Technology* **6**, 915-924.

Makkouk, K.M. and Azzam, O.I. (1986). Detection of broad bean stain virus in lentil seed groups. *LENS Newsletter* **13**, 37-38.

Makkouk, K.M., Bos, L., Azzam, O.I., Katul, L. and Rizkallah, A. (1987). Broadbean stain virus: identification, detectability with ELISA in faba bean leaves and seeds, occurrence in West Asia and North Africa and possible wild hosts. *Netherlands Journal of Plant Pathology* **93**, 97-106.

Malin, E.M., Roth, D.A. and Belden, E.L. (1983). Indirect immunofluorescent staining for detection and identification of *Xanthomonas campestris* pv. *phaseoli* in naturally infected bean seed. *Plant Disease* **67**, 645-647.

Manandhar, J.B. and Cunfer, B.M. (1991). An improved selective medium for the assay of *Septoria nodorum* from wheat seed. *Phytopathology* **81**, 771-773.

Mangan, A. (1983). The use of plain water agar for the detection of *Phoma betae* on beet seed. *Seed Science and Technology* **11**, 607-614.

Masmoudi, K., Duby, C., Suhas, M., Guo, J.Q., Guyot, L., Olivier, V., Taylor, J. and Maury, Y. (1994). Quality control of pea seed for pea seed-borne mosaic virus. *Seed Science and Technology* **22**, 407-414.

Masmoudi, K., Suhas, M., Khetarpal, R.K. and Maury, Y. (1994). Specific serological detection of the transmissible virus in pea seed infected by pea seed-borne mosaic virus. *Phytopathology* **84**, 756-760.

Mathur, S.B. and Cunfer, B.M. (1993). 'Seedborne diseases and seed health testing of wheat'. Danish Government Institute of Seed Pathology for Developing Countries, Copenhagen, Denmark.

Maude, R.B. and Presley, A.H. (1977). Neck rot (*Botrytis allii*) of bulb onions. I. Seed-borne infection and its relationship to the disease in the onion crop. *Annals of Applied Biology* **86**, 163-180.

Maury, Y., Duby, C., Bossennec, J.M. and Boudazin, G. (1985). Group analysis using ELISA: determination of the level of transmission of soybean mosaic virus in soybean seed. *Agronomie* 5, 405-415.

Maury, Y., Bossennec, J.M., Boudazin, G., Hampton, R., Pietersen, G. and Maguire, J. (1987). Factors influencing ELISA evaluation of transmission of pea seed-borne mosaic virus in infected pea seed: seed-group size and seed decortication. *Agronomie* 7, 225-230.

McGee, D.C., Brandt, C.L. and Burris, J.S. (1980). Seed mycoflora of soybeans relative to fungal interactions, seedling emergence and carryover of pathogens for subsequent crops. *Phytopathology* 70, 615-617.

McGuire, R.G. and Jones, J.B. (1989). Detection of *Xanthomonas campestris* pv. *vesicatoria* in tomato. *In* 'Detection of Bacteria in Seed and Other Planting Material (A.W. Saettler, N.W. Schaad and D.A. Roth, eds), pp. 59-62. American Phytopathological Society, St. Paul, Minnesota.

Mehta, Y.R. (1990). Management of *Xanthomonas campestris* pv. *undulosa* and *hordei* through cereal seed testing. *Seed Science and Technology* 18, 467-476.

Mia, M.A.T., Mathur, S.B. and Neergaard, P. (1985). *Gerlachia oryzae* in rice seed. *Transactions of the British Mycological Society* 84, 337-338.

Middleton, J.T. (1944). Seed transmission of squash mosaic virus. *Phytopathology* 34: 405.

Ming, D., Ye, H.Z., Schaad, N.W. and Roth, D.A. (1991). Selective recovery of *Xanthomonas* spp. from rice seed. *Phytopathology* 81, 1358-1363.

Mink, G.I. and Aichele, M.D. (1984). Detection of Prunus necrotic ringspot and prune dwarf viruses in *Prunus* seed and seedlings by enzyme-linked immunosorbent assay. *Plant Disease* 68, 378-381.

Mohan, S.K. and Schaad, N.W. (1987). An improved agar plating assay for detecting *Pseudomonas syringae* pv. *syringae* and *P. s.* pv. *phaseolicola* in contaminated bean seed. *Phytopathology* 77, 1390-1395.

Morrall, R.A.A. and Beauchamp, C.J. (1988). Detection of *Ascochyta fabae* f.sp. *lentis* in lentil seed. *Seed Science and Technology* 16, 383-390.

Naumann, K., Karl, H., Gierz, E. and Griesbach, E. (1986). On the detection of *Pseudomonas syringae* pathovar *tomato*, the causal agent of the bacterial speck disease, on tomato seeds. *Zentralblatt für Mikrobiologie* 141, 301-321.

Neergaard, P. (1977). 'Seed pathology', Vol. I and II. MacMillan Press, London, UK.

Nemeth, J., Laszlo, E. and Emody, L. (1991). *Clavibacter michiganensis* ssp. *insidiosus* in lucerne seeds. *EPPO Bulletin* 21, 713-718.

Noble, M. (1965). Introduction to Series 3 of the Handbook on Seed Health Testing. *Proceedings of the International Seed Testing Association* 30, 1045-1115.

Nolan, P.A. and Campbell, R.N. (1984). Squash mosaic virus detection in individual seeds and seed lots of cucurbits by enzyme-linked immunosorbent assay. *Plant Disease* 68, 971-975.

Paludan, N. (1982). Virus attack in Danish cultures of sweet pepper (*Capsicum annuum* L.) specially concerning tobacco mosaic virus. *Acta Horticulturae* 127, 65-78.

Parashar, R.D. and Leben, C. (1972). Detection of *Pseudomonas glycinea* from soybean seed lots. *Phytopathology* 62, 1075-1077.

Paulsen, M.R. (1990). Using machine vision to inspect oilseeds. *INFORM - Interna-*

*tional News on Fats, Oils and Related Materials* **1**, 50-55.

Pesic, Z. and Hiruki, C. (1986). Differences in the incidence of alfalfa mosaic virus in seed coat and embryo of alfalfa seed. *Canadian Journal of Plant Pathology* **8**, 39-42.

Prabhu, M.S.C., Safeeulla, K.M., Venkatasubbaiah, P. and Shetty, H.S. (1984). Detection of seed-borne inoculum of *Peronosclerospora sorghum* in sorghum. *Phytopathologische Zeitschrift* **111**, 174-178.

Prosen, D., Hatziloukas, E., Schaad, N.W. and Panopoulos, N.J. (1993). Specific detection of *Pseudomonas syringae* pv *phaseolicola* DNA in bean seed by polymerase chain reaction-based amplification of a phaseolotoxin gene region. *Phytopathology* **83**, 965-970.

Pryor, B.M., Davis, R.M. and Gilbertson, R.L. (1994). Detection and eradication of *Alternaria radicina* on carrot seed. *Plant Disease* **78**, 452-456.

Raizada, R.K., Albrechtsen, S.E. and Lange, L. (1990). Detection of bean common mosaic virus in dissected portions of individual bean seeds using immunosorbent electron microscopy. *Seed Science and Technology* **18**, 559-565.

Ranganathaiah, K.G. and Mathur, S.B. (1978). Seed health testing of *Eleusine coracana* with special reference to *Dreschslera nodulosa* and *Pyricularia grisea*. *Seed Science and Technology* **6**, 943-951.

Rao, B.M., Prakash, H.S., Shetty, H.S. and Safeeulla, K.M. (1985). Downy mildew inoculum in maize seeds: techniques to detect seed-borne inoculum of *Peronosclerospora sorghi* in maize. *Seed Science and Technology* **13**, 593-600.

Rasmussen, O.F. and Reeves, J.C. (1992). DNA probes for the detection of plant pathogenic bacteria. *Journal of Biotechnology* **25**, 203-220.

Rasmussen, O.F. and Wulff, B.S. (1990). Identification and use of DNA probes for plant pathogenic bacteria. *In* 'Proceedings of the 5th European Congress on Biotechnology' (C. Christiansen, L. Munck and J. Villadsen, eds), pp. 693-698. Munksgaards, Copenhagen.

Rasmussen, O.F. and Wulff, B.S. (1991). Detection of *Pseudomonas syringae* pv. *pisi* using PCR. *In* 'Proceedings of the 4th International Working Group on *Pseudomonas syringae* Pathovars', (R.D. Durbin, G. Surico and L. Mugnai, eds), pp. 369-376. Stamperia Granducale, Florence.

Rennie, W.J. and Tomlin, M.M. (1984). Repeatability, reproducibility and interrelationship of results of tests on wheat seed samples infected with *Septoria nodorum*. *Seed Science and Technology* **12**, 863-880.

Richardson, M.J. (1990). 'An annotated list of seed-borne diseases - section 1.1 of the Handbook on seed health testing'. International Seed Testing Association, Zürich, Switzerland.

Ridout, M.S. (1994). Three stage designs for seed testing experiments. *Applied Statistics*. In press.

Roberts, S.J., Phelps, K., Taylor, J.D. and Ridout, M.S. (1993). Design and interpretation of seed health assays. *In* 'Proceedings 1st ISTA *Plant Disease* Committee Symposium on Seed Health Testing, Ottawa, Canada' (J.W. Sheppard, ed.), 115-125. International Seed Testing Association, Zürich, Switzerland.

Rohloff, I. and Marrou, J. (1981). Working sheet no. 9 - Lettuce mosaic virus. *In* 'ISTA Handbook on Seed Health Testing, section 2 - Working Sheets, each dealing with one pathogen on one host' (J. Jorgensen, ed.). International Seed

Testing Association, Zürich, Switzerland.

Ryder, E.J. and Johnson, A.S. (1974). A method for indexing lettuce seeds for seedborne lettuce mosaic virus by airstream separation of light from heavy seeds. *Plant Disease Reporter* **58**, 1037-1039.

Saettler, A.W., Schaad, N.W. and Roth, D.A. (1989). 'Detection of bacteria in seed and other planting material'. American Phytopathological Society, St. Paul, Minnesota.

Schaad, N.W. (1988a). 'Laboratory guide for identification of plant pathogenic bacteria'. American Phytopathological Society, St. Paul, Minnesota.

Schaad, N.W. (1988b). Reports of the working groups for comparative tests - Working group on bacterial pathogens. *In* 'Report of the 19th International Seminar on Seed Pathology, Wageningen, 1987', pp. 13-16. International Seed Testing Association, Zürich, Switzerland.

Schaad, N.W. (1989). Detection of *Xanthomonas campestris* pv. *campestris* in crucifers. *In* 'Detection of Bacteria in Seed and Other Planting Material (A.W. Saettler, N.W. Schaad and D.A. Roth, eds), pp. 68-75. American Phytopathological Society, St. Paul, Minnesota.

Schaad, N.W. and Forster, R.L. (1989). Detection of *Xanthomonas campestris* pv. *translucens* in wheat. *In* 'Detection of Bacteria in Seed and Other Planting Material (A.W. Saettler, N.W. Schaad and D.A. Roth, eds), pp. 41-44. American Phytopathological Society, St. Paul, Minnesota.

Schaad, N.W., Azad, H., Peet, R.C. and Panopoulos, N.J. (1989). Identification of *Pseudomonas syringae* pathovar *phaseolicola* by a DNA hybridization probe. *Phytopathology* **79**, 903-907.

Shaarawy, M.A., Abdelmonem, A.A., Soleman, N.K. and Sherif, S. (1991). Evaluation of methods for detecting *Dreschslera* sp. and *Fusarium moniliforme* from wheat seeds. *Assiut Journal of Agricultural Sciences* **22**, 117-128.

Sheppard, J.W. (1993). Diagnostic sensitivity, specificity and predictive values in evaluation of test methods. *In* 'Proceedings 1st ISTA *Plant Disease* Committee Symposium on Seed Health Testing, Ottawa, Canada' (J.W. Sheppard, ed.), 132-142. International Seed Testing Association, Zürich, Switzerland.

Sheppard, J.W., Wright P.F. and DeSavigny, D.H., 1986. Methods for the evaluation of EIA tests for use in the detection of seed-borne diseases. *Seed Science and Technology* **14**, 49-59.

Sheppard, J.W., Roth, D.A. and Saettler, A.W. (1989). Detection of *Xanthomonas campestris* pv. *phaseoli* in bean. *In* 'Detection of Bacteria in Seed and Other Planting Material (A.W. Saettler, N.W. Schaad and D.A. Roth, eds), pp. 17-29. American Phytopathological Society, St. Paul, Minnesota.

Shetty, S.A. and Shetty, H.S. (1988). Development and evaluation of methods for the detection of seed-borne fungi in rice. *Seed Science and Technology* **16**, 693-698.

Shetty, H.S., Khamzada, A.K., Mathur, S.B. and Neergaard, P. (1978). Procedures for detecting seed-borne inoculum of *Sclerospora graminicola* in pearl millet (*Pennisetum typhoides*). *Seed Science and Technology* **6**, 935-941.

Shrivastava, M. and Yadav, S. (1990). Detection of loose smut infection in wheat seeds by ergosterol estimation. *Comparative Physiology and Ecology* **15**, 17-20.

Sijam, K., Chang, C.J. and Gitaitis, R.D. (1991). An agar medium for the isolation and identification of *Xanthomonas campestris* pv. *vesicatoria* from seed.

*Phytopathology* **81**, 831-834.

Sijam, K., Chang, C.J. and Gitaitis, R.D. (1992). A medium for differentiating tomato and pepper strains of *Xanthomonas campestris* pv. *vesicatoria*. *Canadian Journal of Plant Pathology* **14**, 182-184.

Singh, D.V., Mathur, S.B. and Neergaard, P. (1974). Seed health testing of maize. Evaluation of testing techniques with special reference to *Dreschslera maydis*. *Seed Science and Technology* **2**, 349-365.

Slack, S.A. and Shepherd, R.J. (1975). Serological detection of seed-borne barley stripe mosaic virus by a simplified radial-diffusion technique. *Phytopathology* **65**, 948-955.

Stockwell, V.O. and Trione, E.J. (1986). Distinguishing teliospores of *Tilletia controversa* from those of *T. caries* by fluorescence microscopy. *Plant Disease* **70**, 924-926.

Taylor, J.D. (1984). Demonstration of the dilution plating method for the detection of *Pseudomonas phaseolicola* and *Pseudomonas pisi*. Report on the 1st International Workshop on Seed Bacteriology, Angers, France, 1982, pp. 33-35. International Seed Testing Association, Zürich, Switzerland.

Taylor, J.L. (1993). A simple, sensitive and rapid method for detecting seed contaminated wisth highly virulent *Leptosphaeria maculans*. *Applied and Environmental Microbiology* **59**, 3681-3685.

Taylor, J.D. and Phelps, K. (1984). Estimation of percentage seed infection. Report on the 1st International Workshop on Seed Bacteriology, Angers, France, 1982, pp. 12-14. International Seed Testing Association, Zürich, Switzerland.

Taylor, J.D., Phelps, K. and Dudley, C.L. (1979). Epidemiology and strategy for the control of halo blight beans. *Annals of Applied Biology* **93**, 167-172.

Taylor, J.D., Phelps, K. and Roberts, S.J. (1993). Most probable number (MPN) method: origin and application. *In* 'Proceedings 1st ISTA *Plant Disease* Committee Symposium on Seed Health Testing, Ottawa, Canada' (J.W. Sheppard, ed.), 106-114. International Seed Testing Association, Zürich, Switzerland.

Tourte, C. and Manceau, C. (1991). Direct detection of *Pseudomonas syringae* pathovar *phaseolicola* using the polymerase chain reaction (PCR). *In* 'Proceedings of the 4th International Working Group on *Pseudomonas syringae* Pathovars (R.D. Durbin, S. Surico and L. Mugnai, eds). pp. 402-403. Stamperia Granducale, Florence, Italy.

Tourte, C. and Manceau, C. (1994). Evaluation of a digoxygenin-labelled phaseolotoxin gene fragment as a DNA probe for detection of *Pseudomonas syringae* pathovar *phaseolicola* from bean seeds. *Seed Science and Technology* **22**, 449-459.

Trigalet, A., Samson, R. and Coleno, A. (1978). Problems related to the use of serology in phytobacteriology. Proceedings of the 4th International Conference on Plant Pathogenic Bacteria, Angers, pp. 271-288.

Trione, E.J. and Krygier, B.B. (1977). New tests to distinguish teliospores of *Tilletia controversa*, the dwarf bunt fungus, from spores of other *Tilletia* species. *Phytopathology* **67**, 1166-1172.

Unnamalai, N., Mew, T.W. and Gnanamanickam, S.S. (1988). Sensitive methods for detection of *Xanthomonas campestris* pv. *oryzae* in rice seeds. *In* 'Advances in Research on Plant Pathogenic Bacteria - Proceedings of the National Symposium on Phytobacteriology, Madras, India, 1986 (S.S. Gnanamanickam and A. Mahadevan, eds), pp. 73-82. University of Madras, India.

Vannacci, G. (1981). Record of *Fusarium moniliforme* var. *subglutinans* Wr. and Reink on onion seeds and its pathogenicity. Reperimento di *Fusarium moniliforme* var. *subglutinans* Wr. et Reink, su seme dicipolla e sua patogenicita. *Phytopathologia Mediterranea* **20**, 144-148.

van Vaerenbergh, J.P.C. and Chauveau, J.F. (1987). Detection of *Corynebacterium michiganense* in tomato seed lots. *EPPO Bulletin* **17**, 131-138.

van Vuurde, J.W.L. and Maat, D.Z. (1983). Routine application of ELISA for the detection of lettuce mosaic virus in lettuce seeds. *Seed Science and Technology* **11**, 505-513.

van Vuurde, J.W.L. and Maat, D.Z. (1985). Enzyme-linked immunosorbent assay (ELISA) and disperse dye immuno assay (DIA): comparison of simultaneous and separate incubation of sample and conjugate for the routine detection of lettuce mosaic virus and pea early browning virus in seeds. *Netherlands Journal of Plant Pathology* **91**, 3-13.

van Vuurde, J.W.L. and van den Bovenkamp, G.W. (1984). Detection of *Pseudomonas syringae* pv. *phaseolicola*. In 'Report of the 18th International Seminar on Seed Pathology, Washington, 1984', pp. 8-11. International Seed Testing Association, Zürich, Switzerland.

van Vuurde, J.W.L. and van den Bovenkamp, G.W. (1989). Detection of *Pseudomonas syringae* pv. *phaseolicola* in bean. In 'Detection of Bacteria in Seed and Other Planting Material (A.W. Saettler, N.W. Schaad and D.A. Roth, eds), pp. 30-40. American Phytopathological Society, St. Paul, Minnesota.

van Vuurde, J.W.L., van den Bovenkamp, G.W. and Birnbaum, Y. (1983). Immunofluorescence microscopy and enzyme-linked immunosorbent assay as potential routine tests for the detection of *Pseudomonas syringae* pv. *phaseolicola* and *Xanthomonas campestris* pv. *phaseoli* in bean seed. *Seed Science and Technology* **11**, 547-559.

van Vuurde, J.W.L., Maat, D.Z. and Franken, A.A.J.M. (1988). Immunochemical technology in indexing propagative plant parts for viruses and bacteria in The Netherlands. In 'Biotechnology for Crop Protection' (P.A. Hedin, J.J. Menn and R.M. Hollingworth, eds), pp. 338-350. American Chemical Society, Washington DC, USA.

van Vuurde, J.W.L., Franken, A.A.J.M., Birnbaum, Y. and Jochems, G. (1991). Characteristics of immunofluorescence microscopy and of dilution plating to detect *Pseudomonas syringae* pv. *phaseolicola* in bean seed lots and for risk assessment of field incidence of halo blight. *Netherlands Journal of Plant Pathology* **97**, 233-244.

Wang, D., Woods, R.D., Cockbain, A.J., Maule, A.J. and Biddle, A.J. (1993). The susceptibility of pea cultivars to pea seed-borne mosaic virus infection and virus seed transmission in the UK. *Plant Pathology* **42**, 42-47.

Webster, D.M., Atkin, J.D. and Cross, J.E. (1983). Bacterial blights of snap beans and their control. *Plant Disease* **67**, 935-940.

Welty, R.E., Milbrath, G.M., Faulkenberry, D., Azevedo, M.D., Meek, L. and Hall, K. (1986). Endophyte detection in tall fescue seed by staining and ELISA. *Seed Science and Technology* **14**, 105-116.

Wharton, A.L. (1967). Detection of infection by *Pseudomonas phaseolicola* (Burkh.)

Dowson in white-seeded dwarf bean seed stocks. *Annals of Applied Biology* **60**, 29-36.

Wong, W.C. (1991). Methods for recovery and immunodetection of *Xanthomonas campestris* pv. *phaseoli* in navy bean seed. *Journal of Applied Bacteriology* **71**, 124-129.

Wu, W.S. and Lu, J.H. (1984). A semiselective medium for detecting seed-borne *Alternaria brassicicola*. *Plant Protection Bulletin* **26**, 67-72.

Wu, W.S. and Mathur, S.B. (1987). Evaluation of methods for detecting *Fusarium moniliforme* in sorghum seeds. *Seed Science and Technology* **15**, 821-829.

Wylie, S., Wilson, C.R., Jones, R.A.C. and Jones, M.G.K. (1993). A polymerase chain reaction assay for cucumber mosaic virus in lupin seeds Australian *Journal of Agricultural Research* **44**, 41-51.

Yao, C.L., Frederiksen, R.A. and Magill, C.W. (1990). Seed transmission of sorghum downy mildew: detection by DNA hybridisation. *Seed Science and Technology* **18**, 201-207.

Yao, C.L., Magill, C.W., Frederiksen, R.A., Bonde, M.R., Wang, Y. and Wu, P. (1991). Detection and identification of *Peronosclerospora sacchari* in maize by DNA hybridization. *Phytopathology* **81**, 901-905.

Yu, S.H., Mathur, S.B. and Neergaard, P. (1982). Taxonomy and pathogenicity of four seed-borne species of *Alternaria* from sesame. *Transactions of the British Mycological Society* **78**, 447-458.

Zimmer, R.C. (1979). Influence of agar on immuno diffusion serology of pea seed-borne mosaic virus. *Plant Disease Reporter* **63**, 278-282.

# 8

# A ROLE FOR PATHOGEN INDEXING PROCEDURES IN POTATO CERTIFICATION

## S.H. De Boer[1], S.A. Slack[2], G.W. van den Bovenkamp[3] and I. Mastenbroek[3]

[1]*Pacific Agriculture Research Centre, Vancouver, British Columbia, Canada*
[2]*Cornell University, Ithaca, New York, United States*
[3]*Netherlands General Inspection Service for Agricultural Seeds and Seed Potatoes, Ede, The Netherlands*

## I. INTRODUCTION

Seed potato production is a highly sophisticated industry in Western Europe, Canada, and the United States that produces relatively disease-free seed of recognized cultivars. Freedom from pathogens is one of the primary concerns of this

Advances in Botanical Research Vol. 23
Incorporating Advances in Plant Pathology
ISBN 0-12-005923-1

industry and it is an important factor in domestic and international trade and marketing of seed. The use of clean seed increases the quantity and quality of subsequent potato crops used for commercial consumption and processing.

Disease incidence in seed potato crops is minimized by beginning production schemes from micropropagated plantlets and by limiting the number of generations seed potatoes are grown in the field prior to commercial production for table-stock and processing. In Canada, for example, "elite" seed can only be grown in the field for a maximum of five generations before it is down-graded to the "Foundation" level.

Seed potato-producing countries usually have national or regional certification programs under which seed potato crops are visually inspected for the presence of disease (Shepard and Claflin, 1975). Tolerance levels are set for each class of seed and differ according to the risk a disease poses for the potato industry. However, the presence of pathogens in the absence of overt disease symptoms is a threat to the industry which cannot be addressed by certification programs based on visual inspection. Laboratory tests are required to index seed potatoes for symptomless and latent infections of specific disease-causing agents. Indexing for viral, bacterial, or nematode pathogens is currently being carried out in various countries to reduce disease incidence in domestic potato production, to access foreign markets, or to eradicate a particular disease. These indexing procedures invariably involve laboratory procedures applied to samples drawn from consignments of seed potato tubers or from foliage during the growing season. The availability of laboratory protocols for sensitive and specific detection of pathogenic microorganisms has proved to be valuable for addressing specific disease problems. In this chapter we review how indexing technologies have been applied successfully for detecting viral diseases, symptomless ring rot infections, contamination by the blackleg pathogen, and infestations of the potato cyst nematode.

## II. VIRUS INDEXING IN THE UNITED STATES

### A. Potato Viruses in North America

Six potato viruses are known to occur in seed potato stocks in North America. They are potato leafroll virus (PLRV) and the potato viruses A (PVA), M (PVM), S (PVS), X (PVX) and Y (PVY) (Shepard and Claflin, 1975). Alfalfa mosaic virus, which causes the calico disease, sugar beet curly top virus and tobacco rattle virus have been reported but are not considered to be carried in the seed system. Slack (1991) performed a survey to determine the relative importance of potato viruses to the potato industry (Table I) and their relative importance compared to disease problems in the seed industry (Table II) and the potato industry generally (Table III). It is clear that PLRV and PVY are the most important viruses. The potato spindle tuber viroid (PSTV) which is often grouped with the virus diseases, has largely been eliminated through systematic testing over the past two decades.

*Table I.* The most important potato viruses in the United States and Canada

| Virus | Disease | Rank | Mean Value* | % of Total | Ranked First (#) |
|---|---|---|---|---|---|
| Potato leafroll virus | Leafroll | 1 | 5.2 | 86 | 19 |
| Potato Virus Y | Mosaic | 2 | 4.0 | 67 | 10 |
| Potato Virus X | Mosaic/Latent | 3 | 3.2 | 54 | 2 |
| Potato Virus S | Mosaic/Latent | 4 | 2.2 | 36 | 1 |
| Potato Virus A | Mosaic | 5 | 1.3 | 22 | 0 |
| Potato Virus M | Mosaic | 6 | 0.8 | 14 | 0 |
| Alfalfa Mosaic Virus | Calico | 7 | 0.2 | 3 | 0 |
| Sugar Beet Curly Top Virus | Curly Top | 8 | 0.1 | 1 | 0 |
| Tobacco Rattle Virus | Stem Mottle | 9 | 0.1 | 1 | 0 |

*Scale of 1 (low) to 6 (high), 32 of 35 respondents.

*Table II.* The most important disease problems in seed potatoes in the United States and Canada

| Disease | Pathogen | Rank | Mean Value* | Ranked First | # Times Ranked |
|---|---|---|---|---|---|
| Ring Rot | Clavibacter | 1 | 1.2 | 12 | 17 |
| Leafroll | Leafroll Virus | 2 | 1.0 | 3 | 20 |
| Mosaic | PVY | 3 | 1.0 | 8 | 13 |
| Blackleg | Erwinia | 4 | 0.8 | 3 | 13 |
| Dry Rot | Fusarium | 5 | 0.3 | 2 | 6 |
| Scab | Streptomyces | 6-8 | 0.3 | 1 | 5 |
| Soft Rot | Erwinia | 6-8 | 0.3 | 0 | 5 |
| Wilt | Verticillium | 6-8 | 0.3 | 1 | 5 |
| Canker/Black Scurf | Rhizoctonia | 9 | 0.2 | 1 | 4 |
| Late Blight | Phytophthora | 10-11 | 0.1 | 0 | 2 |
| Spindle Tuber | Spindle Tuber Viroid | 10-11 | 0.1 | 0 | 2 |
| Mosaic/Latent | PVX | 12-13 | 0.1 | 0 | 2 |
| Silver Scurf | Helminthosporium | 12-13 | 0.1 | 1 | 1 |
| Root Knot | Meloidogyne | 14 | 0.1 | 0 | 1 |
| Early Blight | Alternaria | 15-17 | | 0 | 1 |
| Mosaic | PVA | 15-17 | | 0 | 1 |
| Mosaic/Latent | PVS | 15-17 | | 0 | 1 |

*Scale of 1 (low) to 3 (high), 33 respondents

Table III. The most important diseases of seed and commercial potatoes in the United States and Canada

| Disease · | Pathogen | Rank | Mean Value* | Ranked First | # Times Ranked |
|---|---|---|---|---|---|
| Early Dying/Wilt | Verticillium (complex) | 1 | 1.2 | 8 | 19 |
| Ring Rot | Clavibacter | 2 | 0.8 | 8 | 12 |
| Blackleg | Erwinia | 3 | 0.6 | 2 | 11 |
| Scab | Streptomyces | 4 | 0.5 | 4 | 9 |
| Leafroll | Leafroll Virus | 5 | 0.4 | 2 | 7 |
| Soft Rot | Erwinia | 6 | 0.3 | 1 | 7 |
| Mosaic | PVY | 7-8 | 0.3 | 1 | 6 |
| Canker/Black Scurf | Rhizoctonia | 7-8 | 0.3 | 1 | 6 |
| Dry Rot | Fusarium | 9-11 | 0.3 | 2 | 4 |
| Late Blight | Phytophthora | 9-11 | 0.3 | 3 | 3 |
| Root Knot | Meloidogyne | 9-11 | 0.3 | 2 | 4 |
| Early Blight | Alternaria | 12 | 0.2 | 1 | 4 |
| Silver Scurf | Helminthosporium | 13 | 0.1 | 0 | 2 |
| Bacterial Wilt | Pseudomonas | 14-15 | 0.1 | 0 | 1 |
| Pink Rot | Phytophthora | 14-15 | 0.1 | 0 | 1 |
| Mosaic | PVA | 16-18 | | 0 | 1 |
| Leak | Pythium | 16-18 | | 0 | 1 |
| White Mold | Sclerotinia | 16-18 | | 0 | 1 |

*Scale of 1 (low) to 3 (high), 35 respondents.

## B. Potato Virus Biology

The importance of these viruses is directly related to the activity of aphid vectors. The green peach aphid, Myzus persicae Sulz., is the most important vector. In most seed production areas, green peach aphid migrations occur in the latter half of the season. The relationship of spread to the green peach aphid is particularly linked for the phloem-limited PLRV. Some states utilize aphid pan traps to alert growers that aphid migrations have started. These alerts trigger spray applications to control aphid populations and vine kill applications to minimize tuber-borne virus incidence. Recent outbreaks of the mosaic disease caused by PVY$^O$ have emphasized that other aphid species can be important in spread of this virus within seed lots, especially early-season spread when plant susceptibility to infection is greater. Since PVX is not aphid-transmitted, systematic elimination of this virus from seed lots has been easier.

The potato tuber is the primary inoculum source for all of these potato viruses, however, PVY has a wide host range and weeds or ornamental crops can be an inoculum source. PSTV is the only one of these pathogens known to be transmitted through botanical seed, a reason that it has traditionally plagued germplasm collections. Although spread within the crop may occur via contact, this mechanism of spread is most important only to PVX and PSTV. Thus, seed certification programs have emphasized the use of "virus-free" or "virus-tested" seed lots. This goal has become more realistic in the last two decades with the widespread application of tissue culture systems to propagate plantlets and of serological tests, especially ELISA, to test these plantlets along with the concomitant adoption of limited generation programs that require the use of these tested stocks as the initial or "pre-nuclear" seed class (Slack, 1980). In addition, all states require that a postharvest test be carried out on a sample, usually 400 tubers, of a seed lot that will be entered for certification the following year.

Symptomless infections are the principal reason for significant disease outbreaks in recent years. This is not to say that virus spread within an improperly managed crop cannot occur, but these seed lots are rapidly eliminated from the seed system. Additionally, virus will be detected in properly sampled seed lots for postharvest tests in cases where infection occurs too late for disease symptoms to develop. More problematic have been cases in which distinct disease symptoms do not occur. In one case, two new cultivars which are symptomless or nearly so to $PVY^O$ infection enabled some early seed lots to be infected at high incidence levels. Rapid multiplication and distribution of these lots to meet industry demands led to significant losses as $PVY^O$ moved into other less tolerant cultivars. Control was achieved by systematic removal of these lots and their replacement with pathogen-tested lots. In another case, $PVY^N$ was introduced into North American seed stocks. It was distributed to several seed lots and cultivars before detection because affected plants were symptomless or nearly so. Once detected, systematic testing of seed lots provided for functional eradication. The existence of cultivars that do not express distinct symptoms upon infection and of virus strains that do not incite distinctive symptoms will continue to provide an opportunity for significant disease outbreaks to develop.

## C. Indexing Strategies

The utilization of serological tests to detect potato virus infections in North American seed lots was pioneered by Shepard (Shepard and Claflin, 1975) and Wright (1988) and targeted the detection of symptomless PVX and PVS infections. These initial programs emphasized the collection and testing of leaves from the growing crop. Several Western states still do summer testing of leaves for PVX and mark seed tags as "PVX-tested". This testing is particularly important for seed lots that are not part of a limited generation program (five of 16 states permit "dual" programs, i.e., a limited generation program and a non-limited generation program based on hill selection and clonal increase). In the United States and Canada the term "Limited Generation" denotes that seed certification agencies limit the number of field generations (majority = 7 years; range = 4-8 years) a seed lot, which originates from tissue culture origin, can be entered for certification (Slack, 1993).

In Montana, for example, testing leaves from the growing crop to detect PVX was initiated in 1972. The Ouchterlony double diffusion serological technique was used until 1981 at which time they switched to ELISA. In 1991, they added PVY to their testing service. Testing for PVX and PVY is now mandatory. They employ two leaf picking crews (6 people each) to collect samples with the Nuclear (their first field generation) and Generation 1 seed lot classes being sampled 2-4 weeks after emergence or transplanting and Generations 2-4 being sampled by August 20. Montana has a five year limited generation system for seed lots. Leaf samples are tested in composites of 10 leaves with the percent of infection calculated as follows: $P = 1 \cdot \sqrt[n]{1-x}$, where n = number of leaves in composite, x = number of positive tests and P = probability of infection. The actual incidence of PVX and PVY by seed generation in Montana is shown in Table IV. Lund and Sun (1985) have addressed the general question of sample size determination for the certification of seed potato lots and Clayton and Slack (1988) have addressed this question in terms of zero tolerance disease situations.

The adoption of limited generation systems based on the multiplication of seed lots originating from a pathogen-tested *in vitro* source enabled virus testing to become systematic and comprehensive. Key elements for testing large sample numbers were the development of the ELISA test (Clark and Adams, 1977) and the ready availability of high quality antisera (suitable specificity and avidity). Commercial antisera and ELISA kits are now readily available to complement efforts by private . and public laboratories. Pre-nuclear class seed stocks must be 100% tested for the six viruses noted earlier. As seed stocks are multiplied, samples from successive generations are also tested. The sample intensity and the number of generations tested by ELISA varies by state. More recently, ELISA variations based on blotting sap extracts onto membranes (Whitworth *et al.*, 1993) and polymerase chain reaction amplification (Hadidi *et al.*, 1993) have been reported. These latter assays, as yet, are not utilized on a wide scale.

PSTV is also tested at the 100% level at the pre-nuclear class. Early diagnostic tests were based on tomato bioassays developed in the 1960s (Raymer and O'Brien, 1962) and later gave way to polyacrylamide gel electrophoresis assays (Morris and Wright, 1972; Pfannenstiel and Slack, 1980b). The development of the hybridization assay (Owens and Diener, 1981; Palukaitis *et al.*, 1985) and the commercial availability of this test, however, have been the key to intensive, sustained testing. Most germplasm collections have also been systematically screened for PSTV over the past decade. PSTV has not been considered a production concern in the United States for two decades and has only been reported sporadically in North America (Singh and Crowley, 1985). Continued testing is considered essential, however, because mild PSTV strains and tolerant cultivars do exist (Pfannenstiel and Slack, 1980a) and discontinuance of testing could enable PSTV to be reintroduced through new germplasm sources.

In the state of New York, Cornell University operates a seed potato farm dedicated to the production of nuclear class seed stocks at Lake Placid. This farm is isolated from other potato production and produces >95% of the seed lots entered for certification by private New York seed growers on an annual basis. All seed lots originate from an *in vitro* pathogen-tested source. Tests include ELISA for the potato viruses A, M, S, X, Y and leafroll; nucleic acid hybridization for potato spindle tuber viroid; ELISA for the bacteria *Erwinia carotovora* subsp. *atroseptica* and *Clavibacter michiganensis* subsp. *sepedonicus*; and cultural tests for general bacteria and fungi. Each year tubers harvested from greenhouse-grown plants are reintroduced into

Table IV. Potato virus X (PVX) and potato virus Y (PVY) incidence in field tested leaf samples in Montana.

| | Number of | | Percentage of Positive Tests In a Potato Generation[*] | | | | | | | | |
| | | | Nuclear | | G1 | | G2 | | G3 | | G4 | |
| Year | Fields | Tests | PVX | PVY | PVX | PVY | PVX | PVY | PVX | PVY | PVX | PVY |
|---|---|---|---|---|---|---|---|---|---|---|---|---|
| 1994 | 434 | 271779 | 0.01 | 0.21 | 0.39 | 0.12 | 0.33 | 0.27 | 1.31 | 1.01 | 0.00 | 0.00 |
| 1993 | 510 | 229147 | 0.00 | 0.01 | 0.01 | 0.14 | 0.21 | 0.87 | 1.06 | 0.81 | 4.25 | 1.76 |
| 1992 | 486 | 222607 | 0.02 | 0.07 | 0.07 | 0.20 | 0.42 | 0.48 | 0.93 | 0.53 | 3.64 | 0.00 |
| 1991 | 455 | 346873 | 0.06 | 0.05 | 0.20 | 0.86 | 0.12 | 2.97 | 1.99 | 3.55 | 1.12 | 1.66 |
| MEAN | 471 | 287602 | 0.02 | 0.09 | 0.17 | 0.33 | 0.27 | 1.15 | 1.32 | 1.48 | 2.25 | 0.86 |
| % Allowable Tolerance | | | 0 | 0 | 0 | 0 | 1 | 0.5 | 2 | 1 | 4 | 2 |

[*]100% of Nuclear and G1 plants tested; minimum of 1000 leaves/lot tested G2-G4
Data courtesy Dr. Mike Sun, Director of Seed Potato Certification, Montana, United States

tissue culture and these tests are repeated. General bacterial and fungal contamination has been low (<0.1% incidence) and PSTV has not been detected for more than a decade. When viruses are detected in cultivars being brought into the program, a combined heat and chemotherapy treatment of the *in vitro* plantlets is utilized to obtain virus-free plantlets prior to introducing the cultivar to the Cornell-Uihlein Farm (Griffiths *et al.*, 1990; Sanchez *et al.*, 1991). Following this protocol, viruses have not been detected following repeated *in vitro* culture and greenhouse grow-outs. The experience of testing greenhouse and field-grown plants for viruses at the Cornell-Uihlein Farm is shown in Table V. In addition, a 1992 field survey of private seed grower fields for PVX and PVY detected the following number of positive plant samples: PVX = 0, $PVY^N$ = 0, $PVY^O$ = 4. A total of 27 seed lots representing 82% of the New York seed farms and 13% of the seed lots (19% of the acreage) were tested at the 1,000 plants/seed lot level. Plants were tested in composites of 10 plants/test with positive samples retested as individual plants. The four $PVY^O$ positive plants represented two plants each from two seed lots, one a G2 seed lot of cv. Russet Norkotah and the other a G4 seed lot of cv. Genesee. The New York data for lack of PVX reinfection is similar to the experience in Wisconsin (Hahm *et al.*, 1981).

Utilization of serological tests at higher generation levels (e.g., Generation 3 and above or Foundation and Certified) is usually on a "need" basis to supplement visual inspections. This is in addition to the PVX and PVY testing mentioned above. ELISA tests are used to confirm the identity or presence of specific viruses in mottled or "atypical" plants and to confirm diagnostic symptoms, especially in new cultivars. These tests may be performed during summer inspections or winter grow outs. The use of ELISA testing during winter grow outs has intensified in recent years because

*Table V.* Virus test results at the Cornell-Uihlein Nuclear Seed Potato Farm at Lake Placid, New York.

| Seed Generation | Cultivars or Plants | Number of cultivars and plants tested* | | | | | |
|---|---|---|---|---|---|---|---|
| | | 1990 | 1991 | 1992 | 1993 | 1994 | Totals |
| Prenuclear | Cultivars | 59 | 53 | 55 | 59 | 58 | 284 |
| (greenhouse) | Plants tested | 1350 | 1400 | 1328 | 1380 | 1640 | 7098 |
| | Plants positive | 0 | 0 | 0 | 0 | 0 | 0 |
| Nuclear 1 | Cultivars | 34 | 27 | 37 | 46 | 49 | 193 |
| (field) | Plants tested | 440 | 370 | 490 | 880 | 880 | 3060 |
| | Plants positive | 0 | 0 | 0 | 0 | 0 | 0 |
| Nuclear 2 | Cultivars | 22 | 48 | 36 | 33 | 37 | 176 |
| (field) | Plants tested | 1760 | 1940 | 20,220 | 25,330 | 12,680 | 61,930 |
| | Plants positive | **1 | **1 | **2 | 0 | 0 | 4 |

*Plants tested for potato viruses A, M, S, X, Y and leafroll in composites of 10 plants/test.
**Represents PVS positive plants, cv. Green Mountain in 1990-92 and cv. Katahdin in 1992.

inspection personnel have found that they provide an important, independent assessment of plant health. In addition, the recognition that some cultivars may be symptomless carriers of PVY has led to prescribed testing of these cultivars (e.g., 100 leaflets from 400 tuber winter sample in composites of 5 with positive composites retested individually). Some states have required such testing on seed lots of PVY$^O$ symptomless cultivars before they will permit seed lots to be entered for certification in their state when the seed lot originates from out-of-state.

## D. Current Status and Future Outlook

ELISA tests are an integral part of current seed potato certification programs to control potato viruses in North America and will remain so for the near future. It is possible that DNA-based tests will replace ELISA, at least in part. This change will depend on test cost and availability as well as test specificity and sensitivity. ELISA tests are suitable for testing individual plants for all of the important viruses, but composites exceeding 5-10 plants are useful only for PVX in lots where a low incidence of virus infection is expected. Lack of test specificity is usually not a problem with antisera now available, rather the concern is more with whether or not antisera have suitable broad specificity to detect all strains of a given virus. The use of monoclonals has been central to this discussion as they permit the elimination of background problems in some tests, yet may be too specific for general seed certification use. The development of a potyvirus-specific antiserum (Jordan and Hammond, 1991) has been one interesting development. The use of a panel of monoclonals to simultaneously detect and strain-type PVY (PVY$^O$ and PVY$^N$) has been suggested as another approach (McDonald *et al.*, 1994).

Current programs have been very successful with 0-10% of the seed lots exceeding tolerances of 0.3-1.0% on an annual basis. It is probable, however, that indexing will even increase in importance in the future. Industry awareness of seed quality and the expectation that a given seed quality will be met on a continuing basis has become the norm. Further, as the interests of science and society become increasingly linked, we can expect qualitative and quantitative changes in the insecticides used to control vectors and an increased dialogue over the appropriateness of selected genetic manipulations to enhance virus resistance in cultivars and germplasm. For example, if virus coat protein-mediated resistance was accepted and implemented, then monitoring the effectiveness of that resistance would be important to seed certification programs. Indexing methods would need to be re-evaluated to enhance effective utilization. It is most likely that there will be a mix of susceptible and resistant cultivars and of available prescription-based insecticides. Indexing to define the nature and severity of virus problems may indeed be a prescription requirement.

## III. INDEXING FOR BACTERIAL RING ROT

### A. The Ring Rot Disease

The bacterial ring rot disease of potato, caused by the bacterium, *Clavibacter michiganensis* subsp. *sepedonicus*, is characterized by interveinal chlorosis and curling of leaves, wilting of foliage, and decay in tubers. Tuber decay in the vascular tissue is particularly characteristic and accounts for serious yield loss (Fig. 1). Symptom development, however, is highly dependent on potato cultivar, bacterial strain, inoculum dose, and growing conditions (Bishop and Slack, 1987a; Westra and Slack, 1994; Westra et al., 1994).

In the 1930's and 1940's this disease became a very serious threat to potato production in Canada and the United States and consequently became largely responsible for uniform adoption of a zero tolerance regulation by seed potato certification programs. Under these programs, seed potatoes are visually inspected for ring rot symptoms during the growing season and after harvest. All seed lots with any symptomatic plants or tubers are eliminated from the certification program (De Boer and Slack, 1984). Reasonable levels of control over ring rot have been achieved in this way. However, the persistence of infections that escaped the notice of inspectors as well as symptomless infections, resulted in occasional outbreaks of the disease. Because the pathogen is readily spread to potato tubers from contaminated equipment and potato storages, even slight levels of infection can cause widespread dispersion of the pathogen among seed lots and to new geographic areas (Stead, 1993). In temperate climates ring rot has become the most dreaded disease in seed potatoes because of the monetary cost associated with elimination of an entire potato crop from a certification program and the social stigma attached to its appearance.

*Fig. 1.* Vascular decay in a potato tuber typical of the bacterial ring rot disease.

## B. Biology of *C. m. sepedonicus*

*C. m. sepedonicus* is a Gram-negative, rod-shaped bacterium that grows slowly on artificial media and has no other known natural host but potato. The pathogen is highly infectious but can infect potato plants without causing them to express symptoms of disease. Although some potato cultivars are far more susceptible than others to the bacterial ring rot disease, all cultivars can serve as symptomless carriers of the bacterium (De Boer and McCann, 1990). The severity of disease symptoms not only varies greatly among cultivars, but is also affected by inoculum load and by environmental conditions such as temperature and light (Nelson, 1982; Nelson and Kozub, 1983).

The primary inoculum source for the ring rot disease is infected potato tubers which in turn may cause contamination of planting and harvesting equipment and storage bins. In Canada and the United States where seed potatoes are usually cut to increase the number of seed pieces, the cutting machines are thought to be a primary means by which the disease is spread (De Boer and Slack, 1984). Persistence of *C. m. sepedonicus* in the field has not been adequately studied, but survival in potato stems and tubers left outdoors for many months has been documented (Nelson, 1979, 1980). In North Dakota, *C. m. sepedonicus* was found to be present in roots of sugar beets grown in rotation with potato and subsequently was also found in sugar beet seed (Bugbee *et al.*, 1987). The importance of sugar beet in providing a haven for *C. m. sepedonicus* inoculum from which potato can be reinfected, however, was not established. Also, the apparent survival of *C. m. sepedonicus* in solanaceous weeds in Colorado and their role in the epidemiology of the disease has not been confirmed (Zizz and Harrison, 1991).

In potato, *C. m. sepedonicus* is translocated from the seed piece to the plant and progeny tubers via xylem tissue. Bacterial populations in potato stems increase during the growing season to very high densities (De Boer and McCann, 1989) but symptoms normally do not develop until some time after high bacterial densities have been established. Consequently, potatoes may be harvested before disease symptoms are expressed and bacteria from such infected plant material are widely disseminated by normal farm activities.

## C. Indexing Strategies for *C. m. sepedonicus*

In 1976, Olsson in Sweden first showed that symptomless ring rot infections could be detected by inoculation of potato tuber extracts into eggplants. Subsequently Lelliott and Sellar (1976) elaborated on the procedure and provided data on its sensitivity. Although the sensitivity achieved by Lelliot and Sellar has not been substantiated by subsequent work (Zeller and Xie, 1985), the procedure is still valuable as often the only process by which the pathogen can be isolated from symptomless ring rot infections. The routine application of the eggplant test for indexing, however, is not practical as eggplant tests require a large amount of greenhouse space and carefully controlled growing conditions, are affected by the presence of other microorganisms in test samples, and may itself give rise to symptomless infections (Bishop and Slack, 1987; Persson and Janse, 1988). Despite the success of the eggplant test in specific instances, routine testing with eggplant yields inconsistent results.

The development of the immunofluorescence procedure provided a means for determining the presence of *C. m. sepedonicus* without the need for cultivation of the bacterium. This procedure formed the basis of the protocol for ring rot indexing developed by the European Union (Anon, 1987). The European scheme involves maceration of combined cores of tissue from 200 tubers sampled at the stolon end, extraction of a bacteria-containing fraction, and application of the immunofluorescence procedure to the fraction (Fig. 2). The scheme requires that all suspect positive samples be inoculated into eggplant for isolation of the pathogen because serological cross-reactions occur commonly with tuber-associated bacteria with antisera produced to *C. m. sepedonicus* (Crowley and De Boer, 1982). Non-specificity of antiserum has been the greatest drawback of the immunofluorescence procedure but can be minimized by appropriate antiserum dilution or use of specific monoclonal antibodies (Miller, 1984a,b). Monoclonal antibody 9A1 reacts quite specifically with a somatic antigen of *C. m. sepedonicus*, but occasional cross-reactions with other bacteria have been observed (De Boer and Wieczorek, 1984).

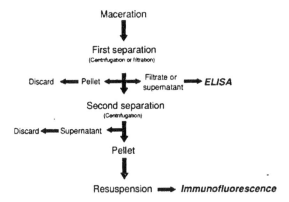

*Fig. 2.*  Preparation of composite potato tuber samples for testing by ELISA and immunofluorescence to detect the presence of *C. m. sepedonicus*. (Adapted from De Boer *et al.*, 1994)

ELISA procedures with polyclonal and monoclonal antibodies have also been used successfully (Corbiere *et al.*, 1987; De Boer *et al.*, 1988). Like other serological tests, cross-reactions may also occur in ELISA but their occurrence is infrequent with the monoclonal antibody 1H3 which is specific for the extracellular polysaccharide produced by *C. m. sepedonicus*. Non-mucoid strains of *C. m. sepedonicus*, however, react poorly in ELISA because they do not produce copious amounts of extracellular polysaccharides as do most wild-type strains (Baer and Gudmestad, 1993). It is not yet known whether non-mucoid strains are important in the epidemiology of the ring rot disease.

The probability of detecting bacterial ring rot in a consignment of seed potatoes depends on the sensitivity and specificity of the laboratory protocols and the number of samples that are tested. In a collaborative study (De Boer *et al.*, 1994) involving six laboratories it was shown that for composite samples, a sensitivity and specificity of 100% could be achieved in individual laboratories but on average both sensitivity and specificity varied from 90-98% (Table VI). When immunofluorescence and ELISA were used as parallel tests such that a positive response was required in only one of the two tests, sensitivity was 100% in almost all laboratories. In contrast, when the two serological tests were used in serial fashion such that a positive response was required in both tests, specificity approached 100%, but sensitivity was reduced.

When the efficiency of laboratory tests is 100%, the probability of detecting ring rot is a function of incidence of infection and sample size (De Boer, 1991). The relationship between these parameters are given by $PD = 1 - (1-p)^n$, where PD is the probability of detection, p is the proportion of infected tubers, and n is the sample size for populations of infinite size. Detection probabilities at several sample sizes and at different disease incidences are given in Table VII. For small potato fields, the number of tubers or plants that are sampled may be a significant percentage of the total population. However, calculations based on all possible combinations of infected and uninfected plants shows that population size has only a small effect on the probability of detection (Clayton and Slack, 1988; De Boer and Hall, unpublished

*Table VI.* Efficacy of the ELISA and immunofluorescence tests for detecting *Clavibacter michiganensis* subsp. *sepedonicus* in composite samples of potato stems and tubers as determined in six different laboratories (De Boer *et al.* 1994)

| Lab | Stems | | | Tubers | | |
|-----|-------------|-------------|------------|-------------|-------------|------------|
|     | Sensitivity | Specificity | Efficiency | Sensitivity | Specificity | Efficiency |
| **ELISA** | | | | | | |
| A   | 100.0 | 94.7  | 96.0  | 92.3  | 90.9  | 91.4 |
| B   | 100.0 | 100.0 | 100.0 | 100.0 | 91.3  | 93.3 |
| C   | 87.5  | 100.0 | 96.7  | 92.3  | 86.4  | 88.5 |
| D   | 100.0 | 91.3  | 93.3  | 100.0 | 90.9  | 94.3 |
| E   | 100.0 | 81.8  | 86.7  | 83.3  | 78.3  | 80.0 |
| F   | 100.0 | 72.7  | 80.0  | 91.7  | 82.6  | 85.7 |
| ALL | 97.7  | 90.1  | 92.0  | 93.3  | 86.7  | 89.0 |
| **Immunofluorescence** | | | | | | |
| A   | 100.0 | 100.0 | 100.0 | 100.0 | 100.0 | 100.0 |
| B   | 100.0 | 100.0 | 100.0 | 100.0 | 100.0 | 100.0 |
| C   | 85.7  | 100.0 | 96.7  | 100.0 | 100.0 | 100.0 |
| D   | 71.4  | 91.3  | 86.7  | 100.0 | 90.9  | 94.3 |
| E   | 100.0 | 100.0 | 100.0 | 83.3  | 100.0 | 94.3 |
| F   | 87.5  | 100.0 | 96.7  | 100.0 | 95.7  | 97.1 |
| ALL | 90.9  | 98.5  | 96.6  | 97.3  | 97.8  | 97.6 |

*Table VII.* Percent probability of detecting bacterial ring rot at several different incidences of infection, with a laboratory test having 100% efficiency, for various sizes of randomly selected samples

| Sample size | \multicolumn{10}{c}{Disease incidence} | | | | | | | | | |
|---|---|---|---|---|---|---|---|---|---|---|
| | 5.0% | 2.0% | 1.5% | 1.0% | 0.5% | 0.2% | 0.1% | 0.05% | 0.02% | 0.01% |
| 50 | 91.792 | 63.212 | 52.763 | 39.347 | 22.120 | 9.516 | 4.877 | 2.469 | 0.995 | 0.499 |
| 100 | 99.326 | 86.466 | 77.687 | 63.212 | 39.347 | 18.127 | 9.516 | 4.877 | 1.980 | 0.995 |
| 200 | 99.995 | 98.168 | 95.021 | 86.466 | 63.212 | 32.968 | 18.127 | 9.516 | 3.921 | 1.980 |
| 300 | 100.000 | 99.752 | 98.889 | 95.021 | 77.687 | 45.119 | 25.918 | 13.929 | 5.824 | 2.955 |
| 400 | 100.000 | 99.966 | 99.752 | 98.168 | 86.466 | 55.067 | 32.968 | 18.127 | 7.688 | 3.921 |
| 500 | 100.000 | 99.995 | 99.945 | 99.326 | 91.792 | 63.212 | 39.347 | 22.120 | 9.516 | 4.877 |
| 600 | 100.000 | 99.999 | 99.988 | 99.752 | 95.021 | 69.881 | 45.119 | 25.918 | 11.308 | 5.824 |
| 700 | 100.000 | 100.000 | 99.997 | 99.909 | 96.980 | 75.340 | 50.341 | 29.531 | 13.064 | 6.761 |
| 800 | 100.000 | 100.000 | 99.999 | 99.966 | 98.168 | 79.810 | 55.067 | 32.968 | 14.786 | 7.688 |
| 900 | 100.000 | 100.000 | 100.000 | 99.988 | 98.889 | 83.470 | 59.343 | 36.237 | 16.473 | 8.607 |
| 1000 | 100.000 | 100.000 | 100.000 | 99.995 | 99.326 | 86.466 | 63.212 | 39.347 | 18.127 | 9.516 |
| 1100 | 100.000 | 100.000 | 100.000 | 99.998 | 99.591 | 88.920 | 66.713 | 42.305 | 19.748 | 10.417 |
| 1200 | 100.000 | 100.000 | 100.000 | 99.999 | 99.752 | 90.928 | 69.881 | 45.119 | 21.337 | 11.308 |
| 1300 | 100.000 | 100.000 | 100.000 | 100.000 | 99.850 | 92.573 | 72.747 | 47.795 | 22.895 | 12.190 |
| 1400 | 100.000 | 100.000 | 100.000 | 100.000 | 99.909 | 93.919 | 75.340 | 50.341 | 24.422 | 13.064 |
| 1500 | 100.000 | 100.000 | 100.000 | 100.000 | 99.945 | 95.021 | 77.687 | 52.763 | 25.918 | 13.929 |
| 1600 | 100.000 | 100.000 | 100.000 | 100.000 | 99.966 | 95.924 | 79.810 | 55.067 | 27.385 | 14.786 |
| 1700 | 100.000 | 100.000 | 100.000 | 100.000 | 99.980 | 96.663 | 81.732 | 57.259 | 28.823 | 15.634 |
| 1800 | 100.000 | 100.000 | 100.000 | 100.000 | 99.988 | 97.268 | 83.470 | 59.343 | 30.232 | 16.473 |
| 1900 | 100.000 | 100.000 | 100.000 | 100.000 | 99.993 | 97.763 | 85.043 | 61.326 | 31.614 | 17.304 |
| 2000 | 100.000 | 100.000 | 100.000 | 100.000 | 99.995 | 98.168 | 86.466 | 63.212 | 32.968 | 18.127 |
| 2100 | 100.000 | 100.000 | 100.000 | 100.000 | 99.997 | 98.500 | 87.754 | 65.006 | 34.295 | 18.942 |
| 2200 | 100.000 | 100.000 | 100.000 | 100.000 | 99.998 | 98.772 | 88.920 | 66.713 | 35.596 | 19.748 |
| 2300 | 100.000 | 100.000 | 100.000 | 100.000 | 99.999 | 98.995 | 89.974 | 68.336 | 36.872 | 20.547 |
| 2400 | 100.000 | 100.000 | 100.000 | 100.000 | 99.999 | 99.177 | 90.928 | 69.881 | 38.122 | 21.337 |
| 2500 | 100.000 | 100.000 | 100.000 | 100.000 | 100.000 | 99.326 | 91.792 | 71.350 | 39.347 | 22.120 |
| 2600 | 100.000 | 100.000 | 100.000 | 100.000 | 100.000 | 99.448 | 92.573 | 72.747 | 40.548 | 22.895 |
| 2700 | 100.000 | 100.000 | 100.000 | 100.000 | 100.000 | 99.548 | 93.279 | 74.076 | 41.725 | 23.662 |
| 2800 | 100.000 | 100.000 | 100.000 | 100.000 | 100.000 | 99.630 | 93.919 | 75.340 | 42.879 | 24.422 |
| 2900 | 100.000 | 100.000 | 100.000 | 100.000 | 100.000 | 99.697 | 94.498 | 76.543 | 44.010 | 25.174 |
| 3000 | 100.000 | 100.000 | 100.000 | 100.000 | 100.000 | 99.752 | 95.021 | 77.687 | 45.119 | 25.918 |
| 4000 | 100.000 | 100.000 | 100.000 | 100.000 | 100.000 | 99.966 | 98.168 | 86.466 | 55.067 | 32.968 |
| 5000 | 100.000 | 100.000 | 100.000 | 100.000 | 100.000 | 99.995 | 99.326 | 91.792 | 63.212 | 39.347 |
| 6000 | 100.000 | 100.000 | 100.000 | 100.000 | 100.000 | 99.999 | 99.752 | 95.021 | 69.881 | 45.119 |
| 7000 | 100.000 | 100.000 | 100.000 | 100.000 | 100.000 | 100.000 | 99.909 | 96.980 | 75.340 | 50.341 |
| 8000 | 100.000 | 100.000 | 100.000 | 100.000 | 100.000 | 100.000 | 99.966 | 98.168 | 79.810 | 55.067 |
| 9000 | 100.000 | 100.000 | 100.000 | 100.000 | 100.000 | 100.000 | 99.988 | 98.889 | 83.470 | 59.343 |
| 10000 | 100.000 | 100.000 | 100.000 | 100.000 | 100.000 | 100.000 | 99.995 | 99.326 | 86.466 | 63.212 |
| 20000 | 100.000 | 100.000 | 100.000 | 100.000 | 100.000 | 100.000 | 100.000 | 99.995 | 98.168 | 86.466 |

data). More precise probabilities, taking into account all available data, can be made using Baysian theory (see Chapter 9).

## D. Indexing for *C. m. sepedonicus* in Canada

Indexing of seed potatoes for bacterial ring rot was initiated in Canada during the late 1970's in response to requirements of off-shore export markets (De Boer, 1987; McDonald and Borrel, 1991). Initially, the Gram stain, the latex agglutination procedure as described by Slack *et al.* (1979), and the immunofluorescence procedure with polyclonal antisera were used. The Gram stain test, however, was soon dropped because it was inappropriate for detecting symptomless infections in plant material that often contained Gram positive endophytic bacteria, although it continued to be a useful confirmatory test for symptomatic plant tissue. In 1986, monoclonal antibody 9A1 was adopted to replace the polyclonal antisera in immunofluorescence and this greatly alleviated the difficulty in distinguishing between typical and atypical fluorescing bacteria by significantly reducing the number of cross-reactants. The ELISA procedure with monoclonal antibody 1H3 was introduced to the laboratories in 1989 and very quickly became an important component of the indexing protocol. From 1989 to 1994, all potato samples submitted for ring rot testing were tested by latex agglutination, immunofluorescence, and ELISA. Advantages and disadvantages of each of these tests as indexing tools for bacterial pathogens are discussed in Chapter 2.

In addition to indexing seed potato lots destined for export, an interest developed in the Canadian potato industry to index domestic seed in order to limit occurrence and spread of the disease. Testing of domestic seed soon was seen as a means by which the disease could perhaps be eradicated from the seed industry, and in 1991 testing became a requirement for certification. Today, a seed lot cannot be certified in Canada unless it "...has been subjected to laboratory tests and not found positive for the presence of *Corynebacterium sepedonicum*" (Anon., 1991).

Currently, 2000 tubers per seed lot destined for export and 400 tubers per domestic seed lot are indexed for ring rot in Canada. Since testing many seed lots is time-consuming and costly, indexing of domestic seed was limited, in 1994, to the ELISA test with confirmatory immunofluorescence applied only on ELISA-positive samples. By limiting indexing requirements to the ELISA test, the need for extracting bacterial cells and conducting the immunofluorescence test, which are tedious and time-consuming when handling many samples, were avoided for most samples. Indexing of all exported lots continues to be done by both ELISA and immunofluorescence to meet foreign requirements.

Identification of ring rot infections, in the absence of plant symptoms, by serological tests alone was an important concept that had to be accepted by the potato industry. Seed lots may be kept from being exported on the basis of a suspect serological test alone, but this does not constitute a positive ring rot diagnosis. A positive ring rot diagnosis of asymptomatic plants in Canada requires a positive result in at least two serological tests using different antibody sources; one of the tests must be immunofluorescence using monoclonal antibody 9A1 or equivalent.

## E. Success of Indexing for Ring Rot

There is only a low probability of detecting ring rot by indexing procedures in consignments of seed potatoes with a low incidence of infection, so some infected seed lots would be expected to escape detection. However, during 1989-92 only 0.19% of 13,570 seed lots grown from seed that tested negative in laboratory tests developed detectable infections the following season. This estimate of false negative test results is high because, in some cases, multiple positive seed lots were derived from a single source lot. Moreover, some of the positive seed lots may have become infected after indexing had correctly identified the lot to be ring rot free.

Over the last decade the percentage of seed lots in which ring rot was detected has decreased steadily in the Canadian provinces of New Brunswick and Prince Edward Island despite the increase in the number of lots that were tested. Elimination of seed lots with symptomless ring rot infections has served effectively to decrease the amount of inoculum present in the industry as a whole. During 1989-1994 all ring rot infections in New Brunswick and Prince Edward Island seed were detected by laboratory tests rather than by visual inspection. These data suggest that ring rot incidence and population levels in stocks have been dramatically reduced by systematic testing. Whether the pathogen can be eradicated depends on the survival characteristics of *C. m. sepedonicus* outside the potato host, but further research is required to determine if the pathogen persists in the environment. If functional eradication, that is, elimination of detectable levels of the disease, can be achieved, indexing for ring rot will need to be continued in Canada only if the pathogen is found to survive outside the host or if it is reintroduced in imported seed lots.

## F. Recent Developments

The advent of DNA-based tests has raised the prospect of even more specific and sensitive indexing procedures. Several DNA probes for *C. m. sepedonicus* have been described but direct probing of potato extracts lacks sensitivity (Drennan *et al.*, 1993). Primers to amplify unique fragments of *C. m. sepedonicus* DNA by the polymerase chain reaction (PCR) may provide for a more sensitive test (Schneider *et al.*, 1993). *In situ* DNA hybridization or PCR of target bacteria on microscope slides may also be possible but has not yet been described for the ring rot pathogen. Ultimately, the DNA-based techniques are more likely to be most useful to confirm positive diagnoses based on serology rather than as primary indexing methodologies because of the cost, labour, and time they require.

## IV. INDEXING FOR BLACKLEG

### A. The Blackleg Disease

The blackleg disease is caused by the bacterium *Erwinia carotovora* subsp. *atroseptica*. Although blackleg-like symptoms can also be caused by *E. carotovora* subsp. *carotovora* and *Erwinia chrysanthemi* under regimes of high field temperatures, the blackleg disease in most temperate potato growing regions is primarily caused by *E. carotovora* subsp. *atroseptica*. The causal bacterium survives well on tuber surfaces particularly within lenticellular tissue, with the result that seed tubers are the main inoculum source for the disease (Perombelon, 1982; Perombelon and Kelman, 1980). Survival of the bacterium in soil and in surface water may also occur at low levels although data on field survival mostly involves *E. carotovora* subsp. *carotovora* (Harrison and Brewer, 1982). *E. carotovora* subsp. *carotovora* is a far more heterogenous subspecies, is considerably more ubiquitous, and generally has a more rapid growth rate than *E. carotovora* subsp. *atroseptica*.

Development of the blackleg disease depends greatly on growing conditions such as temperature and rainfall (Perombelon and Kelman, 1980). However, *E. carotovora* subsp. *atroseptica* may spread from mother tubers to progeny tubers below ground without any visible evidence of disease in the plant (Elphinstone and Perombelon, 1986a; Lapwood and Harris, 1980). Thus highly contaminated seed potatoes may be derived from an apparently healthy crop. The bacterium may also be translocated from the stem to progeny tubers through the stolons via vascular tissue and cause decay in the tubers prior to harvest. During harvest, bacteria from decayed tubers may be widely disseminated throughout the crop leading to high levels of infection (Elphinstone and Perombelon, 1986b).

### B. Indexing Tests for Blackleg in the Netherlands

Seed potato certification rules of the Netherlands General Inspection Service for Agricultural Seeds and Seed Potatoes (NAK; a private foundation responsible for the inspection under government supervision) prescribes a nil tolerance for blackleg symptoms caused by *Erwinia* spp. during field inspection of basic seed (classes S, SE and E). In 1983, the NAK introduced an additional post-harvest laboratory test in order to eliminate the most heavily contaminated seed lots from clonal material, and classes S and SE, that had escaped detection during field inspection due to lack of disease symptoms. In recent years this obligatory laboratory test has been restricted to seed lots of clonal material that can be marketed at the class S and SE levels. For new clonal or pre-basic material, tests are conducted on request at no cost to growers.

To index for blackleg, sap is extracted from the potato peel of 24 tubers per seed lot with a Pollähne press and the sap from each tuber is tested separately by an indirect double-antibody sandwich ELISA. Polyclonal antibodies against *E. carotovora* subsp. *atroseptica* serogroup I *sensu* De Boer *et al.* (1979) produced by the Research Institute for Plant Protection (IPO-DLO) in the Netherlands are used as the primary antibody and alkaline-phosphatase conjugated anti-rabbit antibodies as the second antibody. The antiserum does not react with *E. carotovora* subsp. *atroseptica* serogroups XVIII, XX and XXII which have never been found in the Netherlands in the past nor in recent surveys (Appels, personal communication). The antiserum must, however, be cross-absorbed to prevent reaction with *E. carotovora* subsp. *carotovora* serogroup II strains (Vruggink and Maas Geesteranus, 1975). Recently it was observed that the newest available antiserum cross-reacts with a *Comomonas sp.*, *Janthinobacterium lividium* and an *Enterobacter* sp. (van der Wolf *et al.*, 1994). This recent problem with cross-reactions has been addressed by testing a second group of 72 individual tubers from the same seed lot with three additional antisera when the first sample of 24 tubers is positive in ELISA. By using these three antisera, which differ in specificity, it is possible to recognize the cross-reactions. All ELISA-positive seed lots are down-graded to class E.

## C. Indexing Results

During 1983-1993, 6,800 - 19,500 samples of seed potatoes were tested annually in the Netherlands with ELISA for the presence of the blackleg pathogen. In this period, 0.1 - 3.5% of seed lots representing one-year-old clones were down-graded to a lower seed class for *E. carotovora* subsp. *atroseptica* contamination, and between 0.2% and 14.6% of seed lots in the SE class were downgraded. The highest percentages of positive tests occurred those years in which much rain fell during the growing season and during harvest. Seed lots that were downgraded because of high ELISA values, but which remained within the clonal selection system at the class E level or lower, were carefully monitored during field inspections the year following testing. Over the years, 60% of the four- and five-year-old clones and class SE seed lots that were downgraded, developed blackleg symptoms in the following year. In contrast, only 13% of seed lots from these classes that were negative in ELISA developed blackleg in the field the following year and were subsequently downgraded based on symptom expression.

Low sensitivity of ELISA ($10^5$-$10^6$ cells/ml) and the small sample size permits detection of only the most highly infected seed lots. However, in the Netherlands, the test was always meant to complement visual inspection and to eliminate, from the top of the clonal selection system, the latently infected seed lots carrying a high inoculum load but which escaped detection during field inspections. The correlation of ELISA-positive samples with symptom development in the field the following season indicates the possibility that seed potato quality can be improved by using a relatively insensitive test on a small sample size. The fact that bacteria along with soluble antigens can be spread evenly over a large number of tubers during harvest and are detected by ELISA, probably contributes to the success of the testing despite the small sample size. Use of a test that only detects viable bacteria would probably improve the procedure but would also certainly require testing a larger sample size.

Routine indexing for blackleg on large numbers of samples only became possible with the development of automated ELISA equipment and computerized analysis of test results. Possibly preincubation of samples in selective or semi-selective medium, which would allow multiplication of the target bacterium, used in combination with serology or DNA-based techniques such as amplification by PCR would be helpful to lower the detection threshold. Perhaps these techniques can also be used for detection of *E. chrysanthemi*.

## V. INDEXING FOR POTATO CYST NEMATODES

### A. Certification Guidelines

Guidelines of the European Union require that member states treat the potato cyst nematodes (*Globodera rostochiensis* and *Globodera pallida*) as harmful organisms which are forbidden to be brought into their territory and against which measures must be taken if its presence is established. Also an official inspection on a representative sample must be carried out before transportation of seed potatoes to another member state takes place. In the Netherlands this guideline resulted in legislation that allows production of propagation material only on soil which has been found to be free of potato cyst nematodes.

On the basis of this legislation, the requirements for production and certification of seed potatoes are set in the Netherlands by the NAK (Fig. 3). Seed potatoes can only be brought into circulation after certification by the NAK has taken place, and a seed potato crop is not accepted for inspection by the NAK unless the results of a soil sample analysis for cyst nematodes by a recognized laboratory (of which the NAK laboratory is one) are presented.

At the end of the potato production cycle, before certification, a sample of soil from the surface of potato tubers is also tested for the presence of cysts in the NAK laboratory. If cysts are found, the grower is obligated to brush all soil from the surface of the potato tubers prior to obtaining certification.

### B. Sampling Techniques and Analysis

The minimum sample size required for the analysis of land destined for seed potato production is 200 cc of soil per one third hectare. The samples are taken either by hand using a sample drill or with an automatic sampling device mounted on a vehicle. In both cases samples are taken from the top 5 cm of soil. Subsamples of 4 cc are taken at each intercept of a 7.5 x 7.5 meter grid. Depending on the dimensions of the field this pattern may be adapted to a maximum of 5.5 x 11 meter grid. Furthermore, 200 cc of soil is gathered from under the grading machine for each 10-15 tonnes of seed potatoes and analyzed before certification.

Soil samples are air dried under moderate temperature conditions, not exceeding 28°C. Subsequently the samples are fractionated using an automated aqueous extraction method. The floating fraction is dried and fractionated a second time using acetone as the diluent. The remaining fraction which constitutes 1-5% of the original sample by weight is screened under a microscope at 15 x magnification for the presence of potato nematode cysts. Any cysts that are found are tested for viability

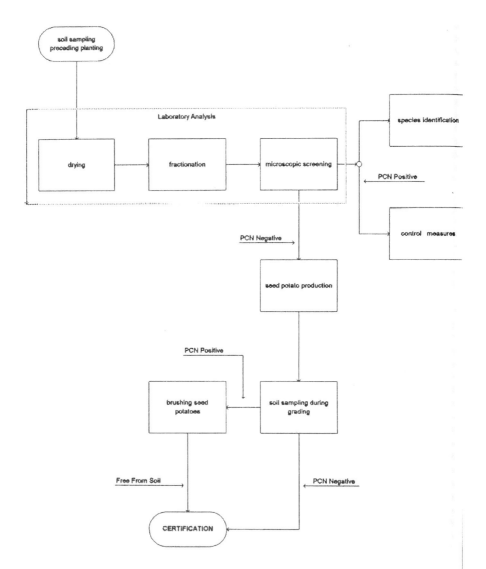

Fig. 3. Flowchart for sampling and testing for the potato cyst nematode (PCN) in the Netherlands.

by crushing them and observing the cyst contents. In some laboratories this step is preceded by a 4-7 day incubation of the cysts in a 0.1 % solution of New Blue R (Michrome #657) which stains nonviable larvae in the cysts. If viable cysts are detected production of seed potatoes (and other propagative plant material) is forbidden in the portion of the field from which the sample originated and in a surrounding buffer zone. If desired by the grower, viable cysts are subjected to further identification to the species level.

## C. Identification of *Globodera* species

To distinguish between G. *rostochiensis* and G. *pallida* a standard indirect enzyme-linked immunosorbent assay (ELISA) is carried out using monoclonal antibodies produced by the Department of Nematology of the Wageningen Agricultural University and the Laboratory for Monoclonal Antibodies (LMA) in the Netherlands (Schots *et al.*, 1989, 1992). The content of individual cysts is homogenized carefully by hand in a small amount of Tris/HCl-buffer (pH 7.4) and heated in boiling water for 10 minutes. After centrifugation for 10 minutes at 15,000 $g$ the soluble thermostable protein antigens present in the supernatant, are tested in duplicate with species-specific monoclonal antibodies in an antigen-coated ELISA plate. Dilution series of standard antigen samples are used as controls. The equivalent of one-tenth of an egg is sufficient to identify the *Globodera* species.

On the basis of the *Globodera* species identified, the grower will be advised which potato varieties to grow. Because almost all G. *rostochiensis* populations in the Netherlands appear to be avirulent for cultivars carrying the $H_1$ resistance gene (Bakker *et al.*, 1993), a large number of resistant varieties is available. Whenever G. *pallida* is found, further characterization of the population based on virulence on a set of differential hosts is necessary to determine which crop management scenario is best to control the particular population of G. *pallida* involved. Growers may use several options including varieties with partial resistance, soil fumigation, and/or crop rotation.

## D. Test results

The number of soil samples tested for the presence of potato cysts depends on the acreage submitted for inspection and has varied from 35,000 ha in 1990 to 38,000 in 1993, resulting in 105,000 to 114,000 soil samples. The number of samples of soil from under grading machines can exceed 1,000 per day per laboratory, and totals about 65,000 annually.

Soil analysis prior to planting and subsequent screening of soil adhering to harvested seed potato tubers has proven to be adequate for reducing the spread of potato cyst nematodes to a minimum. Furthermore, to control the potato cyst nematode, Dutch growers are not permitted to grow potatoes more than once every 3 years on the same field. Before 1993 unlimited, preventive fumigation of soils could be carried out but, due to environmental concerns, fumigation has now been restricted. Potato growers are now forced to control potato cyst nematodes in a different manner.

Studies on the nematode sampling technique described above have shown that the technique does not detect the cysts early enough for control without fumigation. In 1989 a new more intensive sampling technique (AMI) was described (Schomaker and Been, 1993). This method is based on a sample size of 2,660 to 5,320 cc per one-third of a hectare. The 40 cc subsamples must basically be taken in a 5 x 5 meter grid pattern and within the top 10 cm of soil. Growers are now urged to have their soils sampled regularly by the AMI procedure and to have species assessment carried out on cysts. This allows them to locate low population levels of potato cyst nematodes and to take specific, curative action.

## VI.  THE FUTURE OF INDEXING

Indexing of potato seed lots is an expensive proposition and much of the testing that has been done in various countries has been supported by government subsidies. With changing economic conditions it is clear that indexing can only be sustained if the seed potato producers are prepared to pay the costs for such programs. The cost-benefit ratio of including laboratory indexing procedures as part of a seed certification program will vary among countries and even within a country. Costs will depend on the particular test being carried out and on the size and number of samples. Benefits will depend on foreign and domestic market requirements and the particular disease being addressed.   It is clear, however, that competition on international markets will ultimately set the standards required for export of seed potatoes in the future.

## REFERENCES

Anonymous (1987). 'Scheme for the Detection and Diagnosis of the Ring Rot Bacterium *Corynebacterium sepedonicum* in Batches of Potato Tubers'. EUR 11288. Office for Official Publications of the European Communities, Luxembourg.

Anonymous (1991). Seeds Act, Seeds Regulations, amendment. *Canada Gazette Part II. Vol. 125. No. 20*

Baer, D. and Gudmestad, N.C. (1993). Serological detection of nonmucoid strains of *Clavibacter michiganensis* subsp. *sepedonicus* in potato. *Phytopathology* 83, 157-163.

Bakker, J., Folkertsma, R.T., Rouppe van der Voort, J.N.A.M., de Boer, J.M. and Gommers, F.J. (1993). Changing concepts and molecular approaches in the management of virulence genes in potato cyst nematodes. *Annual Review of Phytopathology* 31, 169-190.

Bishop, A.L. and Slack, S.A. (1987a). Effect of cultivar, inoculum dose, and strain of *Clavibacter michiganense* subsp. *sepedonicum* on symptom development in potatoes. *Phytopathology* 77, 1085-1089.

Bishop, A.L. and Slack, S.A. (1987b).  Effect of inoculum dose and preparation, strain variation, and plant growth conditions on the eggplant assay for bacterial ring rot. *American Potato Journal* 64, 227-234.

Bishop, A.L. and Slack, S.A. (1992). Effect of infection with *Clavibacter michiganensis* subsp. *sepedonicus* Davis *et al.* on water relations in potato. *Potato Research* 35, 59-63.

Bugbee, W.M., Gudmestad, N.C., Secor, G.A. and Nolte, P. (1987). Sugar beet as a symptomless host for *Corynebacterium sepedonicum*. *Phytopathology* 77, 765-770.

Clark, M.F. and Adams, A.H. (1977). Characteristics of the microplate method of enzyme-linked immunosorbent assay for the detection of plant viruses. *Journal of General Virology* 34, 475-483.

Clayton, M.K. and Slack, S.A. (1988). Sample size determination in zero tolerance circumstances and the implications of stepwise sampling: bacterial ring rot as a special case. *American Potato Journal* 65, 711-723.

Cobierre, R., Hingand, L. and Jouan, B. (1987). Application des methodes ELISA et immunofluorescence pour la detection de *Corynebacterium sepedonicum*: reponses varietales de la pomme de terre au fletrissement bacterien. *Potato Research* 30, 539-549.

Crowley, C.F. and De Boer, S.H. (1982). Non-pathogenic bacteria associated with potato stems cross-react with *Corynebacterium sepedonicum* antisera in immunofluorescence. *American Potato Journal* 59, 1-8.

De Boer, S.H. (1987). The relationship between bacterial ring rot and North American seed potato export markets. *American Potato Journal* 64, 683- 694.

De Boer, S.H. (1991). Current status and future prospects of bacterial ring rot testing. *American Potato Journal* 68, 107-113.

De Boer, S.H., Copeman, R.J. and Vruggink, H. (1979). Serogroups of *Erwinia carotovora* potato strains determined with diffusible somatic antigens. *Phytopathology* 69, 316-319.

De Boer, S.H. and McCann, M. (1989). Determination of population densities of *Corynebacterium sepedonicum* in potato stems during the growing season. *Phytopathology* 79, 946-951.

De Boer, S.H. and McCann, M. (1990). Detection of *Corynebacterium sepedonicum* in potato cultivars with different propensities to express ring rot symptoms. *American Potato Journal* 67, 685-694.

De Boer, S.H. and Slack, S.A. (1984). Current status and prospects for detecting and controlling bacterial ring rot. *Plant Disease* 68, 841-844.

De Boer, S.H., Stead, D.E., Alivizatos, A.S., Janse, J.D., Van Vaerenbergh, J., De Haan, T.L. and Mawhinney, J. (1994). Evaluation of serological tests for detection of *Clavibacter michiganensis* subsp. *sepedonicus* in composite potato stem and tuber samples. *Plant Disease* 78, 725- 729.

De Boer, S.H., Wieczorek, A. and Kummer, A. (1988). An ELISA test for bacterial ring rot of potato with a new monoclonal antibody. *Plant Disease* 72, 874-878.

De Boer, S.H. and Wieczorek, A. (1984). Production of monoclonal antibodies to *Corynebacterium sepedonicum*. *Phytopathology* 74, 1431-1434.

Drennan, J.L., Westra, A.A.G., Slack, S.A., Delserone, L.M., Collmer, A., Gudmestad, N.C. and Oleson, A.E. (1993). Comparison of a DNA hybridization probe and ELISA for the detection of *Clavibacter michiganensis* subsp. *sepedonicus* in field-grown potatoes. *Plant Disease* 77, 1243-1247.

Elphinstone, J.G. and Perombelon, M.C.M. (1986a). Contamination of progeny tubers of potato plants by seed- and leaf-borne *Erwinia carotovora*. *Potato Research* 29, 77-93.

Elphinstone, J.G. and Perombelon, M.C.M. (1986b). Contamination of potatoes by

*Erwinia carotovora* during grading. *Plant Pathology* **35**, 25-33.

Griffiths, H.M., Slack, S.A. and Dodds, J.H. (1990). Effect of chemical and heat therapy on virus concentrations in *vitro* plantlets. *Canadian Journal of Botany* **68**, 1515-1521.

Hadidi, A., Montasser, M.S., Levy, L., Goth, R.W., Converse, R.H., Madkour, M.A. and Skrzeckowski, L.J. (1993). Detection of potato leafroll and strawberry mild-edge luteoviruses by reverse transcription - polymerase chain reaction amplification. *Plant Disease* **77**, 595-601.

Hahm, Y., Slack, S.A. and Slattery, R.J. (1981). Reinfection of potato seed stocks with potato virus S and potato virus X in Wisconsin. *American Potato Journal* **58**, 117-125.

Harrison, M.D. and Brewer, J.W. (1982). Field dispersal of soft rot bacteria. *In* 'Phytopathogenic Prokaryotes'(M. S. Mount and G. H. Lacy, eds), pp. 31-53, Academic Press, London.

Jordan, R.L. and Hammond, J. (1991). Comparison and differentiation of potyvirus isolates and identification of strain-, virus-, subgroup-specific and potyvirus group-common epitopes using monoclonal antibodies. *Journal of General Virology* **72**, 25-36.

Lapwood, D.H. and Harris, R.I. (1980). The spread of *Erwinia carotovora* var. *atroseptica* (blackleg) and var. *carotovora* (tuber soft rot) from degenerating seed to progeny tubers in soil. *Potato Research* **23**, 385-393.

Lelliott, R.A. and Sellar, P.W. (1976). The detection of latent ring rot (*Corynebacterium sepedonicum* [Spieck. & Kotth.] Skapt. et Burkh.) in potato stocks. *EPPO Bulletin* **6**, 101-106.

Lund, R.E. and Sun, M.K.C. (1985). Sample size determination for seed potato certification. *American Potato Journal* **62**, 347- 353.

McDonald, J. and Borrel, B. (1991). Development of post harvest testing in Canada. *American Potato Journal* **68**, 115-121.

McDonald, J.G., Kristjannson, G.T., Singh, R.P., Ellis, P.J. and McNab, W.B. (1994). Consecutive ELISA screening with monoclonal antibodies to detect potato virus Y. *American Potato Journal* **71**, 175-183.

Miller, H.J. (1984a). Cross-reactions of *Corynebacterium sepedonicum* antisera with soil bacteria associated with potato tubers. *Netherlands Journal of Plant Pathology* **90**, 23-28.

Miller, H.J. (1984b). A method for the detection of latent ring rot in potatoes by immunofluorescence microscopy. *Potato Research* **27**, 33-42.

Morris, T.J. and Wright, N.S. (1975). Detection on polyacrylamide gels of a diagnostic nucleic acid from tissue infected with potato spindle tuber viroid. *American Potato Journal* **52**, 57-63.

Nelson, G.A. (1982). *Corynebacterium sepedonicum* in potato: effect of inoculum concentration on ring rot symptoms and latent infection. *Canadian Journal of Plant Pathology* **4**, 129-133.

Nelson, G.A. (1980). Long-term survival of *Corynebacterium sepedonicum* on contaminated surfaces and infected tubers. *American Potato Journal* **56**, 71-77.

Nelson, G.A. (1979). Persistence of *Corynebacterium sepedonicum* in soil and buried potato stems. *American Potato Journal* **56**, 71- 77.

Nelson, G.A. and Kozub, G.C. (1983). Effect of total light energy on symptoms and growth of ring rot-infected Red Pontiac potato plants. *American Potato Journal* **60**,

461-468.

Olsson, K. (1976). Experience of ring rot caused by *Corynebacterium sepedonicum* (Spieck. et Kotth.) Skapt. et Burkh. in Sweden, particularly detection of the disease in its latent form. *EPPO Bulletin* 6, 209-219.

Owens, R.A. and Diener, T.O. (1981). Sensitivity and rapid diagnosis of potato spindle tuber viroid disease by nucleic acid hybridization. *Science* 213, 670-672.

Palukaitis, P., Cotts, S. and Zaitlin, M. (1985). Detection and identification of viroids and viral nucleic acids by 'dot-blot' hybridization. *Acta Horticulturae* 164, 109-118.

Perombelon, M.C.M. (1982). The impaired host and soft rot bacteria. *In* 'Phytopathogenic Prokaryotes'(M.S. Mount and G.H. Lacy, eds.), pp. 55- 69, Academic Press, London.

Perombelon, M.C.M. and Kelman, A. (1980). Ecology of the soft rot erwinias. *Annual Review of Phytopathology* 18, 361-387.

Persson, P. and Janse, J.D. (1988). Ring rot-like symptoms in *Solanum melongena* caused by *Erwinia chrysanthemi* (potato strain) after artificial inoculation. *EPPO Bulletin* 18, 575-578.

Pfannenstiel, M.A. and Slack, S.A. (1980). Response of potato cultivars to infection by the potato spindle tuber viroid. *Phytopathology* 70, 922-926.

Pfannenstiel, M.A., Slack, S.A. and Lane, L.C. (1980). Detection of potato spindle tuber viroid in field-grown potatoes by an improved electrophoretic assay. *American Potato Journal* 70, 1015-1018.

Raymer, W.B. and O'Brien, M.J. (1962). Transmission of potato spindle tuber virus to tomato. *American Potato Journal* 39, 401- 408.

Sanchez, G.E., Slack, S.A. and Dodds, J.H. (1991). Response of selected *Solanum* species to virus eradication therapy. *American Potato Journal* 68, 299-325.

Schneider, J., Zhao, J.L. and Orser, C. (1993). Detection of *Clavibacter michiganensis* subsp. *sepedonicus* by DNA amplification. *FEMS Microbiology Letters* 109, 207-212.

Schomaker, C.H. and Been, T.H. (1993). Sampling strategies for the detection of potato cyst nematodes: developing and evaluating a model. *In* 'Nematology from Molecule to Ecosystem'(F. J. Gommers and P. W. T. Maas, eds), pp. 182-194, Dekker and Huisman, Wildervank, the Netherlands.

Schots, A., Gommers, F.J. and Egberts, E. (1992). Quantitative ELISA for the detection of potato cyst nematodes in soil samples. *Fundamentals of Applied Nematology* 15, 55-61.

Schots, A., Hermsen, T., Schouten, S., Gommers, F.J. and Egberts, E. (1989). Serological differentiation of the potato- cyst nematodes *Globodera pallida* and *G. rostochiensis*: II. Preparation and characterization of species specific monoclonal antibodies. *Hybridoma* 8, 401-413.

Singh, R.P. and Crowley, C.F. (1985). Evaluation of polyacrylamide gel electrophoresis, bioassay and dot-blot methods for the survey of potato spindle tuber viroid. *Canadian Plant Disease Survey* 65, 61-63.

Shepard, J.F. and Claflin, L.E. (1975). Critical analyses of the principles of seed potato certification. *Annual Review of Phytopathology* 13, 271-293.

Slack, S.A. (1991). A look at potato leafroll and potato virus Y: past, present and future. *The Badger Common'tater* 43, 16-21.

Slack, S.A. (1980). Pathogen-free plants by meristem-tip culture. *Plant Disease* 64, 15-17.

Slack, S.A. (1993). Seed certification and seed improvement programs. *In* 'Potato Health Management'(R. C. Rowe, ed), APS Press, St. Paul, Minnesota.

Slack, S.A., Sanford, H.A. and Manzer, F.E. (1979). The latex agglutination test as a rapid serological assay for *Corynebacterium sepedonicum*. *American Potato Journal* **56**, 441- 446.

Stead, D.E. (1993). Potato ring rot control through detection and certification. *In* 'Plant Health and the European Single Market' (D. L. Ebbels, ed.) pp. 135-144, British Crop Protection Council Monograph 54, Fasham, United Kingdom.

van der Wolf, J.M., van Beckhoven, J.R.C.M., de Vries, P.M. and van Vuurde, J.W.L. (1994). Verification of ELISA results by immunomagnetic isolation of antigens from extracts and analysis with SDS-PAGE and Western blotting, demonstrated for *Erwinia* spp. in potatoes. *Journal of Applied Bacteriology* **77**, 160-174.

Vruggink, H. and Maas Geesteranus, H.P. (1975). Serological recognition of *Erwinia carotovora* var. *atroseptica*, the causal organism of potato blackleg. *Potato Research* **18**, 546-555.

Westra, A.A.G., Arneson, C.P. and Slack, S.A. (1994). Effect of interaction of inoculum dose, cultivar, and geographic location on the development of foliar symptoms of bacterial ring rot of potato. *Phytopathology* **84**, 410-415.

Westra, A.A. and Slack, S.A. (1994). Effect of interaction of inoculum dose, cultivar, and geographic location on magnitude of bacterial ring rot symptom expression in potato. *Phytopathology* **84**, 228-235.

Whitworth, J.L., Samson, R.G., Allen, T.C. and Mosley, A.R. (1993). Detection of potato leafroll virus by visual inspection, direct tissue blotting and ELISA technique. *American Potato Journal* **70**, 497-503.

Wright, N.S. (1988). Assembly, quality control and use of a potato cultivar collection rendered virus-free by heat therapy and tissue culture. *American Potato Journal* **65**, 181-198.

Zeller, W. and Xie, Y. (1985). Studies on the diagnosis of bacterial ring rot of potatoes. I. pathogenicity test on eggplants. *Phytopathologische Zeitschrift* **112**, 198-206.

Zizz, J.D. and Harrison, M.D. (1991). Detection of *Clavibacter michiganense* subsp. *sepedonicum* (Spieck. & Kotth.) (Carlson & Vidaver) in common weed species found in Colorado potato fields. *Phytopathology* **81**, 1348.

# 9

# A DECISION MODELLING APPROACH FOR QUANTIFYING RISK IN PATHOGEN INDEXING

## C. A. Lévesque[1] and D. M. Eaves[2]

[1]*Pacific Agriculture Research Centre,*
*Vancouver, British Columbia, Canada, V6T 1X2*
[2]*Department of Mathematics and Statistics, Simon Fraser University,*
*Burnaby, British Columbia, Canada, V5A 1S6*

## I. INTRODUCTION

Risk can be defined as expected magnitude of loss. All decisions in pathogen indexing are ultimately made to reduce to a minimum the risk due to certain infectious agents. The likelihood that a given proportion of a lot of plants is infected by a pathogen must be determined to make a decision about this given lot. In this chapter, all the examples are related to either entire plants or seeds but the functions can be applied to plant parts, tubers or any other plant propagules. In assessing the presence of a pathogen in a plant population, samples of a predetermined size must be collected and processed.

In keeping with this volume's main emphasis on laboratory techniques to assess the presence of pathogens, in this chapter we emphasise mathematical modelling of costs and risks in support of acceptance-rejection decisions. Most of the chapter will be

Advances in Botanical Research Vol. 23
Incorporating Advances in Plant Pathology
ISBN 0-12-005923-1

devoted to sample size since it is the first and most practical decision that has to be made in pathogen indexing. Three sampling methods and their corresponding analyses will be described, in increasing order of mathematical complexity. (i) Simple direct binomial sampling which requires the analysis of a predetermined number of plants, and assumes that the number of contaminated sampling units in the sample can be determined perfectly. (ii) Simple direct "bulk" sampling which makes the same assumption of determination of the exact number of contaminated plants, but is based on the sampling of a predetermined area, volume or a part of the plant lot. (iii) Assays for detecting the mere presence of contamination in a pooled sample which provide a simple sampling and processing method, assuming that the imperfect detection capability has been properly calibrated. Throughout this chapter we will emphasise the founding of accept-reject decisions upon some knowledge of relative economic costs of the competing decision alternatives. This will be founded upon the summarisation of sample information in a probability distribution over the range of possible incidence of disease. This summarisation is provided in the form of a posterior distribution, by Bayes' theorem. Implicit in our decision-tree illustration is the possibility of repeated sampling. The application of all the methods discussed in this chapter has been kept simple through the use of software. The programs ("functions") that will be introduced were written in Splus language and are listed in the appendix. The functions whose names begin with a lower case letter are those which come as part of the commercial product Splus version 3.2, while names starting with an upper-case letter denote functions which we have newly created. These new functions are available through the public Internet via either Electronic Mail or File Transfer Protocol (FTP) in a directory named "pathogen_indexing". Either send an E-mail to "statlib@lib.stat.cmu.edu" with the message "send pathogen_indexing from S" or get the "pathogen_indexing" directory through anonymous FTP at the address lib.stat.cmu.edu (128.2.241.142).

The decision procedures discussed here are based on a probability distribution over the set of all possible proportions ($\pi$) of infected plants, represented by the interval $0 < \pi < 1$. We believe that many plant pathologists feel comfortable talking about the probability of the mean being within certain boundaries, or about the probability of a disease incidence proportion being below a critical level. This usage of probability ideas to express the degree of a person's belief in a proposition (like "the true population mean is between 19.2 and 21.8") is known as subjective probability. Subjective probability stands on well-developed theoretical foundations, and is an extension or enhancement of the classical repeated-sampling viewpoint of probability. It leads to intuitive interpretation of confidence intervals or power, through the use of Bayes' theorem with its "posterior probabilities" of a statement about the population mean. For many routine statistical analyses, both the Bayesian and traditional frequentist views often yield the same or similar numbers, notwithstanding the different interpretations. We suggest P. M. Lee (1989) for an introduction to the Bayesian view of statistics; however the reader should be able to grasp and apply the models we have developed without such extra reading.

## II. SIMPLE DIRECT SAMPLING OF A NUMBER OF SEEDS

## A. Probability of Detection and Determination of Sample Size

Distributions about y, the number of infected plants, are sometimes the most appropriate to make a final decision on sample size. This traditional approach determining sample size in pathogen indexing is based on the power, i.e., on the determination of a sufficient sample to achieve a desired chance (e.g. 99%) of detecting infection under a stated "maximum acceptable" disease incidence level, when sampling (with that sample size) is repeated many times. In statistical analysis of an experiment for detecting a difference between the actual population mean and a hypothesised mean, the power of a rejection rule (for determining whether to reject the no-difference hypothesis) is the proportion of times the rule will say "reject" if the same procedure is repeated. Power, of course, also depends on the magnitude of the true difference. Similarly, when calculating a 99% confidence interval around a mean we consider that similarly calculated intervals would catch the true mean within their boundaries 99% of the time. Thus, power and confidence intervals are commonly interpreted with respect to the repetition *ad infinitum* of the same sampling or experimental procedure.

A commonly described acceptance procedure for a seedlot (Geng *et al.*, 1983) is to first decide on a high probability $f_{hi} = P[\text{accept} \mid \pi_t]$ and a low probability $f_{lo} = P[\text{accept} \mid \pi_{nt}]$ with which one wants to accept seedlots having, respectively, a tolerably low proportion $\pi_t$ of infection, and an intolerably high proportion $\pi_{nt}$ (t for "tolerable", nt for "not tolerable"). The consequent required size n of a random sample of seeds in the lot, and a critical acceptance range $y \leq y_c$ for the number y of infected seeds found in the sample, may be found by typing, say, Ssize(0.0005, 0.001, 0.95, 0.01, 10000, 100) where the arguments of the function Ssize() are $\pi_t, \pi_{nt}, f_{hi}, f_{lo}$, startn, and incrn. As it searches for n, Ssize() proceeds through 10000 (startn), then 10100 (startn+incr), 10200, etc. The arguments startn and incrn allow control over computing speed and precision of the answer by setting the starting n and the search increments. The results for the above example are n=47,900 and $y_c$=32. The performance of a particular n, 47900 in this example, and $y_c$ may be further judged visually by co-plotting the two acceptance probabilities $P[y \leq y_c \mid \pi = 0.0005, n=47900]$ and $P[y \leq y_c \mid \pi=0.001, n=47900]$ as functions of $y_c$. Ssize() provides the optimal n and $y_c$ and also produces this plot for the optimal n (Fig. 1). In this example, 95% of an infinitude of samples of 47900 seeds each from a lot that has an infection proportion of 0.0005 will have 32 infected seeds or less, i.e., 0.95= $P[y \leq 32 \mid \pi=0.0005, n=47900]$. Simultaneously, only 1% of an infinitude of samples of size 47900 from a lot with an infection proportion of 0.001 will have 32 infected seeds or less, i.e., 0.01= $P[y \leq 32 \mid \pi=0.001, n=47900]$. Therefore, we will achieve a low probability of accepting a seedlot with an unacceptably high disease incidence and a high probability of accepting a lot with a tolerably low disease incidence if we use a 32-infected-seeds cut-off point with a sample size of 47900.

Sometimes there is no interest in controlling the probability of rejecting a good lot. For example in seed testing for lettuce mosaic virus a standard of 0 positive in 30,000 (y=0, n=30,000) seeds was adopted in California (Grogan, 1980). In this kind of "zero tolerance" situation, n had been estimated at 30000 so

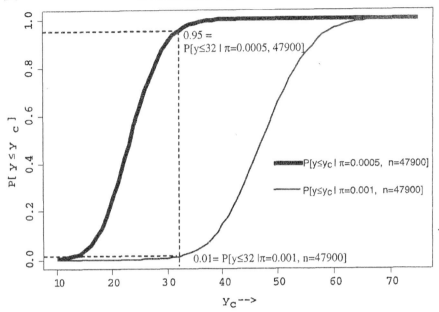

*Fig. 1.* Cumulative probability distributions for P[y≤y$_c$ | π=0.0005, n=47900] and P[y≤y$_c$ | π=0.001, n=47900] obtained by the function Ssize(). The maximum acceptable number of infected seeds in a sample of 47900 is 32 if a lot with 0.0005 disease incidence must be accepted 95% of the time and a lot with 0.001 incidence must be accepted only 1% of the time.

P[y=0 | π=0.00022] = $(1-\pi)^n$ = 0.00136. The 30000 sample size could have been estimated directly with the function Ssize(0,0.00022,1,0.00136,5000,100). Using a similar approach, any sample size in Table VII of Chapter 8 can also be estimated with Ssize(). For example, if one wants a probability of detection of 0.99 with an infection proportion of 0.1%, i.e., P[y>0 | π=0.001,n=4700]=0.99, Ssize(0,0.001,1,0.01,100,100) will specify a sample size of 4700. This also illustrates the sampling cost of trying to reduce the risk of rejecting a lot with a tolerable disease incidence level. We needed ten times more seeds to be able to accept with a probability of 0.95 a lot with 0.05% disease incidence and accept with a probability of 0.01 a lot with 0.1% disease. We can reduce the sample size ten-fold if we only want to minimise the probability of acceptance of a diseased lot.

## B. Bayesian Analysis and the Sample Size

In this subsection, we discuss distributions about π instead of distributions about y as in the previous subsection. The decision-maker's personal views about probable infection levels and their consequences are an integral part of risk assessment. The true value of the infection proportion (0<π<1) is not known and has to be expressed

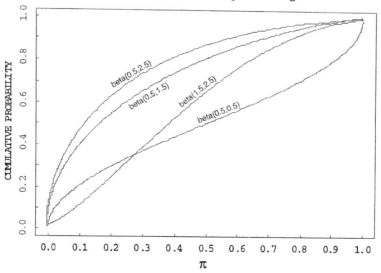

*Fig.* 2. Cumulative density functions for various beta(α,β). The infection proportion is given by π. These curves were obtained with the function betacum. ().

as a probability distribution over the interval $0<\pi<1$, i.e., as a curve representing the probability density (an idealised histogram) over the whole range of π-values. Being able to attach probabilities to π directly is a key in risk assessment. We shall use such probability distributions to express decision-makers' views or partial or vague knowledge about the location of the unknown proportion of infection π. The family of beta(a,b) distributions (a>0, b>0) is sufficiently rich to allow expression of widely different views about the location of π, since there is always a choice of a,b which can express any location of concentrated probability within $0<\pi<1$, and any degree of spread or vagueness. If a particular beta(a,b) prior distribution expresses our knowledge prior to sampling, and if a random sample of n plants in a lot turns out to contain y infected plants, then according to Bayes' theorem our revised or *posterior* distribution of π, actually the conditional distribution of π given y, is the beta(y+a, n-y+b). According to Jeffreys (1961), the beta(0.5, 0.5) is best regarded as the distribution appropriate for expressing absolutely no information about π prior to sampling. Any beta(a,b) probability density function (PDF) can be plotted automatically using the Splus function Betadens(), by typing Betadens(a,b). The cumulative density function (CDF) plots the area under the PDF up to a given π-value and shows the probability that π is not larger than that value (Fig. 2). Any beta(a,b) CDF can be plotted by similarly using the Splus function Betacum(). Typing pbeta($\pi_0$,a,b) gives a subjective probability that π<$\pi_0$. Starting with the beta(0.5,0.5) distribution, if we now examine one plant selected randomly from the field and determine that it is not infected, then the resulting *posterior* distribution, given what we know (n=1, y=0), is the beta(0.5,1.5), whose CDF has been co-plotted along with other CDF in Fig. 2. This expresses conditional probabilities given the data event y=0, for example $P[\pi \leq 0.5 \mid y=0]$; looking at the beta(0.5,1.5) CDF n Fig. 2 we see that it attributes a probability of approximately 0.8 to the proposition π < 0.5.

More precisely, this is found by typing pbeta(0.5,0.5,1.5), getting 0.818. If a second plant is sampled and also tests negative, the updated posterior becomes the beta(0.5, 2.5). If a third plant is collected and gives a positive test in the pathogen detection assay, the new posterior distribution is the beta(1.5,2.5).

A more realistic example was discussed previously for lettuce mosaic virus in seeds where a standard of 0 positive in 30,000 (y=0, n=30,000) was adopted (Grogan, 1980). The cumulative way of expressing the consequent posterior probability function beta(0.5, 30000.5) is given by the function Betacum(0.5,30000.5) and can be plotted to estimate P[$\pi \le$critical $\pi$ | y=0, n=30000] for various critical $\pi$-values (Fig 3). The inputs in this function are based on Jeffreys' prior or a = b = 1/2 and on the fact that no infected seeds were found in a sample of 30000. If one wants to establish the posterior probability that $\pi \le$ 0.00022, the answer 0.9997 can be obtained from the plot on Fig. 3 or by calculating the conditional probability P[$\pi$<0.00022 | y=0, n=30000] directly with the Splus function pbeta(0.00022,0.5,30000.5). This number resembles the 0.999 from Grogan (1980) which was based on the binomial probability P[y=0 | $\pi$=0.00022] = $(1-\pi)^n$ = 0.00136. The probability of detection or P[y>0 | $\pi$=0.00022] = 1-$(1-\pi)^n$ = 0.99864 is commonly used to design sampling schemes (see Chapter 8 for more examples). Note, however, that this is not a probability concerning $\pi$, but one concerning the frequency distribution of imaginary future y-values, i.e., it is the proportion of a very large number of samples of 30000 seeds which will have at least one infected seed given that $\pi$=0.00022.

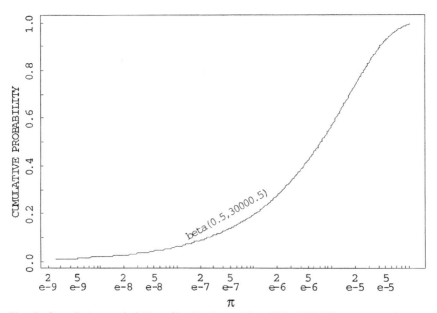

Fig. 3. Cumulative probability distribution of beta(0.5, 30000.5) over a relevant range of infection proportions ($\pi$). This curve was obtained with the function Betacumlog().

Data-informed or *posterior* probabilities concerning $\pi$ were calculated with Bayes' theorem. It is sometimes argued that this is more directly relevant to the actual problem than is a frequency distribution of occurrences of infected seeds for an infinite number of samples of size n. We may determine that the critical $\pi$ =0.00030 will give exactly 0.999 for P[$\pi \leq$critical $\pi$|y=0] with the function qbeta(0.999,0.5,30000.5). We will expand on this lettuce seed example of a "zero tolerance" where a 0-in-30000 standard has been adopted and show how our functions can be used to obtain probabilities about the proportion $\pi$ of infection. As discussed above, Betadens(0.5, n+0.5) portrays the posterior distribution for the zero tolerance case if there is no prior knowledge about the infection proportion, since y=0. As an alternative way of expressing this result, the areas under the density curve can be better visualised with the CDF Betacumlog(0.5, n+0.5), a function similar to Betacum() except that it has a logarithmic scale on the horizontal axis (Fig. 3). From such a graph we can directly get probabilities about $\pi$ which are based on the fact that there were no infected seeds in a sample of 30000. At probability 0.9 on the vertical axis, the proportion $\pi$ on the horizontal axis is 0.000045. Therefore, one can say that P[$\pi$<0.000045|y=0] = 0.90. We have designed Betacum.9(a,b) for the purpose of looking at probabilities in the 0.9 to 1.0 range that $\pi$ is less than a certain value. We believe that this will be useful in "zero tolerance" situations among other scenarios. This function generates a graph that identifies critical infection proportions in P[$\pi$ < critical $\pi$| y] for the probability values P = 0.9, 0.95, 0.99, 0.999 and 0.9999 (Fig. 4).

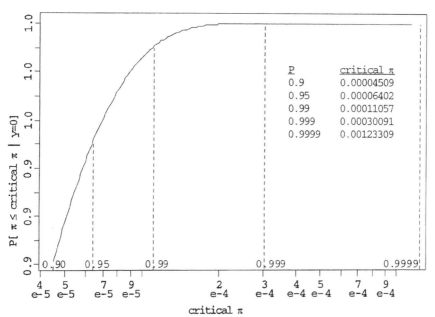

*Fig. 4.* CDF of beta(0.5, 30000.5) showing the probability that P[$\pi$<critical $\pi$| y=0]. This graph and table were produced by the function Betacum.9().

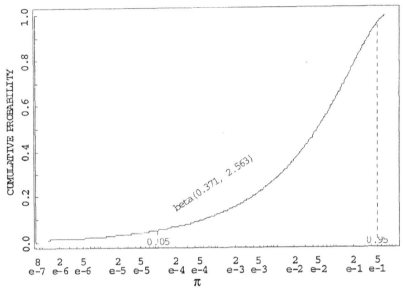

*Fig. 5.* Cumulative probability of a beta distribution that simultaneously give P[$\pi \leq 0.0001$] = 0.05 and P[$\pi \leq 0.5$] = 0.95. This plot was obtained with the function `Betac(0.0001,0.5,0.05,0.95)`.

The family of beta(a,b) distribution is rich enough to express a broad range of views about $\pi$. For example, someone who attributes a 5% chance to each of the two situations $\pi > 0.5$ (i.e. P[$\pi \leq 0.5$] = 0.95) and $\pi < 0.0001$ can discover that it is the beta(0.371, 2.563) distribution which expresses his/her views, by typing `Betad(0.0001,0.5,0.05,0.95)`. The function `Betad(pilo,pihi, flow,fhigh)` also plots the resulting beta(a,b) density to enable the users to judge further whether their choice of the arguments `pilo, pihi, flow, fhigh` satisfactorily expresses their views about $\pi$. Of course this plot could also be made separately with `Betadens()`. Contemplation of the beta(a,b) CDF `Betac(0.0001,0.5,0.05,0.95)` is also helpful (Fig. 5). If one judges from field and empirical experience that $\pi$ has a given probability of being above a certain level and another probability of being below another level, it is possible to use this subjective judgement to determine the prior beta(a,b) with the function `Betad(pilo,pihi,flow,fhigh)`. Such use of past experience will be discussed further in section IIIB.

## C. Modelling Expected Costs of Chance Consequences

In this section, we use our knowledge about the probabilities related to different values of $\pi$ to estimate costs of losses. We will first explain how the functions were derived and use a case study as example. Often the scientist may start with intuitively determined values of $\pi_t$, $\pi_{nt}$, $f_{hi}$, and $f_{lo}$ where an acceptance rule for which $f_{hi}$ = P[accept | $\pi_t$] and $f_{lo}$ = P[accept | $\pi_{nt}$] is required. It is our aim here to bridge the gap between this approach and its economic consequences. We will therefore model expected costs $C_a(\pi)$ of lot acceptance and $C_r(\pi)$ of rejection as functions of the true infection proportion $\pi$. We then find the smallest n for which, for some critical acceptance limit $y_c$, the risk $R(\pi) = C_a(\pi)P[\text{accept} \mid \pi] + C_r(\pi)P[\text{reject} \mid \pi]$ is acceptably low, say R, for both of $\pi=\pi_t$ and $\pi=\pi_{nt}$. Setting $R(\pi)=R$ at both of these $\pi$-values and solving for their respective $P[\text{accept} \mid \pi]$ values $p_{hi}$ and $p_{lo}$ gives approximately $p_{hi} = 1 - R/C_r(0)$ and $p_{lo} = R/C_a(1)$, where $C_r(0)$ is the maximum cost when a lot without any infection is rejected and $C_a(1)$ is the maximum cost when a lot with a 100% infection is accepted. Realistically, from a plant buyer's standpoint for example, $C_r(\pi)$ decreases from its maximum to 0 as we let $\pi$ increase, while $C_a(\pi)$ increases from 0 to its maximum. The exact solutions are $p_{hi} = [(R-C_r(\pi_t))]/[C_a(\pi_t)-C_r(\pi_t)]$ and $p_{lo} = [R-C_r(\pi_{nt})]/[C_a(\pi_{nt})-C_r(\pi_{nt})]$. The subjective inputs $C_r(0)$ and $C_a(1)$ are, respectively, the maximum expected costs of rejecting and accepting and may be calculated by modelling the costs of chance outcomes.

In an attempt to introduce a semblance of economic realism, and incidentally to remove possible inconsistencies in the subjective personal numerical inputs to the determination of n and $y_c$, we will model the expected cost functions $C_a(\pi)$ and $C_r(\pi)$. For this we must identify each of the possible consequences of accepting the lot of plants and of rejecting it. This means we articulate our view of all possible decision/chance structures by means of a *decision tree* (Kaplan, 1993; Smith, 1988; Winkler, 1972). A simple example of a decision tree that we will discuss in more detail is the one that says we must either reject or accept the lot. Possible outcomes after a decision have to be estimated. For example, one could try to estimate that the final acceptance of a lot will result in either a major (national) agricultural disaster (M), or in a crop loss to just a grower or small region (L for local), or in neither of the above (S for satisfactory), and that rejection (R) will have no further consequences. Numerical costs $C_M$, $C_L$, $C_S$, and $C_R$ of these respective situations must also be known in order to be able to model the overall expected cost. Finally, we must specify the outcome probabilities $P[M \mid \pi]$, etc., of each of the chance outcomes, as functions of $\pi$. As a result we have $C_a(\pi) = C_M P[M \mid \pi] + C_L P[L \mid \pi] + C_S P[S \mid \pi] = (C_M-C_S)P[M \mid \pi] + (C_L-C_S)P[L \mid \pi] + C_S$. On the other hand $C_r(\pi) = C_R$ because this simple example envisions no chance *sequelae* to rejection.

First, the probability of a local or major disaster has to be modelled against all possible $\pi$'s. $P[M \mid \pi]/P[M \mid \pi=1]$ and $P[L \mid \pi]/P[L \mid \pi=1]$ are each assumed to be a logistic $\pi$-curve, meaning one of the form $\pi \to \Pi(\sigma(\eta-\lambda))$ where $\Pi(t) = 1/(1+e^{-t})$, $\eta = \log[\pi/(1-\pi)]$, $\lambda$ is a location parameter, and $\sigma$ is a scale parameter. The coefficients $\lambda$ and $\sigma$ govern the location and sharpness of the "break" in the curve, respectively, as $\pi$ is changing from 0 toward 1. The parameter values $\lambda$ and $\sigma$ for the curve of a particular outcome, say $\lambda_M$ and $\sigma_M$ for a major agricultural disaster (M), may be determined with the function Scurvepi (pilo,pihi,flow,fhigh) whose arguments are low values $\pi_{lo}$ and $f_{low}$ for which we want $P[M \mid \pi=\pi_{lo}]/P[M \mid \pi=1] = f_{low}$, and high values $\pi_{hi}$ and $f_{high}$ for which $P[M \mid \pi=\pi_{hi}]/P[M \mid \pi=1] = f_{high}$. The maximum probabilities $P[M \mid \pi=1]$, $P[L \mid \pi=1]$ of the two curves $P[M \mid \pi]$ and $P[L \mid \pi]$ must also be supplied (with sum less than 1) for the cost analysis described further. Of course $P[S \mid \pi] = 1 - P[M \mid \pi] - P[L \mid \pi]$ at any $\pi$. Scurvepi() returns $\lambda_M$ and $\sigma_M$ and creates a graph of $\Pi(\sigma_M(\eta-\lambda_M)) = P[M \mid \pi]/P[M \mid \pi=1]$ vs $\pi$. This graph should be examined for consistency with the scientist's views about the probability of other infection proportions $\pi$ causing a major agricultural disaster (M).

The posterior distribution of $\pi$ provides rational basis for decision-making. The better decision is that which produces the smaller of the two expected costs $E[C_a(\pi)|y]$ of acceptance, and $E[C_r(\pi)|y]$ of rejection. In the following example, these will be calculated using the beta$(y+\alpha, n-y+\beta)$ distribution of $\pi$ and using the facts that $E[C_r(\pi)|y] = C_R$ and $E[C_a(\pi)|y] = (C_M - C_S) P[M|\pi=1] E[\Pi(\sigma_M(\eta - \lambda_M))] + (C_L - C_S) P[L|\pi=1] E[\Pi(\sigma_L(\eta - \lambda_L))] + C_S$. To do this for given values of $\lambda$s and $\sigma$s, we will obtain $E[\Pi(\sigma(\eta - \lambda))]$, where $\eta = \log[\pi/(1-\pi)]$ and $\pi$ is beta$(\alpha,\beta)$-distributed, by using the numerical integration function $Exp.pils(loc, sca, alfa, beta)$. The meaning of, for example, the term $P[M|\pi=1] E[\Pi(\sigma_M(\eta - \lambda_M))]$, is the expected cost of the outcome M, assuming the beta$(alfa, beta)$ distribution is that which expresses our current knowledge about $\pi$. Interactive exploration of sensitivity of decision to inputs may be carried out by calculating $E[C_a(\pi)|y]$ with the function $Exp.cos.bi(costs, loc, sca, apy, nmypb, maxprobs)$. In this example $costs$ is the string $(C_M, C_L, C_S)$, while $loc = (\lambda_M, \lambda_L)$ and $sca = (\sigma_M, \sigma_L)$ are strings obtained from $Scurvepi()$. The values $apy$ and $nmypb$ are inputs to define the beta$(apy, nmypb)$ distribution which is the one which expresses our current knowledge about the infection proportion $\pi$. If we started with the beta$(\alpha,\beta)$ prior distribution of $\pi$ and then obtained a random sample of n plants observing y infected plants, we would use the argument values $apy = \alpha+y$ and $nmypb = n-y+\beta$. Finally, $maxprobs$ is the vector $(P[M|\pi=1], P[L|\pi=1])$, i.e. the probabilities of disasters given that $\pi=1$.

*A case study with Lettuce Mosaic Virus*
The example of the seed-transmitted lettuce mosaic virus (LMV) can be used again to illustrate how we can model what is known about a major and a local disaster with $Scurvepi(pilo, pihi, flow, fhigh)$. Many of the quantitative inputs, e.g., costs, are highly subjective, and must of course be faced if there is any hope for quantitative modelling, and the present machinery allows, indeed, requires, us to face them. To gain a realistic perspective on the adequacy of an acceptance procedure, analyses with a range of subjective inputs should be done. If LMV incidence in seeds is less than 0.022%, the probability of a major outbreak is very low (Grogan, 1980). We could say that if a seed lot with $\pi=0.00022$ is accepted there is a 0.0001 probability of a major disaster given that the probability of a major disaster is 1.0 if $\pi=1$. If we consider that there is only a 0.5 probability of a major disaster if all plants are infected, i.e. $P[M|\pi=1] = 0.5$, then $P[M|\pi = 0.00022] = 0.0001 \times 0.5 = 0.00005$. The $Scurvepi()$ function simply provides the coefficients for a logistic curve from 0 to 1 probability that fits the points we provided. We can look at $P[M|\pi=1]$ as the adjustment factor that will transform this 0 to 1 curve into a curve from 0 to the maximum probability of a disaster when $\pi=1.0$. It has been demonstrated that if the proportion of infected seeds is between 2-5% an outbreak with 25%-96% of all plants infected will almost inevitably occur (Tomlinson, 1962). We could say that there is 0.95 probability of a major disaster with $\pi=0.02$ if the probability of a major disaster is 1.0 with $\pi=1$, i.e. $P[M|\pi=0.02]/P[M|\pi=1] = 0.95$. If $P[M|\pi=1] = 0.5$, then the actual probability of a major disaster with $\pi=0.02$ is $P[M|\pi=0.02] = 0.95 \times 0.5 = 0.475$. This conservative scenario reflects the danger that if a seed lot with more than 2% infection is not rejected, the outbreak resulting from the acceptance of these seeds could result in a trade embargo for lettuce seed and/or severe losses in areas where seed from this lot were planted if the disease is spread by aphids. Using what we

know about the impact of a 0.02% and a 2% LMV seed contamination on crops, we could try to generate a model for major disasters with Scurepi(0.00022,0.02,0.0001,0.95). It provides the location parameter $\lambda$ = -4.998, and the scale parameter $\sigma$ = 2.683 that give a logistic curve from 0 to 1.0. $P[M \mid \pi]/P[M \mid \pi=1]$ is also plotted against a range of $\pi$-values (Fig. 6A) using the function $P[M \mid \pi]/P[M \mid \pi=1] = 1/(1+e^{-\sigma(\eta-\lambda)})$ given that $\eta = \log[\pi/(1-\pi)]$. The relationship between $\pi$ and $\eta$ can be visualised by comparing the two scales at the bottom of Fig. 6. For example $\eta = -3.89$ for $\pi=0.02$ which gives $P[M \mid \pi = 0.02]/P[M \mid \pi=1] = 1/[1+e^{-2.683(-3.89+4.998)}] = 0.95$, one of the inputs in the function (see dash lines in Fig. 6). Using either Fig. 6A or $1/[1+e^{-\sigma(\eta-\lambda)}]$ one can find for example that $P[M \mid 0.003]/P[M \mid \pi=1] \approx 0.10$ which is in agreement with the fact that the 0.3% LMV tolerance limit originally used for certification provided inconsistent control (Grogan, 1980). Experiments done in the United Kingdom have shown that lettuce seeds with infection proportions from 0.00003 to 0.001 will keep aphid spread to a minimum and produce crops with less than 1% infection (Maude, 1988; Tomlinson, 1962), though LMV control can be inadequate with as low as 0.1% seed transmission in other situations (Grogan, 1980). Using the scales on the right of Fig. 6, it is possible to visualise the probability of a major disaster, $P[M \mid \pi]$, for different $P[M \mid \pi=1]$. Given that the spread of LMV by its aphid vectors is mainly local (Tomlinson, 1970) we could say that if a seed lot with $\pi=0.00003$ is accepted the probability of local disaster is 0.0001 with $P[L \mid \pi=1]$ =1.0, i.e. $P[L \mid \pi= 0.00003]/P[L \mid \pi=1] = 0.0001$. Even if there is no seed infection, there is still a possibility of damage caused by LMV introduced in the crop from infected weed hosts. Given the uncertainty about the disease occurrence if seeds with 0.001 infection are planted, we could say that there is 0.5 probability of a local disaster with $\pi=0.001$ if the probability of a local disaster is 1.0 with $\pi=1$, i.e. $P[L \mid \pi=0.001]/P[L \mid \pi=1] = 0.5$. Scurvepi(0.00003,0.001,0.0001,0.5) gives $\lambda=$ -6.907 and $\sigma=$ 2.626, the location and scale parameters, respectively. $P[L \mid \pi]/P[L \mid \pi=1]$ is also plotted against a range of $\pi$-values (Fig. 6B). The effect of various $P[L \mid \pi=1]$ on the probability of a local disaster $P[L \mid \pi]$ can be seen on the left of Fig. 6B.

If we take an example where no infected seeds were found in a sample of 30000 seeds, it is possible to plot the probability density function of beta($y+\alpha,n-y+\beta$) given by Betadens(0.5, 30000.5) along with $P[M \mid \pi]/P[M \mid \pi=1]$. In Fig. 7, the beta density functions are co-plotted with the probability of disaster shown on Fig. 6. The product of these two $\pi$-functions with the length increment $\partial\pi$ gives the integral $E[\Pi(\sigma(\eta-\lambda))]$. The function Exp.pils(-4.99,2.68,0.5,30000.5) plots $P[M \mid \pi]/P[M \mid \pi=1]$ and gives $E[\Pi(\sigma_M(\eta-\lambda_M))] = 8.5 \times 10^{-7}$ (Fig. 7A). Multiplying $E[\Pi(\sigma_M(\eta-\lambda_M))]$ by $P[M \mid \pi=1]$ gives the probability, given any $\pi$, of a major disaster. For example, if $P[M \mid \pi=1]=0.5$, the overall probability of a major disaster given that no infected seeds were found in a sample of 30000 is $(8.5 \times 10^{-7}) \times 0.5 = 4.25 \times 10^{-7}$. The corresponding functions for a local disaster are shown on Fig. 7B.

Based on Maude (1988) and some statistics on lettuce production in United Kingdom, the major disasters in lettuce production due to LMV have been established at £M 3 per year. The cost of a more localised outbreak in a year has been established at £124,444. The annual cost of the seed testing program being carried yearly is approximately £40,000 based on the cost estimate of £10/kg for seed testing,

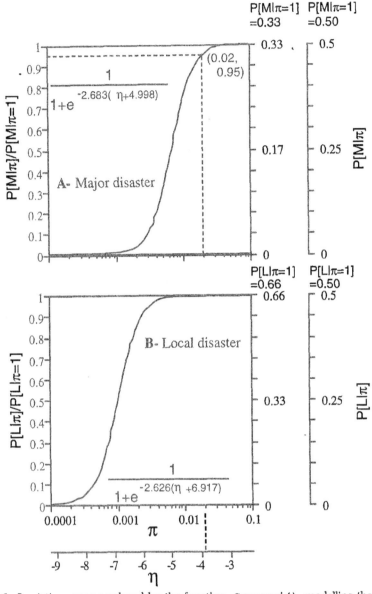

Fig. 6. Logistic curves produced by the function Scurvepi() modelling the probability of a disaster as a function of η which is log [π/(1-π)]. The logistic curve $1/[1+e^{-\sigma(\eta-\lambda)}]$ goes from a probability of 0 to 1 (vertical axis on the left) and can give the actual probability of a disaster for the various π-values once adjusted for the maximum probability of a disaster given by P[M|π=1] or P[L|π=1] (vertical axis on the right).

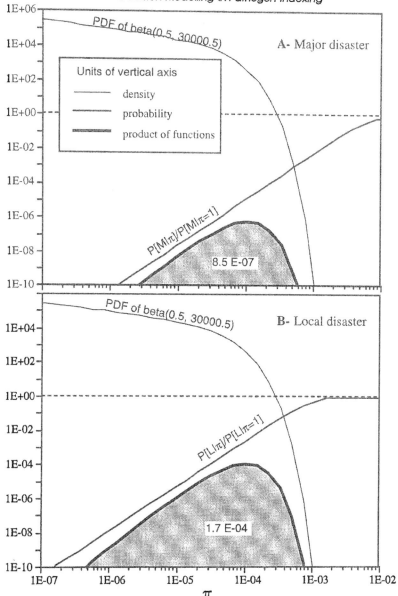

*Fig. 7.* Modelling of probabilities for a major disaster (**A**) and a local disaster (**B**). The integration of the product of the functions (area of shaded area) obtained by the function Exp.pils() gives the overall probability of a major or local disaster for all $\pi$'s. This is based on the fact that no infected seeds were found in a sample of 30000 [beta(0.5,30000.5)] and the probability of a major or local disaster is 1.0 if $\pi = 1.0$ (P[M|$\pi$=1] = P[L|$\pi$=1] = 1).

an unavoidable cost regardless of disease incidence. The vector $(C_M, C_L, C_S)$ was set at (3000000, 124444, 40000) using the Splus command costs <- c(3000000, 124444, 40000). The location and scale parameter for local and major disasters given above were used using the Splus commands loc <- c(-4.989163, -6.906755) and sca <- c(2.683245, 2.625849), respectively. To estimate the cost, the vector maxprobs or $(P[M|\pi=1], P[L|\pi=1])$ also has to be entered. Various pairs of maximum probabilities were tried with a range of infected seeds found in samples using the function Exp.cos.bi(costs,loc,sca,apy,nmypb,maxprobs) where y is the number of infected seeds and apy = $\alpha$+y and nmypb = n-y+$\beta$. The expected costs for various possible combinations of y's and pairs of $(P[M|\pi=1], P[L|\pi=1])$ are plotted on Fig. 8. For example, if 50 infected seeds out of 30000 are found in a seed lot, the cost given $(P[M|\pi=1], P[L|\pi=1]) = (0.33,0.33)$ can be found by typing Exp.cos.bi(costs,loc,sca,50.5,29950.5,c(0.33,0.33)) which estimates the cost at £85360, one of the points of the solid line on Fig. 8. Vectors can be entered directly into the function by typing, for example, c(0.33,0.33) or defined first by typing maxprobs <- c(0.33,0.33) and then entered by typing the name maxprobs into the function. The cost of seed testing is the minimum expected cost and this cost rises in an exponential fashion as the number of infected seeds found in a sample goes up. It is interesting to see that if the probabilities of a major and local disaster are set at 0.5 for both, the expected cost is the same as when the probabilities are set at 0.33 and 0.66 for major and local disasters, respectively, for up to 50 seeds in a sample. If the number of infected seeds found in a sample is 100, then the expected cost will be £60000 more if the probabilities of major and local disasters are both set at 0.5. If the probabilities for disasters given that $\pi=1$ are set at 0.1, the expected cost stays relatively low even with a high number of infected seeds in a sample.

## III. SIMPLE DIRECT "BULK" SAMPLING

### A. Sample Size

In the previous section we discussed decisions based on a number of infected plants in a sample. These analyses applied to the total number y of plants observed to be infected out of n plants randomly sampled from a lot, effectively without replacement. Now consider a sampling scheme where plants in a 100 meter row are each visually inspected or the seeds contained in a scoop of a standard volume are individually tested. How many different rows should be inspected in a field or how many separate scoop sub-samples from a lot should be processed? We define the length of row, the volume, the area, or the weight sampled as a subregion. More precisely, the above examples are described as N non-overlapping subregions $R_i$ (i=1,2,..N) of a homogeneous field or lot that are individually collected, and for which the total number $y_i$ of occurrences of an infected plant or in each subregion is counted. Write $|R_i|$ for the sizes of the subregions. For example, in the context of seedlot acceptance sam-

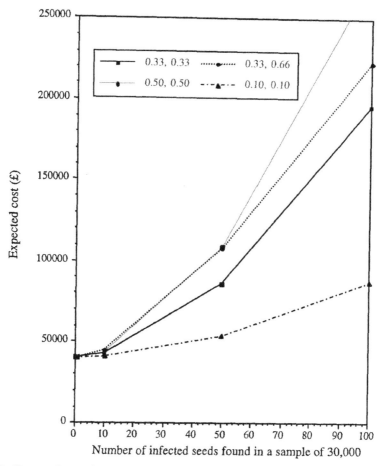

*Fig. 8.* Expected cost due to LMV outbreak based on the number of infected seeds found in a sample of 30000 seeds. Different maximum probabilities of local and major disaster when $\pi=1$ were used as input and gave results that were plotted separately. Each point was estimated by the function Exp.cos.bi().

pling, by taking all $|R_i| = 1$ we could model the sampling of N scoops of seeds and find out how many infected seeds were in each sample. Assuming homogeneity in space of the infected seeds and assuming there are many seeds in each region taken in the sample, and assuming we have perfect detection capability, the $y_i$ have Poisson distributions with means $E[y_i]$ proportional to the region sizes $|R_i|$. We can write $\mu$ infected seeds per unit size for this proportionality constant, so that $E[y_i] = \mu|R_i|$. We are interested in controlling the contamination density $\mu$ of an accepted seedlot. Here we consider only the case where all $|R_i|$ are the same; the more general analysis is slightly more complex. In this case $R_\bullet = \Sigma_{i=1}^{N} |R_i|$, the total area or volume to be sampled, may be considered to be the sample size, and the information in the sample is summarised by the total of infected plants $y_{\bullet} = \Sigma_{i=1}^{N}\, y_i$. Just as with binomial samples, in order to distinguish between $\mu = 4$ and $10$ infected seeds per scoop with probabilities $0.95$ and $0.99$ respectively, we find from Ssize(4, 10, 0.95, 0.01, 1.1, 0.01) that we should sample a total of about $n = 3.06$ scoops and accept the lot when $y_{\bullet} \leq y_c = 18$ (Fig. 9). In this example, 95% of an infinity of samples of 3.06 scoops each from a lot that has an infection proportion of 4 infected seeds per scoop would have a total of 18 infected seeds or less, i.e., $0.95 = P[y_{\bullet} \leq 18 \mid \mu = 4, R_{\bullet} = 3.06 \text{ scoops}]$. Simultaneously, only 1% of an infinity of samples of 3.06 scoops each from a lot with an infection proportion of 10 per scoop would have a total of 18 infected seeds or less, i.e., $0.01 = P[y_{\bullet} \leq 18 \mid \mu = 10, R_{\bullet} = 3.06 \text{ scoops}]$.

## B.   Bayesian Analysis and the Sample Size

In section IIB we used the beta$(\alpha,\beta)$ distribution to express some probabilities about the proportion $\pi$ of infection. Similar to what was done in section IIB, a distribution from which probabilities directly related to $\mu$, the number of seeds or plants per unit size, will be discussed. The distribution family gamma$(\alpha,\beta)$ (a different distribution for each $\alpha > 0$, $\beta > 0$) is sufficiently rich to allow expression of virtually anyone's partial knowledge about the value of $\mu$. This has mean $\alpha/\beta$ and standard deviation $\sqrt{\alpha}/\beta$. Its histogram can be viewed with the function Gamdens(alfa,beta). Furthermore, using a particular gamma$(\alpha,\beta)$ prior distribution gives $\mu$ the posterior distribution gamma$(y_{\bullet} + \alpha, R_{\bullet} + \beta)$, where $y_{\bullet} = \Sigma_{i=1}^{N}\, y_i$ and $R_{\bullet} = \Sigma_{i=1}^{N}\, |R_i|$. What prior

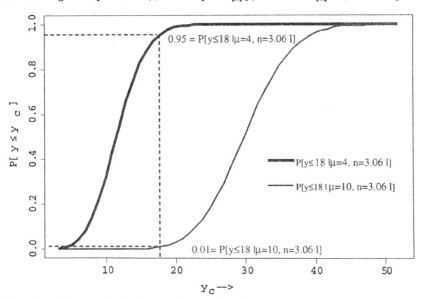

*Fig. 9.* Cumulative probability distributions for $P[y_{\bullet} \leq y_c \mid \mu = 4, R_{\bullet} = 3.06 \text{ scoops}]$ and $P[y \leq y_c \mid \mu = 10, R_{\bullet} = 3.06 \text{ scoops}]$ showing that 18 is the maximum number of infected plants that could be accepted in a lot. These were produced by the function Ssize().

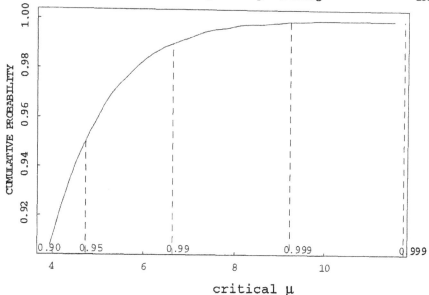

critical μ

*Fig. 10.* Cumulative Density Function for the posterior gamma(2,1) distribution.
P[μ<critical μ | y.=0] can be visually estimated in the 0.9-1.0 probability range. This
distribution was obtained with the function Gamcum.9().

α and β shall we use? The commonly accepted expression of complete prior ignorance about μ is taken to be α=2 and β=0, no matter that this "improper" Jeffrey's
prior distribution has infinite mass. For example, if a scoop of seeds is sampled and
no infected seeds were found, the minimally informative gamma(2,0) prior distribution
of μ together with the data y. = 0 gives the gamma(2,R.) posterior distribution, so
that given that R.=1, the posterior distribution is gamma(2,1). Fig. 10 shows the
gamma(1,2) plotted using the    Gamcum.9(2,1)    function which is like the
Gamcum(2,1) except that it zooms in the portion of the distribution above the prob-
ability of 0.9. It can be estimated from Fig. 10 that P[μ<9 | y.=0] ≈ 0.999 and the
exact answer can be found by typing 1*qgamma(0.999,2), obtaining μ<9.23. If a
region of size 1 holds about 30000 seeds this corresponds to infected seed proportion π < $\frac{9.23}{30000}$ or 0.0003, in rough agreement with the binomial sampling analysis
P[y>0 | π=0.00022] = 1-(1-π)$^n$ = 0.9986 discussed in section IIA.

As opposed to prior ignorance, prior opinion about μ can be expressed through a
gamma(α,β) by specifying two percentage points with two corresponding μ's, an upper point and a lower point. For example, visual inspection of Potato Virus Y (PVY)
or Potato Leaf Roll Virus (PLRV) can be done by counting the number of infected
plants in 100 meters of row. Based on the past experience of several seasons of inspections there were, say, 1 infected plant or less in 5% of the rows counted and 10
infected plants or more in 1% of the rows. In a way similar to the beta distribution,
we can find that particular α and β, which expresses P[μ≤1] = 0.05 and P[μ>10] =
0.01. The function Gammad(1,10,0.05,0.99) tells us that the gamma(3.07,0.85)
expresses our partial knowledge about μ. Gamdens(3.07,0.85) then draws the

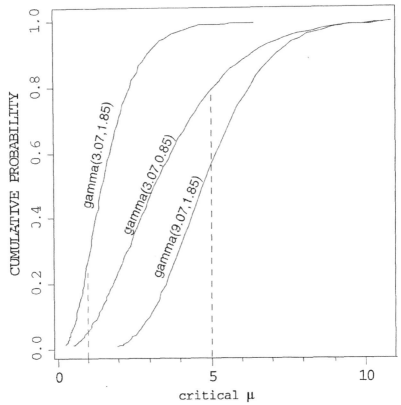

Fig. 11. Cumulative density function of a prior gamma(3.07,0.85) that simultaneously show that P[μ≤1] = 0.05 and P[μ>10] = 0.01. Two posterior gamma distributions based on y. = 0 and R.=1 [gamma(3.07,0.85)] or y. = 6 and R.=1 [gamma(3.07,0.85)] are also shown. Each distribution was obtained with the function Gamcum().

PDF for gamma(3.07,0.85) distribution and Gamcum(3.07,0.85) draws the CDF (Fig. 11). The CDF of gamma(3.07,0.85) shows not only that P[μ≤1] = 0.05 and P[μ≤10] = 0.99 as planned but it can also be used to determine for example that P[μ≤5] ≈ 0.8. With this prior distribution, if one goes into a field from the same area and finds no infected plants in a 100 meter row, i.e., y. = 0 and R.=1, the posterior distribution gamma(3.07,1.85) illustrated in Fig. 11 is obtained. Now P[μ≤1] ≈ 0.25 instead of 0.05. If instead of no infected plants we had found 6 infected plants, the posterior distribution would have been gamma(9.07,1.85). From Fig. 11 it can be seen *a posteriori* that P[μ≤5] ≈ 0.55 instead of the previously obtained 0.8.

## C. Modelling Expected Costs of Chance Consequences

Cost modelling for chance consequences of a decision based on a Poisson-distributed sample proceeds as with a binomial sample, except that the Poisson density $\mu$ now plays the role of the binomial parameter $\pi$: $P[M|\mu]/P[M|\mu=\text{infinity}]$ and $P[L|\mu]/P[L|\mu=\text{infinity}]$ are each assumed to be a logistic $\mu$-curve, meaning one of the form $\mu \rightarrow \Pi(\sigma(\eta-\lambda))$ where $\Pi(t) = 1/(1+e^{-t})$ and $\eta = \log(\mu)$. Such curves are S-shaped with $\mu=0 \rightarrow 0$, $\mu=\exp(\lambda) \rightarrow 1/2$, and they approach 1 as $\mu$ approaches infinity. Appropriate values of $\lambda$ and $\sigma$ can be found with the function Scurvemu(mulo,muhi,flow,fhigh), which works exactly like Scurvepi(pilo,pihi,flow,fhigh).

Again, the decision with lower expected cost is taken. Expected costs $E[C_a(\mu)|y]$ of accepting and $E[C_r(\mu)|y]$ of rejecting are calculated using the gamma(y.+$\alpha$, R.+$\beta$) distribution of $\mu$. The calculations proceed exactly as with binomial samples, using the function Exp.cos.ga(costs, loc, sca, ypa, rspb, maxprobs). As in section IIC, costs is the string $(C_M,C_L,C_S)$, the strings loc = $(\lambda_M,\lambda_L)$ and sca = $(\sigma_M,\sigma_L)$ are obtained from Scurvemu(), and the gamma(ypa, rspb) distribution is the one which expresses our current knowledge about $\mu$, and maxprobs is the vector $(P[M|\pi], P[L|\pi])$.

## IV. INDIRECT BIOASSAYS: DETECTING THE PRESENCE OF CONTAMINATION IN A POOLED SAMPLE

## A. Sensitivity and Probability of Detection

In the foregoing analysis perfect detection was assumed, so the test result was expressed as a number $y$ of detected contaminated plants in a sample of a given size. The resulting mathematical connection between test result and the contamination proportion $\pi$ or density $\mu$ was therefore a simple formula: either $P[y|\pi,n]$ was binomial or $P[y|\mu,R.]$ was Poisson. We now consider the situation in which $y$ is not observed and we have an analytical method which merely detects (or not) the presence of some contamination in a sample of plants. The probability Prob[det| y,n] given that there are $y$ contaminated plants in the sample is called the *sensitivity*. Several examples of such "pooled" sample testing are described in this volume. The functional form by which the conditional detection probability (given $\pi$) P[det| $\pi$,n] = $\sum_{y=0}^{\inf}$ Prob[det| y,n] P[y|$\pi$,n] depends upon $\pi$ is therefore of interest since it allows modelling of sensitivity for a range of $\pi$'s. This function P[det| $\pi$,n] of $\pi$ is calculated and plotted by the Detprob(n,nscount,calprob,scon). The argument nscount is a string of y-values for which Prob[det| y,n] is known presumably through previous calibration, while the string calprob contains their corresponding calibrated Prob[det| y,n] values. The optional argument scon has a default of 0.99 and simply adjusts the scaling of the resulting plot. Prob[det| 1,n] and Prob[det| 0,n] must be included as the first two elements of the string calprob even if they are only guesses, with Prob[det| 0,n] > 0 if the detection method can produce false positives. Then the function Detprob() fills in the missing $(y, \text{Prob}[\det | y,n])$ pairs by fitting a smooth y-curve known as a *spline*, extending from the pair $(1, \text{Prob}[\det | 1,n])$ to the pair $(y_{max}, \text{Prob}[\det | y_{max}, n])$, where $y_{max}$ is the largest y-value in nscount, and assumes that $\text{Prob}[\det | y,n] = \text{Prob}[\det | y_{max}, n]$ constantly over all $y > y_{max}$. This assumption can be relaxed and altered by including a fictitious or uncalibrated $(y, \text{Prob}[\det | y,n])$ for some y beyond $y_{max}$.

$P[\det | \pi, n]$ has a simple closed mathematical expression only under very restrictive assumptions, e.g., if $\text{Prob}[\det | 0,n]=0$ with $\text{Prob}[\det | y,n]$ the same constant sensitivity, say p, for all $y>0$, which gives $P[\det | \pi, n] = (1-e^{-n\pi})p$. For example, Kimble *et al* (1975) reported that the sensitivity of a *Chenopodium* pooled assay for LMV in lettuce seeds was 0.8 if one infected seed was present in a sample of 500. Using the formula $(1-e^{-n\pi})p$ with p=0.8 it is possible to determine that $P[\text{detection} | \pi = 0.00015, n=500] = 0.058$. This example was also used by Geng *et al.* (1983). This estimate, in addition to the whole range of probabilities of detection with n=500 over the entire range $\pi$, can also be obtained from Detprob() by taking nscount=(0,1,2,4) with calprob=(0,0.8,0.8,0.8). The y-values above 1 in nscount are arbitrary but necessary to fit the spline. These extra values do not influence the results once the maximum p=0.8 is reached.

Generally, the sensitivity will not always be at its maximum. $P[\det | \pi, n] \leq (1-e^{-n\pi})\text{Prob}[\det | y_{max}, n]$, especially when n or the $\text{Prob}[\det | y,n]$ values are not large. There may be situations where the sensitivity of the assay has been tested for more than one infected seed in the sample. What sensitivity value should one use in such case? Kimble *et al* (1975) reported that 1 infected seed in 500, 1000 and 2000 healthy seeds could be detected with probabilities of 0.8, 0.6, and 0.5, respectively. If we assume that the virus particles would have the same concentrations after grinding 4 infected seeds in a sample of 2000 as grinding 1 infected seed in a sample of 500, it is possible to make the assumption from Kimble *et al* (1975) that 1, 2, and 4 seeds in a sample of 2000 could be detected with probabilities of 0.5, 0.6, and 0.8, respectively. If these increasing sensitivities are not taken into consideration, i.e. if we assume $P[\det | \pi, 2000] = p = 1.0$ for all $y>0$, then $P[\det | \pi, 2000] = 1-e^{-2000\pi}$ (Fig. 12). However, we do know that the sensitivity with low number of infected seeds in a sample is not 1.0 but we do not know how many infected seeds are in a sample. A possibility would be to take a conservative approach and use p=0.5, which is the sensitivity if there is only one seed in the sample of 2000. In this case $P[\det | \pi, 2000] = 0.5 \times 1-e^{-2000\pi}$ (Fig. 12). If the maximum sensitivity is used then $P[\det | \pi, 2000] = 0.8 \times 1-e^{-2000\pi}$ (Fig. 12). These curves can be obtained with Detprob(2000, nscount, calprob) taking nscount=(0,1,2,4) and calprob= (0,0.5,0.5,0.5) or calprob=(0,0.8,0.8,0.8) for p=0.5 and p=0.8, respectively. These two simplified applications of the function Detprob() use only part of the information that is known about sensitivity. When $\pi$ is in the range of 1/2000 the sensitivity should be around 0.5 and when $\pi$ is around 1/500 it should be approximately 0.8. This is what the function Detprob() can accomplish. It can be seen on Fig. 12 that Detprob(2000, c(0,1,2,4),c(0,0.5,0.6,0.8)) produces a more realistic probability of a detection curve which follows the 0.5 sensitivity curve for low $\pi$-values and gradually ends up following the 0.8 sensitivity curve for higher $\pi$-values. In the following section, we will see how the above results can be used to make decisions about $\pi$.

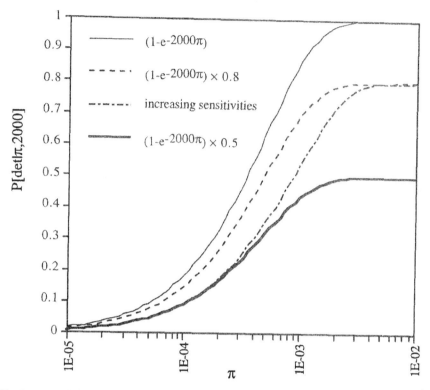

*Fig. 12.* Probabilities of detection of LMV for a range of π-values using a sample of size 2000. If the assay has a constant sensitivity, the probability of detection is determined by multiplying sensitivity by $1-e^{2000\pi}$. The different curves were produced with the function Detprob() using either a single sensitivity or the increasing sensitivities extrapolated from Kimble *et al.* (1975).

## B. Number of Pooled Samples of Size n

We mentioned in section IIA and IIB that acceptance of lettuce seed lots for LMV has been set at π≤0.00022 which is implemented by enforcing a zero tolerance in a sample of 30000 seeds tested individually. With a pooled sample of 2000 seeds used for the *Chenopodium* test (Kimble *et al.*, 1975) where one can only ascertain presence or absence of the disease in the batch of seeds, the probability of detection is 0.1898 when π=0.00022 as determined in the previous section by the Detprob() function using a range of sensitivities (Fig. 13). How many samples of 2000 seeds should be processed if we want to use the *Chenopodium* test and have 0.1% chance of accepting a seed lot with π=0.00022 given the probability of detection p=0.1898? The probability of obtaining k samples that do not show any evidence of LMV through the *Chenopodium* assay given that p=0.1898 can be determined by the formula $(1-p)^k$ which gives a probability of 0.001 with 33 samples. See Geng *et al.* (1983) for more details on this approach.

Given n, to distinguish between $\pi_t$ and $\pi_{nt}$ with required acceptance probabilities as in section IIB we will need k samples of n plants to accept with a high probability a lot with $\pi_t$ disease incidence and to accept with a low probability a lot with $\pi_{nt}$ incidence. To determine the required number k of times we may again use the function Ssize(). To find both k and the critical acceptance region $z < z_c$ for the observed number z of positive detections, the first two arguments of Ssize() should be $\delta(\pi_t,n)$ and $\delta(\pi_{nt},n)$, which can be found on the graph of $\delta(\pi,n)$ = P[detection | $\pi$, n] discussed in the previous section. Let us expand on the example from Geng *et al.* (1983) that we already discussed in section IIB using the pooled sample of 2000 seeds. We want the high probability $f_{hi}$ = P[accept | $\pi_t$ = 0.0005, n=2000] = 0.95 and the low probability $f_{lo}$ = P[accept | $\pi_{nt}$ = 0.001, n=2000] = 0.01. If sensitivity is perfect (i.e. p=1.0), then $\delta(\pi_t,n) = 1-e^{-2000\times0.0005} = 0.632$ and $\delta(\pi_{nt},n) = 1-e^{-2000\times0.001} = 0.865$. Some values of P[det | $\pi$=0.0005,n=2000] and of P[det | $\pi$=0.001,n=2000] are listed in Table I. If sensitivities derived from Kimble *et al.* (1975) are used with the Detprob() function, then $\delta(\pi_t,n) = 0.353$ and $\delta(\pi_{nt},n) = 0.540$. The required size k of random samples of 2000 seeds in the lot, and a critical

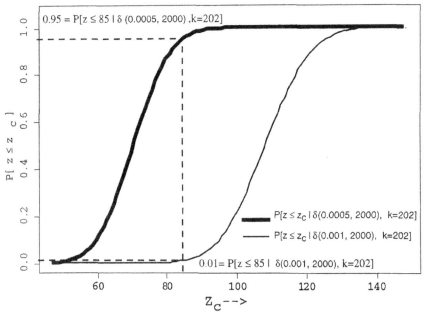

*Fig. 13.* Cumulative probabilities for P[z≤$z_c$ | $\delta$(0.0005, 2000)= 0.353, k=202] and P[z≤$z_c$ | $\delta$(0.001, 2000) = 0.540, k=202] obtained by the function Ssize(). The sensitivity calibrations extrapolated from Kimble *et al.* (1975) were used to determine $\delta(\pi, n)$. The maximum acceptable number z of samples of 2000 seeds out of a total of 202 is 85 if a lot with 0.0005 disease incidence must be accepted 95% of the time and a lot with 0.001 incidence must be accepted only 1% of the time.

*Table I.* Number (k) of samples of 2000 seeds needed and critical number ($z_c$) of infected samples to accept 95% and 1% of the time lots with less than 0.0005 disease incidence ($p_t = 0.0005$) or more than 0.001 ($p_{nt} = 0.001$), respectively. The sample sizes and critical number $z_c$ of lots testing positive were determined using either a constant or variable sensitivity. Number (k*) of samples of 2000 seeds to accept 1% of the time lots with 0.001 disease incidence without any interest in minimising rejection of lots with tolerable disease incidence are compared.

| | Sensitivity of 2000 seed assay | | | |
| --- | --- | --- | --- | --- |
| | Constant[‡] 1.0 | Constant[‡] 0.8 | Variable[†] 0.5 to 0.8 | Constant[‡] 0.5 |
| d($p_t$=0.0005,n=2000)[§] | 0.632 | 0.506 | 0.353 | 0.316 |
| d($p_{nt}$=0.001,n=2000)[§] | 0.865 | 0.692 | 0.540 | 0.433 |
| k | 221 | 276 | 202 | 436 |
| $z_c$ | 159 | 159 | 85 | 157 |
| k* ("zero tolerance") | 2.3 | 4 | 6 | 8 |

[*] number of samples needed if there is no interest in minimising the rejection of lots with a tolerable disease incidence, i.e., 0.01 = P(z=0| d($p_{nt}$=0.001,n=2000), k). A lot with a positive test is immediately rejected.

[‡] Determined by $(1-e^{-np})p$ where the sensitivity p is constant

[†] Determined with the function Detprob() using the sensitivities increasing from 0.5 to 0.8 as derived from Kimble *et al.* (1975).

[§] d(p,n) = P(detection | p, n).

acceptance range $z \leq z_c$ for the observed number $z$ of detections may be found by typing, for example, Ssize(0.353, 0.540, 0.95, 0.01, 2, 1) where Ssize() is the same function as discussed before with arguments $\delta(\pi_t,n)$, $\delta(\pi_{nt},n)$, $f_{hi}$, $f_{lo}$, startn, and incrm (Fig. 13). Assuming the range in sensitivity extrapolated from Kimble *et al* (1975), the lot should be rejected if more than 85 of the 202 samples of 2000 pooled seeds test positive. With this sampling scheme, lots with more than 0.001 disease incidence would be accepted only 1% of the time and lots with less than 0.0005 would be accepted 95% of the time. It is interesting to see that the more realistic scenario, where sensitivity increases as the number of infected seeds in the sample increases, gives the cheaper sampling scheme. However, like we saw in section IIB for individually tested plants, the cost of being able to accept a lot with the tolerable disease level is very high for any of the sensitivity levels. If we simply want to have a 0.01 probability of accepting a lot with the non tolerable disease incidence level of 0.1% without any interest in minimising the rejection of a lot with less than the 0.05% tolerable level, only 3 to 8 samples of 2000 seeds would be needed depending on the sensitivity (Table I).

There is of course a trade-off between n and k, since a bigger n gives a steeper graph and therefore a smaller k. A judicious n,k combination can be arrived at by taking costs of tests and local economics into consideration.

## C. Bayesian Analysis and the Optimal Decision

The probability $\delta$ for detecting contamination in the above size-n test can be expressed with a beta$(\alpha,\beta)$ distribution over $0<\delta<1$. To get the $\alpha$ and $\beta$ for the prior beta$(\alpha,\beta)$ distribution of $\delta$, use $\delta_{lo} = \delta(\pi_{lo},n)$ and $\delta_{hi} = \delta(\pi_{hi},n)$ for the first two arguments of betad() (see section IIA). This is done because we are now using a beta-distribution to express opinion about $\delta$ (not about $\pi$ as in previous sections), and because $\delta$ and $\pi$ are directly linked through the function $\pi$->$\delta(\pi,n)$. The implied cumulative distribution function (CDF) of $\pi$ is B[$\delta(\pi,n)$ | $\alpha,\beta$] for $\pi<1$ and 1 - B[Prob[det | $y_{max}$,n] | $\alpha,\beta$] for $\pi=1$, where B[$\bullet$ | $\alpha,\beta$] is the CDF of beta$(\alpha,\beta)$. The scientist should examine this CDF to confirm that the $\alpha$ and $\beta$ are consistent with his/her overall views about $\delta$. It can be visualised with Probpic(alfa,beta,delta), where the 2-column matrix delta contains the data of the graph of $\delta(\pi,n)$ as returned by Detprob(n,nscount,calprob). Its density over $\pi<1$ can also be viewed, with Probpid(). Because the posterior distribution of $\delta$ is the beta$(z+\alpha,k-z+\beta)$, we may view the posterior distribution of $\pi$ by using z+alfa and k-z+beta for the first two arguments of Probpic() or Probpid(). Subsequent decision analysis proceeds exactly as in the case of direct binomial sampling, based on the beta$(z+\alpha,k-z+\beta)$ posterior distribution of $\delta$.

The sensitivity calibrations for the 2000 lettuce seed samples extrapolated from Kimble et al. (1975) can be used again. First, suppose we look at just k=1 sample of size n=2000 and suppose we get negative result, so z=0. The 2-column matrix delta as returned by Detprob(n,nscount,calprob) contains the pairs of values used for the increasing sensitivities curve shown in Fig. 13. The different 2-column matrices of $\pi,\delta(\pi,n)$ were used successively in the Probpic(0.5,1.5,delta) function using the posterior beta(0.5,1.5) distribution for $\delta$. Fig. 14 shows the different relationships between P[$\pi\leq$ critical $\pi$ | z=0, k=1] and $\pi$ ($0<\pi<1$) depending on the sensitivity. For example, using the range of sensitivities found when 1, 2, and 4 seeds in 2000 were tested, P[$\pi\leq0.001$ | z=0, k=1] = 0.843, i.e., the probability that $\pi$ is less than or equal to 0.001 is 0.843 if a pooled sample of 2000 seeds gives a negative result in an assay that has the range of sensitivities determined by Detprob(2000,c(0,1,2,4),c(0,0.5,0.6,0.8)). This probability can be found in Fig. 14 or by looking at the matrix of values ($\pi$, $\delta$, cdf) produced by Probpic(). If the sensitivity is perfect, then P[$\pi\leq0.001$ | z=0, k=1] = 0.978 (again from Fig. 14 or from Probpic()). If the sensitivity is constantly 0.5 over all y>0 for a 2000 seed sample, then P[$\pi\leq0.001$ | z=0, k=1] = 0.773. Similar to what was observed in the previous section, the CDF P[$\pi\leq$critical $\pi$| z=0, k=1] using the range of sensitivity tested follows the CDF of the low sensitivity at low $\pi$-values and gradually move towards the higher sensitivity (0.8) for higher $\pi$-values. It is interesting to point out that the beta(0.5,2000.5) has a CDF (curve not shown) almost exactly like the CDF for the perfect sensitivity in Fig. 15 (p=1.0) demonstrating that in such a case, it is as good to test all the seeds in one pooled assay as it is to test each seed individually.

In seed testing for LMV (see section IIA), we obtained P[$\pi\leq0.00022$ | y=0, n=30000] as pbeta(0.00022, 0.5, 30000.5) = 0.999723, for the probability that the disease incidence is below 0.00022 if 30000 seeds individually tested were all free of LMV. Considering an indirect bioassay method, what number k of repeated tests of samples of size n=2000, given that there were no detections in any of the k samples,

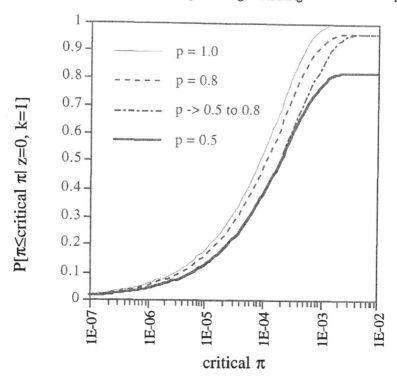

*Fig. 14.* Cumulative density functions assuming different sensitivities and a sampling scenario where no LMV was detected in a sample of 2000 seeds. For three of these curves the sensitivity (p) was set as constant over all $y>0$ and either 1.0 (perfect), 0.8 or 0.5, while for an intermediate curve the increasing sensitivity values 0.5, 0.6, and 0.8 were used.

would give a similarly high probability that $\pi \leq 0.00022$? The answer of course depends on how we model sensitivity, e.g., on whether we assume Prob[det | y,n=2000] is 0 at $y=0$ and the sensitivity is constant over all $y>0$ (say, p=1.0, or 0.8, or 0.5), or whether it is, say, 0.5, 0.6 at $y=1, 2$ and constantly 0.8 over all $y \geq 4$. Under this last assumption, if 33 samples of 2000 seeds each are processed by the *Chenopodium* test, we learn from Detprob(2000,c(0,1,2,4),c(0,0.5,0.6,0.8)) that $\delta(0.00022,2000) \approx 0.189$ and from pbeta(0.189,0.5,33.5) that P[$\pi \leq 0.00022$ | z=0, k=33]= 0.9998. So in this instance k=33 is the required number of samples. The result is influenced by the choice of a sensitivity model, for we see that the other three "constant" models for Prob[det | y,n=2000] give P[$\pi \leq 0.00022$ | z=0, k=33] = 0.99999995, 0.999998, and 0.9997 for sensitivity of 1.0 (perfect), 0.8, and 0.5, respectively (Fig. 15). Continuing with P[$\pi \leq 0.00022$ | z=0, k]= 0.9998, the cumulative posterior probability functions for either P[$\pi \leq$critical $\pi$ | z=0, k=16, p=1.0], P[$\pi \leq$critical $\pi$ | z=0, k=20, p=0.8] or P[$\pi \leq$critical $\pi$ | z=0, k=35, p=0.5] are the same as the P[$\pi \leq 0.00022$ | z=0, k=33] with the variable

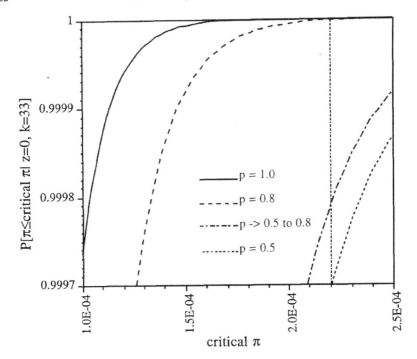

*Fig. 15.* Cumulative posterior probability functions P[π≤critical π|z=0,k=33] re-
sulting from different sensitivity models using samples of n=2000 seeds. Sensitivity
was modelled with Detprob() and the curves were each obtained with Probpic().
The vertical dashed line shows where these different functions intercept with
π=0.00022.

sensitivity shown on Fig. 15. In other words, 33 samples of 2000 seeds with a sensitivity ranging from 0.5 to 0.8 are the equivalent of either 16 samples with sensitivity
of 1.0, 20 samples with sensitivity of 0.8, or 35 samples with sensitivity of 0.5 if no
infected samples are found in any of the *Chenopodium* assays.

## D.  Modelling Expected Costs of Chance Consequences

The analysis of contamination proportion $\pi$ presented in section II was really the
analysis of the binomial(n) parameter $\pi$. The analysis of the binomial(k) parameter
$\delta$, the probability of indirect detection, is exactly the same, but some attention is
needed to connect this analysis to $\pi$ since this is the quantity of interest. Consider
for example the chance of a major agricultural disaster outcome (M; see section IIC).
P[M|$\delta$]/P[M|$\delta$=1]    will be modelled as a logistic    $\delta$-curve, of the form
$\delta \to \Pi(\sigma(\eta-\lambda))$ where $\eta = \log[\delta/(1-\delta)]$. The coefficients $\lambda$ and $\sigma$ are again determined with Scurvepi(), using $\delta(\pi_{lo},n)$ and $\delta(\pi_{hi},n)$ for the first two arguments.

## V. CONCLUSIONS

With an appropriate high-level mathematical computing language like Splus it is possible to model quite complex situations and to facilitate decisions about sample size and acceptance or rejection of lots. We have shown how customised functions can be used to perform heretofore tedious procedures involving complex graphs or tables. Even problems with intuitively simple Bayesian decision modelling were, in the past, hindered by the computing requirement. Inspectors assessing the disease incidence for seed certification programs often have a fairly good idea of what to expect before performing the prescribed sampling procedure. This kind of experience can now be integrated in the prior distribution. If one is still uncomfortable with the idea of using the subjective knowledge prior to sampling, the more neutral prior distributions we have discussed can be used. We hope that this chapter will be a starting point for a wider usage of formal decision modelling.

There is a need for more calibration studies where a detection technique is tested with a range of known infected units in a sample. Data on tolerable disease incidence levels are often absent leading to probably too many "zero tolerance" situations. These studies are labour intensive and not necessarily glamorous but they should be done to assess the risks related to most certification programs. Some of the assumptions that are made in many sampling schemes need to be verified too. For example, uniform spatial distribution of disease in a field is a rare situation, and might therefore also be rare in seedlots, depending on harvesting and transportation processes. Spatial non-uniformity in disease epidemics makes deliberate randomisation of the sampling especially important. Proper random selection of the plants to be tested at harvest time can also help to compensate for the clumping typically seen in disease epidemics.

Ultimately cost will be the driving force behind most decisions. Once the sensitivity of a test with pooled plants has been established, it is quite straightforward to optimise the number of plants in a sample and the total number of samples needed to achieve the required detection limit. In most instances, the cost of accepting a lot with a non-tolerable disease incidence is mere speculation. Accepting such a lot can have socio-political ramifications in cases where trade barriers or new pest introduction could result. It is unlikely that there will ever be a general model for determining economic impact of accept/reject decisions for any disease. Each disease of international importance will need particular attention and require its own calibration and determination of parameters necessary to make rational decisions. Many of these parameters were discussed in this chapter, they include tolerable and non-tolerable disease incidence (in propagation materials, for importation and exportation of edible plant parts, etc.); sensitivity of detection methods for various sample sizes; conditional probabilities of disasters, local or national, given various disease incidence levels; and the use of prior opinion. As subjective as many of these inputs may be, they must be faced explicitly, so that their influence on final decision can be assessed on a "what if" basis. We have tried to provide easy-to-use tools that will allow decision makers to examine a wide range of possibilities for parameters that are not well understood.

## ACKNOWLEDGEMENTS

We wish to thank J. W. Hall and R. A. Lockhart for comments and critical review of this manuscript.

## REFERENCES

Geng, S., Campbell, R. N., Carter, M., and Hills, F. J. (1983). Quality-control programs for seedborne pathogens. *Plant Disease* **67**, 236-242.

Grogan, R. G. (1980). Control of lettuce mosaic with virus-free seeds. *Plant Disease* **64**, 446-449.

Jeffreys, H. S. (1961). 'Theory of Probability.' Oxford University Press, London.

Kaplan, S. (1993). Proceedings of the APHIS/NAPPO International Workshop on the Identification, Assessment, and Management of Risks due to Exotic Agricultural Pests, pp. 123-146.

Kimble, K. A., Grogan, R. G., Greathead, A. S., Paulus, A. O., and House, J. K. (1975). Development, application, and comparison of methods for indexing lettuce seed for mosaic virus in California. *Plant Disease Reporter* **59**, 461-464.

Lee, P. M. (1989). 'Bayesian Statistics: an Introduction.' Oxford University Press, New York.

Maude, R. B. (1988). The value of seed health testing and rotational practices in the control of vegetable diseases. *In* 'Control of Plant Diseases: Costs and Benefits.' (B. C. Clifford and E. Lester, eds.), pp. 263. Blackwell Scientific Publications, Oxford.

Smith, J. Q. (1988). 'Decision Analysis : a Bayesian Approach.' Chapman and Hall, London.

Tomlinson, J. A. (1962). Control of lettuce mosaic by the use of healthy seed. *Plant Pathology* **11**, 61-64..

Tomlinson, J. A. (1970). Lettuce mosaic virus. *In* 'No. 9. Description of Plant Viruses'. Comm. Mycol. Inst., Kew, Surrey, England.

Winkler, R. L. (1972). 'Introduction to Bayesian Inference and Decision.' Holt, Rinehart and Winston, New York.

# Appendix

Splus language functions developed for this chapter.

## Ssize

Calculate the sample size $n$ and critical point $cr$ for either a large binomial(mu) or Poisson(mu)-distributed sample with observed total number $y$ of occurences, so that $y \leq cr$ with probability $prat$ when $mu=mut$ while $y <= cr$ with probability $prant$ when $mu=munt$. For a binomial sample, $mu$ is an infection rate. For a Poisson sample, $n$ is total size of all sampled regions. If the returned optimal $n$ is nstart, use a lower nstart. Use a large nincr for speed, use a low nincr for accuracy. The returned $cr$ should be used only with the optimal $n$. Plot the two cumulative distributions of $y$, under mut and munt.

```
function(mut, munt, prat, prant, nstart, nincr)
{
    if(mut > munt) return("munt must exceed mut")
    if(nstart <= 1)
        return("nstart must exceed 1")
    if(mut > 0)
        crincr <- 1
    else {
        crincr <- 0.1
        mut <- 1e-16
    }
    if(prat == 1)
        prat <- (1 - 1e-16)
    n <- nstart
    pr <- 1
    while(pr > prant + 1e-06) {
        probt <- cumsum(exp( - n * mut + log(n * mut) * (0:170))/gamma(1:171))
        probnt <- cumsum(exp( - n * munt + log(n * n + nincr
    }
    n <- n - nincr
    cr <- cr - crincr
    probt <- cumsum(exp( - n * mut + log(n * mut) * (0:170))/gamma(1:171))
    probnt <- cumsum(exp( - n * munt + log(n * munt) * (0:170))/gamma(1:171))
    nn <- 0:170
    lb <- probt > 0]", xlab =
        "cr -->\nTWO CUM PROB DISTRS OF NR y OF POSITIVES")
    return(n, cr)
}
```

## Betadens
Plot the beta(alfa, beta) density.

```
function(alfa, beta)
{
    eta <- (-2000:2000)/50
    pi <- 1/(1 + exp( - eta))
    dens <- dbeta(pi, alfa, beta)
    max <- max(dens)
    plot(pi[dens > max/100], dens[dens > max/100], type = "l", tck = 0.01,
        ylab = "PROBABILITY DENSITY", xlab = "pi", lab = c(10, 10, 10))
}
```

**Betacum**

Plot the cumulative beta(alfa,beta) distribution.

```
function(alfa, beta)
{
    eta <-`(-2000:2000)/50
    pi <- 1/(1 + exp( - eta))
    cum <- pbeta(pi, alfa, beta)
    plot(pi[(cum > 0.01) & (cum < 0.99)], cum[(cum > 0.01) & (cum < 0.99)],
        type = "l", ylab = "CUMULATIVE PROBABILITY", xlab = "pi", tck
        = 0.01, lab = c(10, 10, 10))
}
```

**Betacumlog**

Plot the cumulative beta(alfa,beta) distribution. Sama as Betacum() except that the x-axis is in logarithmic scale.

```
function(alfa, beta)
{
    eta <- (-2000:2000)/50
    pi <- 1/(1 + exp( - eta))
    cum <- pbeta(pi, alfa, beta)
    plot(pi[(cum > 0.01) & (cum < 0.99)], cum[(cum > 0.01) & (cum < 0.99)],
        type = "l", ylab = "CUMULATIVE PROBABILITY", xlab = "pi", log
        = "x", exp = 0, tck = 0.01, lab = c(40, 10, 3))
}
```

**Betacum.9**

Plot the cumulative beta(alfa,beta) distribution. Same as Betacum() except that that y-axis range is from 0.9 to 1.0.

```
function(alfa, beta)
{
    eta <- (-2000:2000)/50
    pi <- 1/(1 + exp( - eta))
    cum <- pbeta(pi, alfa, beta)
    y <- c(0.9, 0.95, 0.99, 0.999, 0.9999)
    x <- qbeta(y, alfa, beta)
    labels <- c("0.90", " 0.95", " 0.99", " 0.999", "  0.9999")
    plot(pi[(cum > 0.9) & (cum < 1)], cum[(cum > 0.9) & (cum < 1)], log =
        "x", type = "l", ylab = "P[pi <= cr]", xlab =
        "critical infectiolines(x, y, type = "h", col = 2)
    Cumprob <- tapply(x, y, mean)
    return(Cumprob)
}
```

**Betad**

It is thought that pi < pilo with probability flo and pi > pihi with probability 1-fhi. Find alfa and beta for which the beta(alfa,beta) distribution satisfies these conditions, and plot the histogram. The values as and bs are starting points in the search for alfa and beta.

```
function(pilo, pihi, flo, fhi, as = 0.1, bs = 0.1)
{
    alfa <- as
    beta <- bs
    h <- 1e-10
    eps <- 1
```

```
    it <- 0
    while((eps > 1e-12) & (it < 100)) {
        if((alfa < 0) || (beta < 0))
          return("not convergent", it)
        f1 <- pbeta(pilo, alfa, beta) - flo
        f2 <- pbeta(pihi, alfa, beta) - fhi
        d11 <- (pbeta(pilo, alfa + h, beta) - pbeta(pilo, alfa - h, beta))/h/2
        d12 <- (pbeta(pilo, alfa, beta + h) - pbeta(pilo, alfa, beta - h))/h/2
        d21 <- (pbeta(pihi, alfa + h, beta) - pbeta(pihi, alfa - h, beta))/h/2
        d22 <- (pbeta(pihi, alfa, beta + h) - pbeta(pihi, alfa, beta - h))/h/2
        d <- d11 * d22 - d12 * d21
        if(abs(d) < 1e-10)
          return("not convergent", d)
        deltaa <- (d22 * f1 - d12 * f2)/d
        deltab <- ( - d21 * f1 + d11 * f2)/d
        eps <- sqrt(deltaa^2 + deltab^2)
        it <- it + 1
        alfa <- alfa - deltaa
        beta <- beta - deltab
    }
    eta <- (-2000:2ens > max/100], type = "1")
    return(it, alfa, beta, eps)
}
```

**Betac**

It is thought that pi < pilo with probability flo and pi > pihi with probability 1-
fhi. Find alfa and beta for which the beta(alfa,beta) distribution satisfies these
conditions, and plot the cumulative distribution function (CDF). The values as and
bs are starting points in the search for alfa and beta.

```
function(pilo, pihi, flo, fhi, as = 0.1, bs = 0.1)
{
    alfa <- as
    beta <- bs
    h <- 1e-10
    eps <- 1
    it <- 0
    while((eps > 1e-12) & (it < 100)) {
        if((alfa < 0) || (beta < 0))
          return("not convergent", it)
        f1 <- pbeta(pilo, alfa, beta) - flo
        f2 <- pbeta(pihi, alfa, beta) - fhi
        d11 <- (pbeta(pilo, alfa + h, beta) - pbeta(pilo, alfa - h, beta))/h/2
        d12 <- (pbeta(pilo, alfa, beta + h) - pbeta(pilo, alfa, beta - h))/h/2
        d21 <- (pbeta(pihi, alfa + h, beta) - pbeta(pihi, alfa - h, beta))/h/2
        d22 <- (pbeta(pihi, alfa, beta + h) - pbeta(pihi, alfa, beta - h))/h/2
        d <- d11 * d22 - d12 * d21
        if(abs(d) < 1e-10)
          return("not convergent", d)
        deltaa <- (d22 * f1 - d12 * f2)/d
        deltab <- ( - d21 * f1 + d11 * f2)/d
        eps <- sqrt(deltaa^2 + deltab^2)
        it <- it + 1
        alfa <- alfa - deltaa
        beta <- beta - deltab
    }
    eta <- (-2000:2000)/50
    pi <- 1/(1 + exp( - eta))
    y <- c(flo, fhi)
    x <- qbeta(y, alfa, beta)
    labels <- y
    cum <- pbeta(pi, alfa, beta)
```

```
plot(pi[(cum > 0.01) & (cum < 0.99)], cum[(cum > 0.01) & (cum < 0.99)],
    type = "1", ylab = "CUMULATIVE PROBABILITY", xlab = "pi", log
    = "x", exp = 0, tck = 0.01, lab = c(40, 10, 3))
text(x, 0, labels)
lines(x, y, type = "h", col = 2)
return(it, alfa, beta, eps)
}
```

## Scurvepi

Plot Scurve(pi) = 1/{1+exp[-sca*(eta-loc)]} where pi=1/{1+exp(-eta)} using loc and sca such that Scurve(pilo)=frlo and Scurve(pihi)=frhi. For a plot of the Scurve(pi) with given sca and loc use Exp.pils().

```
function(pilo, pihi, frlo, frhi)
{
    lodlo <- log(pilo/(1 - pilo))
    lodhi <- log(pihi/(1 - pihi))
    sca <- (log(frhi/(1 - frhi)) - log(frlo/(1 - frlo)))/(lodhi - lodlo)
    loc <- lodhi - log(frhi/(1 - frhi))/sca
    eta <- (-2000:2000)/50
    pi <- 1/(1 + exp( - eta))
    logit <- 1/(1 + exp( - sca * (eta - loc)))
    plot(pi[(0.01 < logit) & (logit < 0.99)], logit[(0.01 < logit) & (logit
        <0.99)], type = "1")
    return(loc, sca)
}
```

## Exp.pils

Calculate overall expected probability of a disaster, using the beta(alfa,beta) distribution for pi, of the quantity Scurvepi(pi) = 1/{1+exp[-sca*(eta-loc)]} where pi=1/{1+exp(-eta)}. Appropriate loc and sca can be obtained with Scurvepi(). Also plot Scurvepi(pi).

```
function(loc, sca, alfa, beta)
{
    eta <- (-2000:2000)/50
    pi <- 1/(1 + exp( - eta))
    func <- ((1/(1 + exp( - sca * (eta - loc)))))) * dbeta(pi, alfa, beta) *
        pi * (1 - pi)
    expval <- sum(func)/50
    logit <- 1/(1 + exp( - sca * (eta - loc)))
    plot(pi[(0.01 < logit) & (logit < 0.99)], logit[(0.01 < logit) & (logit
        <0.99)], type = "1")
    return(expval)
}
```

## Exp.cos.bi

Calculate expected cost based on a binomial (direct) sample. The vector "costs" is a string of costs of the k chance outcomes of one decision. The vector "loc" contains the location parameters of the s-curves which model the probabilities of all but the last-listed of the outcomes, while "sca" contains their scale parameters. These can be obtained from Scurvepi(). The beta(ypa,nmypb) distribution describes the state of our knowledge about the infection rate pi of the seed lot. "maxprobs" contains the maximum probabilities (ie, prob when pi=1). The function "exp.pils" must be present.

```
function(costs, loc, sca, ypa, nmypb, maxprobs)
```

```
{
    expv <- rep(0, k - 1)
    for(i in 1:(k - 1)) {
        expv[i] <- maxprobs[i] * exp.pils(loc[i], sca[i], ypa, nmypb)
    }
    expcost <- sum((costs[1:(k - 1)] - costs[k]) * expv) + costs[k]
    return(expcost)
}
```

## Gamdens
Plot density of the Gamm(alfa,beta) distribution.

```
function(alfa, beta)
{
    lam <- exp(log(25) * ((-1000:1000)/100))
    dens <- beta * dgamma(beta * lam, alfa)
    md <- max(dens)
    plot(lam[dens > md * 0.01], dens[dens > md * 0.01], type = "1")
}
```

## Gamcum
Plot cumulative probability density of the Gamm(alfa,beta) distribution.

```
function(alfa, beta)
{
    lam <- exp(log(25) * ((-1000:1000)/100))
    cum <- pgamma(beta * lam, alfa)
    md <- max(cum)
    plot(lam[(cum > md * 0.01) & (cum < md * 0.9999)], cum[(cum > md * 0.01)
        & (cum < md * 0.9999)], type = "1", ylab =
        "CUMULATIVE PROBABILITY", xlab = "lambda")
}
```

## Gamcum.9
Plot density of the Gamm(alfa,beta) distribution.  Same as Gamcum except that the y-axis range is from 0.9 to 1.0.

```
function(alfa, beta)
{
    lam <- exp(log(25) * ((-1000:1000)/100))
    cum <- beta * pgamma(beta * lam, alfa)
    md <- max(cum)
    y <- c(0.9, 0.95, 0.99, 0.999, 0.9999)
    x <- beta * qgamma(y, alfa)
    labels <- c("0.90", " 0.95", " 0.99", " 0.999", "  0.9999")
    plot(lam[(cum > 0.9) & (cum < 0.9999)], cum[(cum > 0.9) & (cum < 0.9999)],
        type = "1", ylab = "CUMULAT- tapply(x, y, mean)
    return(Cumprob)
}
```

## Gammad
Find alfa and beta for which the Gamma(alfa,beta) distribution has  lamlo  for its plo-fractile and  lamhi  for its phi-quantile.

```
function(lamlo, lamhi, plo, phi, beta = (lamhi * (1 - phi) + lamlo * plo)/(1 +
    plo - phi)/(((lamhi - lamlo) * (phi - plo))/2/0.6)^2, alfa = (beta * (
    lamhi * (1 - phi) + lamlo * plo))/(1 + plo - phi))
{
    . h <- 1e-1ues. Try some others.")
        f1 <- pgamma(lamlo * beta, alfa) - plo
```

```
      f2 <- pgamma(lamhi * beta, alfa) - phi
      d11 <- (pgamma((beta + h) * lamlo, alfa) - pgamma(lamlo * (beta -
        h), alfa))/h/2
      d12 <- (pgamma(lamlo * beta, alfa + h) - pgamma(lamlo * beta, alfa - h))/h/2
      d21 <- (pgamma(lamhi * (h + beta), alfa) - pgamma(lamhi * ( - h + beta),
        alfa))/h/2
      d22 <- (pg)
        return("Bad initial value, Try some other values.", d)
      deltab <- (d22 * f1 - d12 * f2)/d
      deltaa <- ( - d21 * f1 + d11 * f2)/d
      eps <- sqrt(deltaa^2 + deltab^2)          .
      n <- n + 1
      alfa <- alfa - deltaa
      beta <- beta - deltab
    }
    return(n, alfa, beta, eps)
}
```

## Scurvemu

Plot Scurve(mu) = 1/{1+exp[-sca*(eta-loc)]} against mu=exp(eta) using loc and sca such that Scurve(mulo)=flo and Scurve(muhi)=fhi. For a plot of Scurve(mu) with given sca and loc use Exp.muls().

```
function(mulo, muhi, flo, fhi)
{
    lmulo <- log(mulo)
    lmuhi <- log(muhi)
    sca <- (log(fhi/(1 - fhi)) - log(flo/(1 - flo)))/(lmuhi - lmulo)
    loc <- lmuhi - log(fhi/(1 - fhi))/sca
    eta <- (-3000:3000)/50
    mu <- exp(eta)
    logit <- 1/(1 + exp( - sca * (eta - loc)))
    plot(mu[(0.01 < logit) & (logit < 0.99)], logit[(0.01 < logit) & (logit
      <0.99)], type = "l")
    return(loc, sca)
}
```

## Exp.cos.ga

Calculate expected cost based on a Poisson (indirect) sample. "costs" is a string of costs of the k chance outcomes of one decision. "loc" contains the location parameters of the s-curves which model the probabilities of all but the last-listed of the outcomes, while "sca" contains their scale parameters. These can be obtained from Scurvemu(). The Gamma(ypa,rspb) distribution describes the state of our knowledge about the contamination density mu of the seed lot. "maxprobs" contains the maximum probabilities (ie, limiting prob as mu -> infinity). The function "exp.muls" must be present.

```
function(costs, loc, sca, ypa, rspb, maxprobs)
{
    if(sum(maxprobs) >= 1) return("sum(maxprobs) must  < 1")
    k <- length(costs)
    expv <- rep(0, ks[1:(k - 1)] - costs[k]) * expv) + costs[k]
    return(expcost)
}
```

## Detprob

The values ssiz = number of seeds sampled and nscont = string of possible numbers y of contaminated seeds in the sample for which detection probability Py is known.

Ideally contains only y for which Py has been calibrated but should include y=0 and 1, even if P0 and P1 are only guesses. calprobs = string of Py for y in nscont. The model used for all Py is a fitted spline from y=1 to max(nscont) and then constant. Constancy beyond the highest calibrated y can be altered by extending nscont to include y with fictitious Py. Plot and return graph of delta(pi) = detection probability as a function of contamination proportion pi, evaluated at npv pi-points between 0 and 1. Plot only that part of graph with delta(pi) < scon*max(delta(pi)).

```
function(ssiz, nscont, calprob, scon = 0.99, npv = 200)
{
    nyv <- max(nscont) + 1
    ym <- as.matrix(rep(1, npv)) %*% (0:(nyv - 1))
    ppi <- exp(as.matrix((-20 * (npv:1))/npv))
    npi <- ssiz * ppi
    npim <- npi %*% rep(1, nyv)
    ynpim <- array(0, c(npv, nyv, 2))
    ynpim[, , 1] <- ym
    ynpim[, , 2] <- npim
    binom <- apply(ynpim[, 1:nyv, ], c(1, 2), poipro)
    binomial <- cbind(binom, (1 - apply(binom, 1, sum)))
    detprobs <- spline(nscont[-1], calprob[-1], n = nyv - 1, xmin = 1, xmax =
        nyv - 1)$y
    cpm <- calprob[length(calprob)]
    detprobs <- c(calprob[1], detpro<- cbind(ppi, detpr))
    delta <<- detpro
    return(cpm, detpro)
}
```

## Probpid
Plot the density of pi assuming delta=delta(pi) is beta(alfa,beta)-distributed. The 2-column matrix delta is the set of points of the graph of delta(pi) as returned by Detprob().

```
function(alfa, beta, delta)
{
    lenpi <- nrow(delta) - 1
    ddetpr <- delta[-1, 2] - delta[1:lenpi, 2]
    dpi <- delta[-1, 1] - delta[1:lenpi, 1]
    ddetdpi <- ddetpr/dpi
    deriv <- (ddetdpi[-1] + ddetdpi[1:(lenpi - 1)])/2
    d, dens[dens > 0.01 * max(dens)], type = "l")
}
```

## Probpic
Plot the cumulative dist function of pi assuming delta = delta(pi) is beta(alfa,beta)-distributed. The 2-column matrix delta is the set of points of the graph of delta(pi) as returned in Detprob().

```
function(alfa, beta, delta)
{
    cdf <- pbeta(as.vector(delta[, 2]), alfa, beta)
    ma <- max(delta[, 2])
    plot(delta[(cdf > 0.01) & (cdf < 0.99 * ma), 1], cdf[(cdf > 0.01) &
        (cdf < 0.99 * ma)], type = "l")
} .
```

# 10

# QUALITY CONTROL AND COST EFFECTIVENESS OF INDEXING PROCEDURES

## C. L. Sutula

*Agdia Incorporated, Elkhart, Indiana, United States*

## I. INTRODUCTION

Managed propagation of quality plants and plant products requires more than horticultural excellence. It also requires tests or indexing procedures of controlled quality to ensure freedom from specific pests and pathogens. These tests need to be performed within a technical and business strategy that intends to avoid the introduction or passage of plant pathogens while selecting for desirable factors in the finished product. At the time they are applied, the tests may be the only indicators that propagation of the desired plants should be continued. It is reasonable to expect that such tests should perform reliably and in a predictable way, that is, under quality control, each time they are used. Since many of the tests can cost more than the commercial value of a single plant, when and how they are applied directly impacts on their cost effectiveness.

There are many indexing procedures and tests that can be used to signal the

Advances in Botanical Research Vol. 23
Incorporating Advances in Plant Pathology
ISBN 0-12-005923-1

presence of a pathogen. It is beyond the scope of this chapter to survey the many tests used. Instead, the very popular enzyme linked immunosorbent assay (ELISA) will be used as the model test system in this chapter to illustrate important elements of quality control and cost components. The principles developed using this model are applicable to other types of tests.

Laboratories often examine plant samples to confirm a disease diagnosis or to indicate the absence of a specific plant pathogen. In the first case, the sample is usually examined to provide information in the context of an horticultural crisis. In the second case, the sample is tested to avoid or prevent a crisis. The second case is the focus of this chapter. Using ELISA as the model system the factors that must be under control will be examined and examples provided. Also, the cost effective use of indexing by ELISA will be discussed.

## II. WHAT NEEDS TO BE CONTROLLED?

ELISA is a method by which one or more antibodies are used to signal or measure the presence of an entity by generating color. This can be done using a variety of arrangements and protocols. One common protocol, often called double antibody sandwich ELISA (DAS-ELISA), involves trapping of antigen in wells of microtitre plates with an antibody bound to the plate and subsequent detection of the trapped antigen with an enzyme-antibody conjugate. Addition of an enzyme substrate, which develops a colored product on conversion, permits identification of wells in which antigen was present. However, the color generated is not always directly related to the presence of what we wish to detect. If one considers the entire test process, a large number of possibilities exist that could explain the absence or presence of color development.

### A. Analysis of Possible Test Outcomes

When color develops in a well of a DAS-ELISA plate, it may be the result of normal, uniform color development, or non-homogenous color formation. In a properly performed DAS-ELISA the color in wells of positive samples should first appear as a uniform ring located at the inner surface of the well. As color generation continues the ring of color mixes producing a uniform color in all the liquid in the well. Frequently, the color may generate from a single point in a well, or from patches or even from bathtub-like ring deposits. Wells producing color non-homogeneously do not produce reliable results.

Careful analysis of various possible plate outcomes highlights the various ELISA steps that need to be controlled and what can go wrong. The following paragraphs depict such an analysis. Although some of the observations may seem trivial, the analysis emphasises that all aspects of all steps must be considered in quality control indexing protocols conducted as part of plant disease control programs.

When color is present in every well and all wells have the same intensity, this may be due to the following circumstances: (1) the plate was coated with conjugate, (2) the plate was coated with antibody contaminated with conjugate, (3) the plate was not washed after the conjugate step, (4) the substrate solution already was colored, (5) the plate was left in sunlight or very bright light causing spontaneous substrate conversion, or (6) the enzyme-antibody conjugate was not compatible with the plate coating antibody, i.e. in the indirect format the conjugate antibody binds to the coating antibody rather than the secondary antibody.

When color is present in every well but the wells have different intensities of color this may be caused by: (1) all wells having a positive sample, (2) poor washing of the ELISA plate between steps, (3) inadequate mixing of conjugate, (4) or again exposure to high light intensity.

Distinct and varying patterns of color development may be due to the successful completion of ELISA showing different levels and absence of pathogen among the samples. However if color radiates from positive controls to adjacent wells, this would suggest that positive sample washed over into wells with negative samples while carrying out the procedure. A problem with the washing steps between addition of reagents may also be indicated when only the perimeter wells are positive or more strongly so. The so-called "edge-effect" also results in non-specific positive reactions in perimeter wells probably due to characteristics of or defects in the microtitre plate used for the test. Color gradients in a row or column in the plate could result from sample carry-over during extraction or pipeting, or by sample drying out in one portion of the plate. Pipeting errors could also account for a single row or column appearing different from the rest of the plate.

Non-specific color development in ELISA can also result from (1) interaction of the antibody-enzyme conjugate with healthy plant tissue, (2) presence of enzyme activity in sample, (3) microbial contamination of buffers, reagents, or wash water used in the procedure, or (4) incorrectly prepared reagents.

Lack of any color development in the plate could be due to operator errors at any step in the ELISA procedure such as use of the wrong antibodies for coating or as secondary antibodies, failure to add substrate, use of incorrect buffers, use of incorrect concentrations, or use of the wrong substrate for the particular enzyme-antibody conjugate used. It could also be due to the conjugate or antibody preparations being inactive because they have exceeded the storage time or by having become contaminated.

The above paragraphs illustrate the approach used to gain an understanding of the elements which need to be controlled. One simply lists and prioritizes all of the plausible reasons for certain observations while performing a test. The analysis need not be exhaustive at the beginning. Even a preliminary analysis produces useful directions and initiatives for quality control and new items can be added later. The outline given is realistic and presents items that have been communicated by hundreds of persons performing ELISA worldwide over the past 13 years. Of the 38 items we have identified, 23 of the failures in the integrity of the test are the result of operator error, nine of them deal with reagent quality, and six of them involve technique while performing the test.

The importance of positive and negative controls on each ELISA plate cannot be overemphasized. The reaction in control wells will often quickly help to recognize and identify problems with an ELISA test. Of course, possible inactivity of positive control samples and operator error in loading control samples must also be considered in analysis of possible failures in ELISA.

## B. Major Items to Consider

If the analysis is extended to the idea of performing an ELISA that is reliable and predictable in its performance, on demand or when desired, then the following general and specific items need attention:
- people
- materials, including plasticware, reagents, water quality, etc.
- test protocol
- interpretation of results
- method of reporting results
- record keeping
- workplace and facilities
- documenting of all processes
- audit

Clearly, people prepare, perform, interpret, report, file, maintain and inspect the performance of any indexing procedure. The prior analysis of ELISA indicated about 60% of the errors can be caused by the persons performing the test. Another 16% of the errors can result from the technique they used.

The workplace can enhance or detract from the desired test performance. Even the smallest of items can affect performance. ELISA requires a clean, organized workplace with properly operating equipment to obtain consistent, quality results. For example, plant sap can be diluted about 1 million fold before it is not responsive in most ELISA tests. Thus, a small amount of pathogen positive plant sap can be a serious cause of contamination in the work space.

The test must be interpreted and reported before it can become the basis for action. A test cannot be interpreted reliably if it is not performed in a consistent, reproducible way. Reporting results and keeping records are the ways the process of testing is completed. Documenting all the procedures used in the testing process and auditing are proven, effective tools to enhance the performance of personnel performing ELISA.

## III. QUALITY CONTROL OF ELISA

Quality control of ELISA is based on items identified in the previous section. Minimum requirements are presented here which have been effective in a commercial laboratory setting over the last ten years.

## A. Training of Personnel

Training must be provided on safety, the proper care of the environment, good laboratory practices (GLP), and the indexing procedures. This is a considerable undertaking that should never stop. Records should be kept of the training performed, the program goals for continuing training, updating, and the skill audits of personnel.

## B. Documentation of Material Transfers

It is very useful to document all transfers of material to the laboratory. For each item received a record must be entered in the materials received log. This log should be a bound notebook or computerized spreadsheet or database, in which the receipt of materials is recorded, day by day, as it occurs. Each item received must be given a thorough inspection for damage and conformity of labelled goods to what was ordered. Each item should be assigned a unique number, for example A6678, which is written on or permanently attached to the material. It is convenient to have the number increase sequentially as items are received. Each entry follows the field list itemized in Table I.

The materials received log is not used for recording the receipt of plant samples for testing. These are recorded in another log with a different set of information requirements. The materials received log becomes an invaluable aid as it is maintained over time. Being able to track every container of every item enables one to specify exactly what materials were used in preparing components for a test.

*Table 1.* The list of items that would normally be included in a Materials Received Log

Date
Item Number
Description of Item
Purchase Order Number
Quantity
Vendor
Vendor Order Number
Storage Location
Checked by: (initials)
Comments

## C. Test Optimization

An ELISA will perform in a more consistent way if the test is optimized. Important elements in test optimization include selecting and validating test parameters, selecting appropriate thresholds, and tracking positive and negative control samples.

*1. Selection and validation of test parameters*
One should begin by selecting superior materials including microtitre plate, antibodies, enzyme conjugates, chemicals for buffers, and distilled water. Antibodies and conjugates able to detect an ever growing list of pathogens are available from commercial suppliers. High quality, controlled materials ought to be chosen, otherwise the test may change each time another order of materials is obtained.

When assembling materials for a new test, it is necessary to establish and confirm the optimum concentration of all reagents. This is necessary even when purchasing reagents from a commercial supplier. The data and protocol, if provided with the reagents, may not produce the optimum test on *your* plates with *your* buffers in *your* work space when performed by *your* staff. If a kit is being used that contains an assembled test then the manufacturer should have optimized the test.

It is common to search for optima in ELISA with 'checkerboard' type experiments which vary one or more variables in a systematic way, as in a serial dilution of coating antibody, sample extract or enzyme conjugate. For example, many ELISAs follow the family of curves illustrated in Figure 1 where test response of a single positive sample is shown at various concentrations of coating antibody at several dilutions of enzyme conjugate. Note that performing the test at coating levels less than 2 µg/mL places the test in a condition that is more sensitive to small changes in concentration of coating antibody. Performing the test above 2 µg/mL places the test on a response plateau that is less sensitive to even large changes in the concentration of coating antibody.

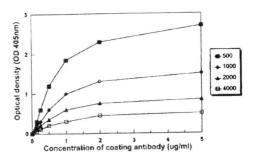

*Figure 1.* Response of an ELISA for prunus necrotic ringspot virus to changes in the concentration of coating antibody at four dilutions (1:500 - 1:4000) of enzyme-antibody conjugate.

Figure 1 also demonstrates that test response is proportional to the concentration of enzyme conjugate. Doubling the concentration of enzyme conjugate often doubles the test response. The time enzyme-antibody conjugate is incubated in the test has a similar effect. Doubling incubation time is like using twice the concentration of enzyme conjugate. Since more time or a higher concentration just produces a stronger test response other conditions must be introduced to optimize incubation time and concentration. This can be done by testing select, known positive and negative samples at a series of convenient incubation times. The condition producing the best positive versus negative response, i.e., the highest optical density ratio, in a relatively convenient time is selected. The basic procedure for optimizing ELISA is discussed by Voller *et al.*, (1979). An early example of this process applied to plant virus ELISA appears in the paper by Clark and Adams (1977) following the first demonstration of ELISA applied to plant viruses (Voller *et al.*, 1976).

The type and brand of microtitre plate that is used has a major effect on ELISA. Since 1976 many investigators have collected data on the effect of the microtitre plate on the assay, but very little of this experience has been published. Early plates were quite variable between and within individual plates. Current plates are much improved; however, very large variations in performance still can be observed between plates from different manufacturers. The performance often depends on the antibody and incubation times used. Thus, one plate may not work well for all ELISA tests of interest.

### 2. Selection of threshold

Once the ELISA is running in a repeatable manner a deliberate examination and selection of the positive/negative threshold must be made. This is done by determining the response of the ELISA with many known negative and positive samples. From these data a positive/negative threshold is chosen. As discussed by Sutula *et al.*, (1986) this process is simply empirical. It helps to chart the results in the form of a histogram or in tables in which assay response is increasing. The net outcome can be a simple rule, such as, positive samples are those with absorbances >0.300. Often, the data obtained with the ELISA may allow a more complete rule, such as when negatives are <0.100, positives are >0.200, and borderline results are in the range of 0.100 - 0.200. Later, as the performance of the test becomes more predictable, the borderline range may be shortened and the positive threshold lowered.

ELISA used in indexing procedures present a special case. Most of the samples during indexing will be negative and thus they can be considered a negative population. Any samples falling outside of the population are examined for the possibility that they are positive. This is illustrated in Figures 2 and 3. Figure 2 shows the frequency distribution of absorbance values when the entire plate contains negative samples. Note that negatives are not just one point on the chart. Instead, the negatives in Figure 2 define a population in which 95 samples produce absorbance values between 0 and 0.080. The positive control in this figure is at 0.960. Figure 3 shows a frequency distribution of samples when some of them are positive. In this case the negative population still occupies the same space as in Figure 2. Positive samples are those with absorbance values >0.200.

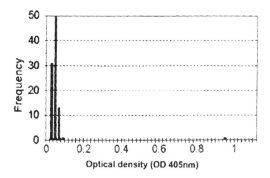

*Figure 2.* Frequency of absorbance (optical density at 405 nm) values in an ELISA test for potato virus Y on extracts from potato sprouts that tested negative for the virus (except for the positive control).

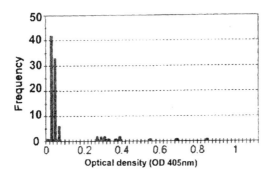

*Figure 3.* Frequency of absorbance (optical density at 405 nm) values in an ELISA test for potato virus Y on extracts from potato sprouts some of which tested positive for the virus.

The point of this illustration is that the entire negative population can serve as the ELISA 'negative'. It is easy to make interpretative errors when only a single negative sample is used to represent a negative population. You can avoid this by examining the shape and position of the negative population for each plate. Most of the time, this is easily done by visual inspection of the histogram of the data from each plate. After deciding whether the positive control and the negative population are behaving in an expected manner, the positive samples are identified.

### 3. Tracking of Control Samples
The everyday performance of the ELISA is monitored using a set of 'in-house' controls. These are usually a collection of negative and positive samples whose quantitative response in ELISA is already known from previous experiments. These controls are aliquoted or prepared in a form and volume convenient for everyday use and then preserved by freezing or drying.

Each time the ELISA is performed the results obtained for the controls are recorded, inspected and charted. The emphasis is on the absorbance value of the control and not only on whether it is positive or negative. For example, the expected value for the positive control with potato virus Y in Figure 2 was 0.900 with a range of 0.700 - 1.100. The charting and maintenance of database records are important to establish the history and expectation of performance. ELISA can be reproduced plate to plate within a 10% coefficient of variance. Day to day results for a particular control that varies more than 20% should receive attention to make certain the ELISA is not declining and that everything is working properly.

## D.  Maintenance of Work Environment

Some indexing tests such as ELISA do not necessarily have to be done in a laboratory setting. Quality ELISA results can be obtained in farm sheds, basements, kitchens, and extension offices as well as in modern laboratories. However, the work space should be clean and organized. In addition, the work space in which an indexing program is carried out, should be arranged to allow for an easy flow through of samples, controls, and other materials required for the tests.

Extraneous material that causes problems in ELISA occurs largely from four sources, namely, plant sap and tissues, fungal and bacterial contamination of solutions, dirt or dust from the environment, and body fluids. For example, human saliva contains alkaline phosphatase and talking over an unprotected ELISA plate can produce false positive readings. Droplets of plant sap or pieces of plant tissue from a positive sample that drop into adjacent wells of the microtitre plate can be a source of antigen resulting in a false positive reaction. Soil and dust accumulate in work spaces that are not frequently cleaned and their presence can affect the final results. Elimination of contamination sources is helpful for achieving consistent ELISA results.

## E. Documentation of Procedures

The procedures used to perform tests that constitute the indexing protocol should be documented. Even in the single person laboratory, documented procedures can help to explain to customers and regulatory personnel exactly what is done to perform each test. Minimal documentation must at least include details of test protocols including the preparation of buffers and other test components, as well as sources of reagents such as antibodies and conjugates. The documentation should also include the maintenance and calibration of all equipment that is used and even the cleaning, opening, and closing of the laboratory. There are also safety, GLP, and other regulatory requirements that must be satisfied. The specific procedures for these requirements as well as documentation demonstrating their completion should be included.

## F. Auditing

It is important to audit indexing protocols regularly. Audits are proven, effective tools that enhance the performance of personnel performing ELISA and other procedures in any organization. Simply put, people respect what you inspect. In indexing laboratories there ought to be a regular process of auditing that ranges over all aspects of the organization. Attention should be paid to documentation, skills, performance, controls, records, that is, every part of the process that can affect the results. It is useful to combine regularly scheduled audits with unannounced ones and to include knowledgeable visitors in the process. Audits, findings, and follow-up actions should all be documented as well. The audits will help demonstrate and document long term, reproducible performance.

## IV.  COST ANALYSIS

The cost of indexing procedures at any facility varies considerably and can be difficult to compute.  In common with many activities, indexing procedures may include start-up, continuation, and maintenance costs.  Start-up costs include the hiring and training of personnel, and the acquisition of facilities, materials and equipment.  Continual costs include the purchase of reagents, kits, and disposables; compensation for staff and management; and office expenses.  Maintenance may include cleaning, repair, training and attendance at professional meetings.

If several assumptions are made, an approximate cost of indexing programs can be estimated.  At many facilities in the United States an estimate of $20/hr/person would reflect staff and management costs with another $10/hr/person for overhead, facilities, and office costs.  Laboratories process about 50 to 200 samples/person/day on a continuing basis.  These assumptions produce an average sample processing cost of about $1.50 to $5.00 per sample.

Next, the cost of materials including reagents, kits and accessory disposables needs to be considered.  Reagents are the basic materials that are used to prepare an ELISA, i.e., antibodies and antibody-enzyme conjugates.  Typically, the supplier of these materials provides a protocol for use but limits responsibility to the identity, bioactivity and amount of materials provided.  Kits often include all the components required to perform the test which arrives ready-for-use.  Most often key accessory disposables and laboratory equipment are not provided.  The supplier furnishes test instructions and extends responsibility for a stated level of performance if the test is properly done.

Prices for ELISA materials cover a broad range (Table II).  The low end prices are often quoted for reagents used in programs with large sample volumes, as in potato seed certification.  The high end prices reflect reagents that are very difficult to prepare, quantitative tests, and tests with small annual sales.

Disposable accessories are pipette tips, sample extraction pouches, test tubes and all of the various devices that are used to hold, process and dispense extracts of samples.  They usually represent a small fraction of the total test cost.  The cost of kits varies primarily by the number of tests that can be performed and by the ease-of-use provided.  Kits supplying only 1-70 tests cost about $2.00 to $16.00 per test.  By adding sample processing, test performance, and cost of disposables an estimate of $2.00 to $6.00 per sample is obtained when performing one test on 50-200 samples/person/day.  If additional tests are done on the same sample extract the cost would increase by about $0.25 to $0.50/test at the high sample rate and about $1.00 at 50 samples/person/day.

The reasonableness of the estimates can be examined by considering the cost of ELISA services from a variety of laboratories and diagnostic clinics (Table II).  The rates for single samples are included for comparison but are outside the initial assumptions.  These rates are higher because the reporting and business costs are applied to a single sample.  As the number of samples and or the number of tests increases, the fees charged compare favourably with the above estimates.

Price ranges for additional tests and services that are frequently done in indexing procedures are also given in Table II.  Since these tests are labour intensive the top of the range cannot be set when the tests fail to perform or when the problem is not

*Table II.* An estimate of the price range of materials and services for conducting ELISA tests in the United States in 1995.

| Type | Item | Price range/test |
|------|------|------------------|
| ELISA materials | Reagents | $0.10 - $1.00 |
| | Disposables | $0.10 - $0.35 |
| | Kits (200 - 500 tests) | $0.50 - $1.00 |
| | Kits (1 - 70 tests) | $2.00 - $16.00 |
| ELISA service | 1 sample, 1 test | $4.00 - $50.00 |
| | 1 sample, 6 tests | $20.00 - $65.00 |
| | 10 samples, 1 test | $25.00 - $100.00 |
| | 10 samples, 6 tests | $150.00 - $170.00 |
| | 100 samples, 1 test | $125.00 - $600.00 |
| | 100 samples, 6 tests | $600.00 - $1100.00 |
| Other service tests | Bacterial culture | $5.00 - $50.00 |
| | Plant bioassay | $50.00 - ? |
| | Electron microscopy | $50.00 - ? |
| | Nucleic acid hybridization | $2.00 - $12.00 |
| | Polymerase chain reaction | $20.00 - $120.00 |

routine.

The requirement for indexing is not uniform throughout the year. The need for indexing procedures is seasonal by crop and the number of samples and tests performed varies dramatically during the year. This produces special problems in the management of test quality and test cost. The assumptions used above imply an annual cost of about $75,000.00 and a sample volume of 12,500 - 50,000. Many laboratories and indexing programs sustain this level of testing only for a fraction of the year.

Diagnostic services are evolving from a time when the examination of a single or small set of samples for a disease was provided free. Now, only several clinics and laboratories in the United States still do not charge growers in their state for such services. Although, the indexing of plants and seed for the presence of pathogens is a demanding process that requires talent, time, and money, the procedures play a crucial role in propagation of disease-free planting material. Continued, routine use of indexing procedures is part of the cost of propagating plants and should be included in all business and financial plans.

## V.  COST EFFECTIVENESS

## A.  Business Objectives

The business objectives of any propagation program include growing the crop free from certain pathogens to a desired volume and quality at harvest. It is expected to achieve the objective in a certain interval of time at or below a particular cost. There is considerable business risk in any program just because so many variables can affect the outcome when plants are being nurtured. There is special risk too. Often, the crop will have no value if certain pathogens are present in the mother or nuclear stock and are propagated in the harvested crop. Common examples of this are potato leaf roll virus in tissue cultured potato, chrysanthemum stunt viroid in chrysanthemums and *Xanthomonas campestris* pv. *pelargonii*, the causal agent of bacterial blight in geranium.

For many crops, the possibility of a pathogen being present dominates the business risk until it is clear that the pathogens of concern are absent. As propagation continues and the potential harvest volume of the crop increases, the objectives regarding pathogen presence may change. As long as the plants are in a vector free environment, for example in tissue culture containers, growth chambers, and isolation greenhouses, they may continue to be free of select pathogens. Once the plants have been introduced to the field or to vector containing environments, a low level of pathogen may be acceptable in the crop. This depends very much on the crop, pathogen, and vector.

At any end-point of this process the harvested crop is not 'disease free' or 'virus free' just because it was propagated through tissue culture or a rigidly controlled stem cutting program. Most indexing procedures detect pathogens and not disease. Thus the harvested crop is 'pathogen tested', 'virus tested', or 'bacteria tested' only. There is nothing special about most propagation methods that eliminates pathogens by meristem culture, heat treatment, and/or chemical therapy. Whatever survives this combination of treatments usually is propagated to the harvested crop and may appear again in the next or later generations.

## B. Pathogen Avoidance

Since it is not possible to 'cure' plants that are infected with certain pathogens, strategies have been developed for avoiding infection. Without direct genetic intervention, a popular strategy is locating or generating a mother plant free from the pathogens of concern, and then propagating the mother plant in tissue culture or in an environment free of the pathogen and its vectors. Maintenance of a pathogen- and vector-free environment is required as the plants are propagated to large numbers. Avoiding the recycling of plant material without starting the process anew, that is, one flow through of material, makes this strategy particularly effective.

In plant propagative systems the success of a pathogen avoidance strategy depends on the efficacy of indexing tests to determine whether the mother plants or the first several generations of plants propagated from mother plants are pathogen free. If a pathogen is detected, the mother plant may be returned to therapy or discarded in favor of another plant that tested negative for the presence of pathogens. In this phase the sensitivity of the test is particularly important and other tests besides ELISA such as indicator bioassay, polymerase chain reaction, nucleic acid hybridization, and electron microscopy should be considered. While some of these tests may be quite expensive, they only need to be applied to a few plants and the cost is a small fraction of the total cost of propagating large numbers of plants.

As the number of plants increase to large numbers, the pathogen status can be monitored only by selective testing. Testing of early generations is generally more important than later generations since the pathogen-free status of the mother plant is confirmed by negative test results on the first generation of propagules. Further testing in the process should be reactive rather than mechanically repetitive, that is, it must be in response to problems that arise in the propagation process. For example a breakdown in procedures or environmental controls could lead to the possible reintroduction of pathogens. Ultimately, it is not possible or desirable to test all plants. In programs following pathogen avoidance strategies, freedom from pathogens is assured by environmental controls and well kept records, if starting plant material is known to be clean.

## C. Cost Benefits

No test for a plant pathogen is always correct and no test done on a continuing basis is free. So, indexing procedures also contribute to the ever present risk and cost when propagating plants. If risk were to be minimized by testing all plants the process would be very expensive and in the end there would be nothing to sell. Insufficient or improper testing places a plant propagation program at great risk. The greatest benefit from indexing is achieved by testing parent plants intensively, monitoring whatever is key to final production, and confirming product specifications.

Even when operator and clerical errors are absent a test can become increasingly unreliable if applied to plants containing levels of pathogen close to the limit of test sensitivity. The problem is aggravated when plants fail to show symptoms of pathogen presence early in the propagation process. In addition to problems with test sensitivity, other errors can occur during propagation that may introduce mother stock plants to vectors or to infected or contaminated plants.

Since indexing procedures are at the heart of strategies to produce pathogen-free plants, they are cost effective in an essential way. It is the failure to perform properly the entire propagation process, including indexing assays, which results in economic disasters. For example, multimillion dollar losses of geranium occur in greenhouse operations due to bacterial blight infections; and certification failure of entire potato crops occur due to bacterial ring rot in seed potato crops. The cost of well performed and controlled indexing procedures is insignificant compared to such large crop losses. The benefit of indexing can best be estimated on the basis of potential crop loss.

First hand experience is required to determine the cost effectiveness of various options and choices among indexing procedures. The costs can be estimated for various hypothetical scenarios but only really become meaningful by trial and error. One cost effective approach to indexing is a three step process: (1) test parent plants intensively, (2) monitor and maintain environmental controls which ensure exclusion of pathogens, and (3) ensure that plants meet customer's product specifications at the end of the propagation process. The tests performed in steps 1 and 2 are generally determined by the propagator, whereas, the tasks performed in step 3 are primarily determined by the client purchasing the finished product or an organization setting standards for the commodity. For example, a propagator may test for twenty pathogens in step 1, two or three pathogens in step 2, and nine pathogens in step 3 for producing an exportable crop of potato minitubers at the nuclear level of seed classification. The propagator determines the business risk by the way steps 1 and 2 are performed - the methods used for propagating, environmental controls, and record keeping all enhance or detract from benefit of the indexing procedures. In general, however, correctly performed and controlled indexing procedures are a very cost effective part of plant propagation.

## REFERENCES

Clark, M.F. and Adams, A.N. (1977). Characteristics of the microplate method of enzyme-linked immunosorbent assay for the detection of plant viruses. *Journal of General Virology* **34**, 475-483.

Sutula, C.L., Gillett, J.M., Morrissey, S.M. and Ramsdell, D.C. (1986). Interpreting ELISA data and establishing the positive-negative threshold. *Plant Disease* **70**, 722-726.

Voller, A., Bartlett, A., Bidwell, D.E., Clark, M.F. and Adams, A.W. (1976). The detection of viruses by enzyme-linked immunosorbent assay (ELISA). *Journal of General Virology* **33**, 165-167.

Voller, A. Bidwell, D.E., and Bartlett, A. (1979). 'The Enzyme Linked Immunosorbent Assay (ELISA)'. Dynatech Laboratories, Inc., Virginia.

# AUTHOR INDEX

Numbers in *italics* refer to pages on which full references are listed at the end of each chapter.

## A

Abad, P. 123, *127, 129, 134*
Abbott, A.G. *129*
Abdelmonem, A.A. *212*
Adams, A.H. 222, *239*
Adams, A.N. 18, 23, 62, 65, 66, *69*, 155, *165, 169*, 284, *292*
Adams, S.S. 11, *21*
Agarwal, P.C. 172, 190, *202*
Agarwal, V.K. 171, 172, *202*
Ahohara, K. *56*
Ahrens, U. 42, *46*
Aiche, M.D. *210*
Akkermans, A.D.L. *53*
Alarcon, B. 10, *21*
Alarcon, G. *51*
Albouy, J. *170*
Albrechtsen, S.E. *211*
Alcorn, S.M. 139, *165*
Alderman, S.C. *202*
Alderson, P.G. *166, 167*
Alivizatos, A.S. 23, *49, 239*
Allan, E. 34, *46*
Allen, T.C. *242*
Allison, D.A. *100*
Alvarez, A.M. 31, 33, 35, *46, 47, 53*
Amann, R. *55*
Amica, A. *100*
Amouzou-Alladye, E. 76, *93*
Anderson, M.J. 33, *46*, 154, *164*
Anderson, R.L. *207*
Annis, S.L. 92, *96, 102*
Anselme, C. 184, *202*
Antoniw, J.F. *94*
Appels, N. 88, *93*
Archer, S.A. *167*
Arneson, C.P. *242*
Arnheim, N. *54, 100*
Aron, L.M. *130*

Asay, G.M. *128*
Assad, S. *204*
Asselin, A. *56, 93*
Assouline, I. 190, *203*
Atkin, J.D. *214*
Atkinson, H.J. 121, *127, 128, 133*
Atlas, R.M. 38, 39, 40, 41, 42, *46, 55, 94*
Audy, P. 86, *93*
Avgelis, A. *202*
Avila, F. *102*
Ayers, A.R. 79, 80, *102*
Azad, H. 55, *212*
Azevedo, M.D. *214*
Azzam, O.I. *209*

## B

Bachelier, J.C. *167*
Baer, D. *51*, 228, *238*
Bailey, J.A. *99*
Bailiss, K.W. *165, 202*
Baillie, D.L. *136*
Bain, R.A. 5, *23*
Bajaj, Y.P.S. *168*
Baker, J. *135*
Bakker, J. 119, *128*, 237, *238*
Bakker, P.A.H.M. *56*
Baldini, A. *54*
Ball, E. 24, *51, 167, 169, 207*
Ball, S. 4, 5, 23, *202, 203*
Banowetz, G.M. 76, 79, 80, *93*
Banville, G. *56*
Bar-Joseph, M. *52, 62, 68, 82, 94*, 156, *165*
Barash, I. *53*
Barba, M. *54, 202*
Barbara, D.J. 82, *94*
Bardin, R. *99*
Barfield, C.S. *128*
Barham, R.O. 108, *128*

# SUBJECT INDEX

# DATE DUE / DATE DE RETOUR

| | | | |
|---|---|---|---|
| | | | |
| | | | |
| | | | |
| | | | |
| | | | |
| | | | |
| | | | |
| | | | |
| | | | |
| | | | |
| | | | |
| | | | |
| | | | |
| | | | |
| | | | |

CPSIA information can be obtained
at www.ICGtesting.com
Printed in the USA
BVHW040004141221
623917BV00005B/117

9 781015 063495